Interrelationship Between Insects and Plants

Interrelationship Between Insects and Plants

Pierre Jolivet

CRC Press

Boca Raton   Boston   London   New York   Washington, D.C.

Photo:

t paleozoic Myriapod Arthropleura in the Upper Devonian forest is feeding on
rest floor from vegetal matter and lycopod debris. From an animated diorama
California Academy of Sciences, San Francisco.

Library of Congress Cataloging-in-Publication Data

Jolivet, Pierre, 1922–
    Interrelation between insects and plants / by Pierre Jolivet.
        p. cm.
    Includes bibliographical references and index.
    ISBN 1-57444-052-7 (alk. paper)
    1. Insect-plant relationships. I. Title.
QL496.J64   1998
577.8 ' 5 -- dc21   98-14846
                                                                                      CIP

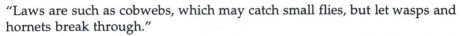

"Laws are such as cobwebs, which may catch small flies, but let wasps and hornets break through."

– Jonathan Swift,
A Critical Essay Upon the Faculties of the Mind

Dedication

To Erika and Julia, studying nature, in the land of the Rising Sun.

Foreword

An incredible number of papers and books have been written on the relationships between insects and plants during the last 20 years. More and more may be expected — for example, some excellent works such as the book by Bernays and Chapman (1994). These give an up-to-date synthesis of the physiology and chemistry of host-plant selection. It is very difficult to select certain works among thousands of papers and nearly one hundred books and monographs dealing sometimes with only one aspect of the question or one group of plant-eating insects.

This book does not pretend to be exhaustive or even to be always up to date. Some group of thinkers, such as those of the University of Wageningen in Holland, have made wonderful studies of the reactions of the phytophagous insects to chemical or visual attractants using their locomotive reactions, as well as data from electro-antennograms and electro-palpograms. Many U.S. laboratories and several pioneer workers (for example, Dethier) have done remarkable work in this field.

The aim of this book is more modest: to summarize in an understandable and simple way the basis of food selection among insects and to review the various aspects of their relationships with plants. The author has previously written a book on ants and plants, *Les Fourmis et les Plantes* (1986) and, in collaboration, several books on the leaf beetles, including works on their host plants (Jolivet et al., 1988, 1994; Jolivet and Cox, 1996; Jolivet and Hawkeswood, 1995). More data were published recently (Jolivet and Cox, 1996).

Ants and beetles have a special relationship with plants, using plants not only as a source of food but also (mostly for the ants), as a place of lodging. Many other insects (for example, Hymenoptera, Lepidoptera, Diptera, etc.) feed on or pollinate plants. Carnivorous plants also feed on insects. This results in special plant adaptations and even plant "behavior". These aspects also are reviewed here. We can see from the Table of Contents that many other aspects, including homochromy and mimetism, are also mentioned here, even paleomimetism, which means that this is the result of a very long evolutionary period. This is not a book on insect physiology but an essay about all aspects of the interactions between insects and plants. Its aim is to be different from the other texts in covering more topics.

There are one thousand and one ways to deal with the relationships between insects and plants. I beg the indulgence of my readers if mine is not the one they expected.

A final point, I treat here (following the old classification) algae, fungi, and cyanobacteria all as plants. I know they can be divided into separate kingdoms, but traditions here prevail over DNA sequences!

Preface

At the risk of being regarded as an eighteenth-century naturalist, I have considered here various aspects of food selection as found among the phytophagous insects. I had a famous predecessor, the "leaf mine" specialist, Martin Hering (1951). It is he who invented a sophisticated classification of mines based on his wide experience. His work is still regarded as the authority all over the world.

I remember in 1959 when I met Hering in Addis-Ababa, Ethiopia. I invited him to my house and we spent the evening discussing mines and their classification. Back in Germany, that same year, he published a paper on the various mines he found in the Land of the Queen of Sheba. I was fascinated by his vast knowledge and his great enthusiasm.

The present book contains topics related to insects and plants generally absent from ordinary books. There are plenty of books on physiology, chemistry, and genetics of plant selection by insects, but none includes carnivorous or myrmecophilous plants, pollination, etc. You have to find that in specialized treatises.

I did my best to cover as many subjects as possible and to explain them in an understandable way. However, it is impossible to be complete in such a broad topic as this. The interrelationships between arthropods and plants have many sides, and some aspects must be deliberately rejected. I hope that my colleagues who wrote books more erudite than mine will be indulgent of this provision.

Pierre Jolivet
Paris, France

The Author

Pierre Jolivet, Ph.D., a Frenchman by birth, was previously entomologist at the Royal Institute of Natural Sciences in Brussels, Belgium, from 1950 to 1957. He then spent most of his life abroad traveling as project director and entomologist for various entomological and plant protection projects for such United Nations specialized agencies as FAO and WHO. He visited most of the tropical countries of southeast Asia, Papua New Guinea, Australia, Africa, and Latin America teaching or doing research for various universities and colleges, and he participated in many tropical research expeditions. Dr. Jolivet was for a short while entomologist at the Museum of Natural History in Paris (1962) and scientific director of the Environmental Unit (EID) in Montpelier, France, with a position in the zoological department of the university (1971 to 1972).

Throughout his career, Dr. Jolivet's interest has been focused on leaf-beetles, taxonomy, biology, plant-insect relationships, ant-plants, and the study of several *Sporozoa*, parasitic insects of the order Gregarina. He has written 16 books, dealing with topics such as ants and plants, carnivorous plants, insects and man, host plants of the Chrysomelidae, and others in collaboration with several specialists, including a series of five volumes on the biology of Chrysomelidae. A recent publication on the biology of leaf-beetles summarizes the knowledge accumulated thus far on the family.

Pierre Jolivet obtained his Ph.D. from the antique and historical Sorbonne in Paris in 1954, just one year before the splitting of Paris University into 12 separate entities. He is now a corresponding member of the National Museum of Natural History in Paris, France, and is a Research Associate at the Florida State Collection of Arthropods, Gainesville. He continues to study beetles and the Chrysomelids and to travel abroad in various missions, mostly in Central and South America and southeast Asia, as a consultant in entomology, biological control, and plant protection.

Contents

chapter one

A brief and simple review of plant-arthropod relationships

The actual interrelationships of arthropods and plants are diverse. Many arthropods — for example, millipedes, centipedes, crustacea, and cockroaches — inhabit forest litter composed of decaying leaves, fungi, and algae, while others live under the bark and dead wood. Still others live on or inside plant tissue, chewing externally on different plant organs, or they feed on or inside roots, leaves, flowers, and stems. Some produce galls or mines, or they even inhabit plant cavities. Sapsucking insects move indifferently from one plant to another. Many, for example, bees and butterflies, are nectar feeders. Bees and other insects also collect pollen while feeding on nectar.

The diet of insects, and of arthropods in general, is based on the structure of their mouth parts. Some have mandibulate or chewing mouth parts, such as most beetles and Orthoptera; others have piercing and suctorial mouth parts. Examples include the present-day Siphonaptera, Hemiptera, and Anoplura; the mosquitoes and those with haustellate mouth parts adapted for sucking liquids without piercing; and the Lepidoptera, Hymenoptera, and most Diptera. These were not prevalent until the Paleozoic and early Mesozoic era. Ephemeroptera, certain Lepidoptera, and some Diptera have mouth parts that are reduced and nonfunctional. These insects do not feed at the adult stage — for example, Saturniidae (Lepidoptera), some genera of Chironomidae, and others. These are termed *aphagous*.

I want to review here generalized plant/arthropod relationships. These are generally simple but sometimes become quite complicated if a complex life cycle is involved, for example that of aphids or parasitic wasps.

1.1 Sapsuckers, fruit eaters, and nectar eaters

Piercing plant tissue for sap sucking is one of the most primitive ways of feeding on plants. This is very common as far back as the Paleozoic, namely among Palaeodictyoptera and Dictyoptera. Actually, many Hemiptera, Thysanoptera, Homoptera, Lepidoptera, and others suck sap from plant

1

stems and leaves or juice from fruit. Female leaf beetles of the subfamily Megalopodinae have the habit, before depositing eggs into the stem of a plant, of cutting the apex of the leaf and sucking the exuded juice. Leafcutting ants always feed on juice running out from the leaves and stalks after cutting.

Aphids normally move freely on young shoots and foliage of plants. A few species live below the surface of the ground on roots, while still others feed on woody trees and shrubs, and a certain number are gall inducers. The physiological changes and the polymorphism of aphids are in some cases the results of chemical changes in the physiology of the growing host plant. In these cases, the influence of the host plant on the behavior and cycle is a very important factor in these relationships.

Hemiptera and Homoptera transmit viruses both directly and indirectly, resulting in damage to their host plants. Most of these insects feed on the contents of the phloem vessels, i.e., on the complex sugar solutions or sap. Aphids penetrate mainly the intracellular leaf or stem tissues. Saliva is often injected into the plant by the aphid's salivary pump, which extends to the end of the sucking tube, along with the other mouth parts, in a tubular sheath around the stylets. This is analogous to the bloodsucking flies. Many sapsuckers (for example, some Reduviidae) have evolved into bloodsuckers (for example, the kissing bugs of North and South America). Some other bugs, such as bed bugs, also pierce the skin of mammals to absorb blood.

Most Thysanoptera feed by penetrating the tissues of plants, as opposed to the xylem and phloem, with their piercing mouth parts and imbibe the sap. Thrips feeding on protoplasts, in liquid suspension, absorb spherical chloroplasts. These may be larger than the narrow food channel, but they are elongated by suction pressure created by the cibarial pump and thus pass up the very narrow maxillary canal (Lewis, 1987). When thrips feed on fresh pollen, they rapidly pierce and suck the contents of many individual pollen grains per minute and discard the empty exine. Only one other insect, the ceratopogonid fly, *Atrichopogon pollenivorous*, removes the contents of individual grains rather than absorbing them whole as do beetles and most flies (Downes, 1955).

The imagoes of butterflies and moths live entirely (with several exceptions) on the nectar of flowers, overripe fruit (hence the habit of some tropical moths to feed on eye secretions and even blood of cattle), honeydew, or other liquid substances such as extrafloral nectaries or animal urine. Few really pierce the skin of fresh fruits or feed on sap.

As we well know, *Atta* ant species, the leafcutter ants, feed on plant sap exuding from the leaves they cut. In addition, they cultivate fungus grown on leaf compost formed from the leaves they bring into the nest. They feed their larvae and themselves with the cryptogam. Even, in Europe, the small ant *Lasius niger* (L.) bites at the edges of the leaves of various plants and feeds on flowing, rather than crystallized, sap (Jones, 1994).

Several beetles feed on both ripe and unripe fruit. Eumolpine leaf beetles (Chrysomelidae) are often found on tropical fruits, mostly in tropical America.

Among the beetles, only the species *Nemognatha* and *Leptopalpus* (Meloidae) evolved an adapted proboscis to absorb nectar. Nectar is absorbed by the Hymenoptera, butterflies, moths, and several other insects with adapted sucking tubes. *Thrips* spp. normally probe vegetation with their stylets and suck sap from pierced cells using their muscular pump. Many adult Diptera, such as ribionids, syrphids, muscids, tachinids, calliphorids, and others, are anthophilous and explore corollae with their narrow, elongated proboscis.

Feeding on nectar from extrafloral nectaries is rather common among ants, as hosts or only visitors of myrmecophilous or ordinary plants. Many other insects such as the parasitoids (Hymenoptera or Diptera), several species of moths, some species of beetles (Elateridae, Chrysomelidae, etc.), mites, and even Thysanoptera feed on extrafloral nectaries, often at a setose place where mites congregate. Galerucine leaf beetles, particularly many species of *Coelomera* or related genera, feed on trophosomes, the food bodies of their host plants, *Cecropia* spp. They also feed on extrafloral nectaries; these food bodies are only modified nectaries. Species of *Pseudomyrmex* ants, living on the Central American genus *Acacia*, feed on the trophosomes (Belt bodies) of the plant and also absorb sugar from the extrafloral nectaries. Many other ants from various plants (species of *Piper* in America and *Macaranga* in Southeast Asia) also feed on food bodies. Generally the nectar from the flowers and the nectar from the extrafloral nectaries are different in composition, the latter generally being richer in proteins and amino acids. While many other insects and mites feed on extrafloral nectaries, ants are the main customers.

1.2   Pollen- and spore-eaters and floral feeding

Bees, many Lepidoptera (Heliconiinae, Micropterygidae, and others), and many Coleoptera (Chrysomelidae, Oedemeridae, Nitidulidae, Melyridae, Cantharidae, Scarabaeidae, etc.) feed on pollen and digest it. Pollen-eating Lepidoptera (Micropterygidae) and Hymenoptera usually feed as both adults and larvae. Some Thysanoptera also feed on pollen. The long adult life of the *Heliconius* spp. (6 months in the wild) is attributed to their pollen feeding. As we know, caterpillars feed on leaves of species of *Passiflora*. The adults eat pollen of various plants and can digest it to support a prolonged oviposition period (Gilbert, 1972). The nitrogenous compounds of pollen (amino acids and proteins) absorbed by the butterfly influence its reproductive strategy.

Arnett (1968) was the first to observe pollen grains germinating in the gut of Oedemeridae, facilitating the digestion of otherwise undigested food. Several Galerucinae (species of *Aulacophora* and *Diabrotica*), which feed on petals of the flowers of the Cucurbitaceae, also ingest pollen. Some other beetles use plant trichomes as gastroliths to help pollen digestion; however, beetles generally do not participate directly in pollination except on some primitive flowers, but with several exceptions. Ants also are poor pollinators. Some pollen-feeders, such as *Anthonomus grandis*, the boll weevil, consume pollen from a wide range of plant species (Benedict et al., 1991).

Kuschel and May (1990) studied Palophaginae (Chrysomelidae). These chrysomelids feed exclusively on coniferous pollen, both as larvae and adults. The beetles have the same range as species of *Araucaria* in Australia, Chile, and probably elsewhere. They share the pollen of species of *Araucaria* with the larvae of the primitive Nemonychidae (formerly Curculionidae). *Palophagoides vargasorum*, from Chile, crush the pollen grains from flowers of species of *Araucaria*. The uncrushed grains probably pass out of the gut undigested (Kuschel and May, 1996).

Pollen will be treated further in Chapter 12 on pollination. A good summary of the topic is given by Samuelson (1994). According to this author, feeding of leaf beetles on pollen started at the end of the Jurassic period about when the separation of the cerambycid and chrysomelid lines occurred; the cerambycids probably specialized on Coniferales, and the early chrysomelids on Cycadales (Willemstein, 1987). A species of *Rhagium* (Cerambycidae) has a mixed diet of fungal spores and pollen. This is probably similar to the behavior of many insects during the Mesozoic era period before the diversification of the angiosperms.

The digestion of pollen is not easy. As we have seen before (Arnett, 1968; Haslett, 1989), pollen germinates inside the insect gut stimulated by digestive juices. The rupturing of the pollen test (a hard, indigestible covering of the pollen grain) enables the contents to be digested. Also, other mechanisms involving plant trichome cells, as in the scarab *Cyclocephala amazona*, are sometimes used. Trichomes are crushed with the pollen grains of *Bactris gasipaes* H.B.K. (Palmae) in the gut of the beetle and act as gastroliths (grinding stones) (Rickson et al., 1990). Most of the pollen grains are so digested and few of them pass through undigested. Birds use gastroliths to break open seeds. Dinosaurs also swallowed gizzard stones for the same purpose. Floral feeding is not necessarily linked with pollen feeding (Samuelson, 1994). Insects that collect nectar do not necessarily collect pollen. It is surprising that among species of Alticinae (Chrysomelidae) only a few collect and eat pollen. Pollen-feeding beetles (Cantharidae, Melyridae, Cleridae, Oedemeridae, Orsodacninae) generally have special adaptations (Crowson, 1981), mostly found in the structure and length of the alimentary tract. Pollen-eating beetles have short guts such as those of carnivorous species.

Actually, many different insects frequent the flowers: Thysanoptera, Lepidoptera, Diptera, Hymenoptera, Coleoptera. Again, this is treated in Chapter 12 on pollination. Only a few species of Coleoptera suck nectar. They do this through a specialized aspirator tube formed from the maxillary laciniae and sometimes the labial palpi. Species of *Nemognatha* and *Leptopalpus*, for instance, suck nectar from a tubiform corolla with a pump similar to those of Lepidoptera. It is a suctorial proboscis, but the beetle's tube cannot be spirally coiled. When species of *Thrips* feed on pollen, they suck out the contents of the individual pollen grain. An individual thrip can destroy 2 to 7% of a flower's pollen in a day (Kirk, 1987).

1.3 Chewers

One tactic "devised" by the plants to protect their leaves is to produce tough leaves. Chemicals such as cellulose and lignin deposited between the veins make the leaves difficult to chew and less nutritious (Raupp, 1985).

Many insects, such as a great many beetles, most caterpillars, and sawfly larvae have chewing mouth parts. These insects generally eat leaves, stems, and roots of plants. Many may feed on seeds (Bruchidae) and, incidentally, the flowers themselves, as we have seen above. Most pollen-feeders also have chewing mouth parts.

There are various ways of feeding on plants, either externally or internally. Thus we can recognize external feeders, leaf miners, borers, and gall inducers. There are those that feed on the visible part of the plant and others that feed on or inside the roots. Such a classification is rather arbitrary, but I will describe the main types of plant feeding. Caterpillars have powerful mandibles well adapted for mastication. Those of leaf-mining caterpillars may be modified for their specialized feeding method. Species of Micropterygidae, as more advanced Lepidoptera, have functional mandibles and laciniae in the adult stage. These moths are pollen-feeders, as shown above, and their maxillae are adapted for the purpose. Adult and larval Chrysomelidae have chewing mouth parts. Some Scarabaeidae beetles, specially adult Passalidae, have mandibles that are not specially modified but are capable of tearing apart decaying wood and chewing it into a condition suitable for consumption by their larvae. Chewing mouth parts are very diverse among insects and are adapted to many different situations.

The way in which different species of insects feed on leaves is not uniform. We already know that the spines surrounding a holly leaf function as a deterrent to most caterpillars, but if we cut around the edges of the leaf, the insect is able to feed voraciously. Where on a leaf that a folivore feeds depends on the ultimate area of the leaf that can be reached and on the overall physical condition of the plant (Coleman and Leonard, 1995). In addition, not all chewers chew in the same manner. Saturniid larvae tend to have a variety of cutting methods, while sphingid larvae tear and crush the leaf to masticate it thoroughly (Bernays and Janzen, 1988). The mandibles of these two families are different in shape which modifies the quality of food eaten.

Chewing insects may feed on or inside the roots, particularly species of Alticinae, Eumolpinae, Galerucinae (leaf beetles), and Curculionidae (weevils). Others, both as adults and larvae, feed inside fungi, on fungal spores, and on various species of algae. Even some thrips (Thysanoptera), normally pollen-feeders, can be mycophagous. Some leaf beetles feed on conidia of species of the genus *Venturia*, a fungus. Which is also true of many specialized beetles and flies. Many cerambycid (longhorn beetles) larvae, larvae of Megalopodinae or Sagrinae (leaf beetles), larvae of some weevils, and caterpillars of some

Lepidoptera dig galleries inside the stems of shrubs as well as herbaceous plants. Some alticine larvae mine the stems of the host plant at the junction between the root and stem and eventually may move indifferently from one to the other. Scolytid and bostrychid beetles are wood borers, as are many other insects such as some ants and caterpillars. Bostrychid beetles make cylindrical burrows in felled timber or dried wood and attack unhealthy trees. Heavy infestation can happen sometimes in forests, and wood-boring beetles have killed many pine trees (as for example, in 1994 in Florida).

Hering (1951) gives the following definition of mines: feeding channels caused by insect larvae inside the parenchyma or epidermis tissues of plants, in which the epidermis, or at least its outer wall, remains undamaged, thus shutting off the mine cavity from outside. The mine provides both living and feeding quarters for the larva.

Mines are a good protection for insects, and some species of Hispinae feeding on *Cecropia* spp. are protected this way from the aggressive species of *Azteca* ants. However, the protection is not perfect, and some parasitoids are experts in finding the larvae under the leaf cuticle, either by smell or by allelochemicals emitted by the injured plant or by the insects themselves. It is the third trophic level of host-plant relationships. Anyway, the balance seems to be favorable to the insect since it persists and survives.

Fossil mines are known, as are fossil galls, with insect frass and eggs inside. Mines and galls will be discussed in another chapter. Often the chewing mouth parts of this insects are slightly modified.

Several weevil larvae, bruchid larvae, and some caterpillars feed on seeds. Seeds are dry and rich in proteins, and a special adaptation is necessary to pierce the contents of the seed and to compensate for the lack of water. The same problem exists for the insects that live in dry foodstuffs. Many of these are of tropical origin. Bruchidae, in many respects, are related to Sagrinae. Some recent Australian classifiers have made them (wrongly, I believe) a subfamily of the Chrysomelidae. John Kingsolver (1995), one of our best specialists of the group, and many other entomologists do not agree with the Australian classification. Most bruchid larvae live in the seeds of Fabaceae and Caesalpinaceae (Leguminosae, *sensu lato*) but they also attack coconuts, palm nuts, and seeds of many other plants. The larvae are very similar to the larvae of weevils, with atrophied legs during the boring stage. These larvae feed on endosperm of the seed. When there are several larvae in the same seed, destruction is almost complete.

Those insects that suck sap from plants or nectar from flowers or chew leaves — the phytophagous insects — are very well adapted to their diet. Feeding on fungi, lichens, and algae requires further adaptation. The fungus eaters will be discussed in Chapter 6.

Cryptogam herbivory (feeding on algae; fungi, especially bryophytes; and lichens) is well known among weevils from the subantarctic islands. There they do not have much choice anyway. Many species of Lepidoptera feed on lichens in the larval stage. They are fully camouflaged, resembling their hosts. Some Coleoptera in the Juan Fernandez islands near Chile feed only on fern spores. Cryptogam feeding remains a necessity in remote and isolated islands where plant diversity is greatly reduced. Many caterpillars are scavengers, feeding on dead plant material along with millipedes or some terrestrial crustaceans.

Fern and tree fern feeding is rather common among some tropical flea beetles. It is a secondary adaptation among a differentiated group; some even are leaf miners. Mining on ferns was probably common during the Paleozoic or the Mesozoic era, but those insects were more primitive than those mining today.

1.4   Carnivory in phytophagous insects

Phytophagous insects and arthropods in general show few exceptions to their normal diet: plants. Detriticolous or saprophagous millipedes, for instance, are somewhat omnivorous, but generally when plant material is the normal food, few species deviate from their standard food source.

Several sucking-piercing insects, including some rare Hemiptera, such as Lygaeidae, Anthocoridae, Tingidae, Coreidae, and Capsidae, and Homoptera, such as Cicadellidae, Cicadidae, Membracidae, Jassidae, and Typhlocybidae, can actually suck blood from mammals, including man. Among the blood-sucking Homoptera, this habit seems to be linked to special atmospheric conditions (Jolivet, 1980). Such is the case of one species of coccinellid in England attracted to human blood. This happened in a dry year, and probably the beetle sucked blood from an open wound.

Among the leaf-eating Chrysomelidae, there are also exceptions (Mafra-Neto and Jolivet, 1994; Mafra-Neto and Jolivet, 1996). Phytophagous insects, including leaf beetles when the adults or larvae are in contact with the eggs, often feed on their own or conspecific eggs. Adults and larvae may eventually feed on larvae. This happens mostly when the eggs are not protected. Also, cannibalism is very widely reported among beetles. Saurophagy was reported for the galerucine *Aplosonyx nigripennis* feeding in Sulawesi on the wounds of a coluber snake. Coprophagy is known for *Oomorphus floridanus*, a lamprosomid feeding on rodent excreta in Florida. Nematophagy is reported for the Colorado potato beetle in Russia, specimens that feed in winter on potatoes rich in saprophagous nematodes. Entomophagy has been reported for *Aristobrotica angulicollis*, a galerucine preying in the Amazon region on living adult meloid beetle.

Among the weevils, the most famous case of coprophagy is found in the species of the genus *Tentegia* in Australia. The species of *Tentegia* feed as larvae in the naturally very dry pellets of dung dropped by marsupials. *T. ingrata* females collect pellets of kangaroo and wallaby dung and make caches of them. The eggs are laid singly in the pellets. *T. bisagata* feeds in possum dung (Morris, 1989). Also, the South American weevil *Ludovix fasciatus* is parasitizing the brood of some species of semi-aquatic grasshoppers

(Acrididae) of the genus *Cornops* which lay egg cases in the stem of the water hyacinth. Adults and larvae of the weevil feed on eggs of the Orthoptera. *Brachytarsus* spp. (anthribids) feed on scale insects (Coccoidea) and are therefore also carnivorous. Some Australian scarabs (*Cephalodesmus*) have reverted to phytophagy.

Also, we can cite adult butterflies, which feed willingly on excreta and urine, and the numerous exceptions to phytophagy by moths in general. Many caterpillars devour foliage, bore into the stems of plants, and attack the roots and timber. Others attack carpets and clothing or stored products. Some are predators on other insects or they feed in beehives, not only on honey, but also on bee grubs. These are all adaptations from the primitive plant-feeding habit.

It is well known that butterflies and moths, in addition to flower nectar feed on a variety of other substances (Turner, 1986): fruit, pollen (heliconiids), bird droppings (ithomiids and heliconiids), mammalian carrion (*Caligo* spp. and *Callicore* spp.), and water contaminated with mammalian droppings and urine. Some nymphalids drink human sweat. Moths are known to feed on fruit, sweat, tears, and even blood (Bänziger, 1971; Buttiker, 1961, 1967, 1973). Even the eye secretions of reptiles can be added to the list. Several observations made in Peru and Brazil show that the heliconiid *Dryas iulia* often drinks caiman tears by inserting its tongue into the corners of the eyes. This butterfly does the same in Peru on turtle eyes, which seem to be its normal food in the area, as the insect probably obtains necessary electrolytes, those not found in the existing puddles but which are present in reptile eyes. It is probable that at the end of the Mesozoic era the great reptiles were bitten by the precursors of fleas and ceratopogonids, and perhaps the tear-drinking habit was already established.

Species of *Melipone* and *Trigona* bees feed on the nectar of flowers, along with honey bees, but several of these species in South and Central America feed on carrion which they mix with pollen (Roubik, 1989). Normally dung beetles feed on excreta, but *Phaneus* spp. in Brazil, for instance, also are carrion feeders.

A comprehensive review of carnivorous herbivores among insects was written by Whitman, Blum, and Slansky (1994). The role of these species in nutritional ecology is important, as they are an additional food source and a key element of survival.

A non-arthropod example of carnivory among invertebrates is the species *Deroceras hilbrandi*, a Spanish slug, which is more carnivorous than herbivorous and feeds on animal carcasses trapped in the carnivorous plant *Pinguicula vallisneriifolia* (Zamora et al., 1996). Kleptoparasitic behavior is risk-free for the slug because it secretes mucilage and is able to crawl freely on the leaves.

Cannibalism is more widespread than usually believed: rabbits and sheep, in particular, can feed on meat. We know the result of feeding cattle animal flours — the Mad Cow disease in England was certainly unfortunate.

But vegetarians, such as cows, adapt very well to that artificial diet. Insects, however, probably react to attractants derived from plants included in their animal food.

chapter two

Plant-feeding insects and arthropods of the geological past

Evidence shows that as soon as the first arthropods crawled out of the sea, they started to feed on any available plant, probably at that time mostly green and blue (Cyanophyta or Cyanobacteria) algae. Then, some primitive green plants evolved, but it seems that at the beginning, after the first chewing arthropods evolved, the sapsucking forms dominated in what was then the Paleozoic forest. (See Figure 2.1.)

2.1 The plant kingdom

If we accept the findings of Stewart and Rothwell (1993), the first vascular plants appeared mid-Silurian, i.e., 435 million years ago. Green algae and cyanobacteria probably colonized swamps and shores of primeval seas even earlier, during the Ordovician period, i.e., early Silurian. It seems that the land plants are derived from progenitors that would be classified with modern Charophyceae (Graham, 1993). According to Graham, the land-adapted plants arrived 450 to 470 million years ago (late Ordovician). The first vascular plant known is the *Cooksonia* sp. It seems even that the first flowering plant (*Pannaulika* Triassic) emerged from the Triassic mud 220 million years ago (Crane, 1993).

Diversification of vascular plants started during the Devonian, and all major groups except the flowering plants were present by the end of the Devonian, i.e., 350 million years ago. Heterospory and seeds appeared in the early seed ferns. Liverworts, mosses, and fungi were already food for many arthropods (see Figure 4.2E).

Chaloner and MacDonald (1980) mentioned that fossil Collembola, species of *Rhyniella*, may have lived on the dead fungus infecting remains of species of *Rhynia* and possibly sucked sap from living *Rhynia* spp. These injuries, probably from insects, are preserved as dark exudations. Trigonotarbids

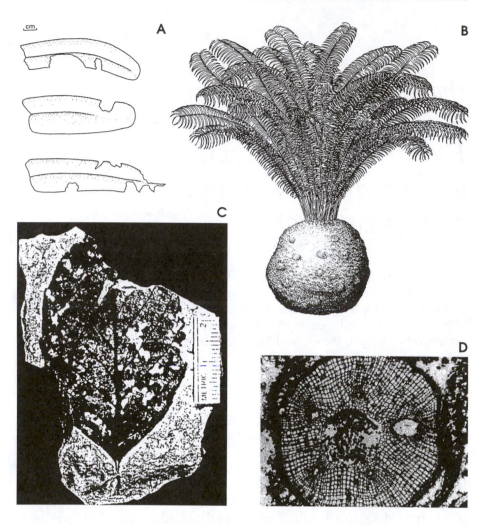

Figure 2.1   Fossil arthropods, fossil plants, and traces of insect feeding. **(A)** *Neuropteris* sp. pinnules, a Paleozoic gymnosperm with marginal bite marks. (Adapted from Scott and Taylor, 1983.) **(B)** Recent reconstruction of a (Cretaceous) Benettitale, *Cycadeoidea* sp., slightly different from previous interpretations. Such plants had bisporangiate cones and probably were self-pollinated or pollinated by boring insects. (Adapted from Delevoryas, 1971.) **(C)** Flea beetle egg deposition on alder leaves (*Alnus parvifolia*; (Middle Eocene, WA). (Adapted from Lewis and Carroll, 1991.) **(D)** Transverse section of *Johnhallia lacunosa* stem containing coprolites. (Adapted from Stidd and Phillips, 1982.) **(E)** Trigonotarbid (*Gelasinotarbus bonanoae*; Devonian, NY); these arachnids supposedly feed mainly on spores. (Adapted from Shear and Kukalova-Peck, 1990.) **(F)** Reconstruction of *Arthropleura armata* (original magnification 0.1×; Carboniferous). Some of these arthropods were very large (up to nearly 2 m in length); supposedly they fed on lycopods and forest litter. The head is not well known, and the mouthparts are still enigmatic. (Adapted from Rolfe and Ingham, 1967.)

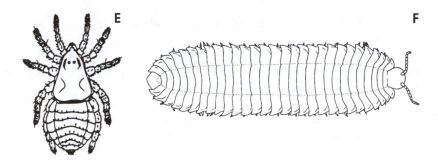

Figure 2.1   (continued)

probably consumed the contents of the sporangia of *Rhynia* spp. Also, these primitive Arachnids may have been the accidental pollinators of other primitive, lower-Devonian vascular cryptograms.

During the Carboniferous, a great diversity of insects populated an already rich forest of primitive ferns, seed ferns, and arborescent lycopods. Species of *Calamites* lived in the extensive swamps of that time. On the forest floor lived a lot of insects and also some big myriapods, present since the Devonian. These were eating lycopod spores, stems, leaf litter, and other types of detritus. Amphibians were numerous, feeding on insects and various small arthropods. At the end of the Carboniferous era, mosses, lycopods, sphenopsids, ferns, seed ferns, and cordaites formed, along with the first conifers, the tropical forests of that period. The abundance of insects, both flying and crawling, was characteristic of the late Carboniferous, or Permian, period.

During the Permian period, changes in climate and glaciation started in the Southern hemisphere, with cooler and drier climates everywhere. In the South, species of *Glossopteris* were abundant after lycopsid and sphenopsid trees became extinct. Reptiles started their diversification.

During the Mesozoic era, the beginning of a semiarid climate, Cycadophytes, Ginkgoales, and Benettitales appeared. Their prototypes of "flowers" were visited by many insects, including the newly evolved beetles which appeared during the Permian period. Recent cladistic analysis seem to suggest that angiosperms are the sister group of the Benettitales and the Gnetales. Most Benettitales have concealed flowers, but some had opened "flowers" which may have been white in color and with scents that attracted insects.

It is certain that the ancestors of cycads formed a major part of the world vegetation during the Mesozoic era. Actually, cycads now number 185 species representing 11 genera (Rothschild et al., 1986). When cycad material is ingested by vertebrate animals, the toxins liberated kill them, even the multistomach ruminants such as cows. Still, a lycaenid butterfly actually feeds on the young foliage of the cycad, *Zamia floridana*, in southeastern United States. This butterfly, *Eumaeus atala*, is aposematic and sequesters the toxins. This seems to be a new adaptation, not an old one, but it shows that in the past all the insects feeding on the cycad group had already developed an immunity against the alkaloids.

During the Mesozoic era, conifers and ferns became abundant, and, as mentioned previously, they are the earliest plants found in Triassic sediments, 226 million years old. This means that flowering plants evolved almost 100 million years earlier than generally thought. The association of leaf beetles with flowering plants is certainly earlier than was recorded before, and fossil pollen has been found in Triassic depots. During the Triassic, it is to be noted, species of *Glossopteris* declined in the Southern continents (see Figure 4.2E).

Under the uniform, mild climate of the Jurassic period, ginkgoes, conifers, ferns, cycads, and cycadoids were abundant. Some pteridosperms survived. The higher insects, such as ants, appeared at the same time as birds. Angiosperms, or flowering plants (monocots and dicots), were abundant in the lower Cretaceous period. Angiosperms rose to dominance in the Upper Cretaceous period. Many insects feed on such plants using many different feeding methods.

The Cenozoic (Tertiary and Quaternary) extends from 65 million years ago to the present. We consider as recent the last 5000 years, with the Pleistocene starting about 2.5 million years ago. Many recent extinctions are due to human interference, and plants and animals are equally affected. As mentioned by Stewart and Rothwell (1993), the floras of the Paleocene (early tertiary) are more in common with the Upper Cretaceous period than the recent eras. The primitive angiosperms, Magnoliaceae, Lauraceae, and Juglandaceae, prospered during this period.

Many of these flowers are cantharophilous, i.e., fertilized by beetles. The new and modern types of angiosperms first appeared during the Eocene (54 million years ago). At the Oligocene, *Metasequoia* spp., actual living relics of that period, were abundant in forests. Only during the Miocene (26 million years ago) did the present-day associations become predominant in those ancient forests.

The Tertiary boundary (called K/T) of the Cretaceous period and its presumed catastrophic events did not affect the insects and many other living forms very much (Whalley, 1987, 1988). Perhaps this is because many insects could fly and find refuge in other unaffected parts of the planet. If the Jurassic species of *Timarchopsis* were really the ancestors of today's species of wingless *Timarcha* (Coleoptera, Chrysomelidae), it is possible that they had wings at that time. Perhaps the brachypterism and apterism started during the Tertiary, or the already apterous species were eliminated at the K/T boundary. As for the plants, several botanists believe a mass killing, similar to that observed after the eruption of Krakatau (Collinson, 1986), took place. According to Stewart and Rothwell (1993), a mass kill does not necessarily mean that there was a mass extinction; the post K/T changes reflect a normal ecological succession and recovery after a mass killing has gradually changed the environment. Nothing is really clear, but there does seem to have been some kind of extinctions as indicated by the K/T layer.

Finally, if Darwin's "abominable mystery" of the origin of flowering plants has not been properly solved, at least now it is better understood. Cladistic analysis links the angiosperms to the Benettitales, an idea previously rejected (Crane, 1993). It is probable that the angiosperms appeared during the Triassic (Cornet, 1993). The number of vascular plants species was estimated to be 500 species in the late Carboniferous, increasing to 3000 species at the beginning of the Cretaceous period. By the end of the Cretaceous period, it was estimated to be 22,500 species followed by the explosive diversification that produced the contemporary land flora, estimated to be about 300,000 vascular plants species (Hughes, 1976; Burger, 1981). One can imagine that the diversification of the insects on those flowers, however, is not necessarily correlated with the multiplication of angiosperms. Lawton and Price (1979), looking at larval agromyzid mining species of the family Apiaceae (Umbelliferae) in the British Isles, state that agromyzid species are considerably below the number that could evolve to exploit these plants. For the British specialists, guilds of specialized leaf miners are not equal assemblages. That means that many more insects could have evolved on the 300,000 species of plants.

After these Cretaceous period catastrophic events, angiosperm dominance was reestablished in layers within a few centimeters of the famous boundary but with a very different composition (Tschudy et al., 1984). Flanders (1962) proposes the hypothesis (one more) that the abrupt disappearance of the dinosaurs at the end of the Cretaceous period was due to a lack of food killing the giant reptiles. Caterpillars could have multiplied so much that they ate all the available food. Did the caterpillar exterminate the dinosaurs? To Flanders, this idea is worth considering. Also, it has been said that the changes in flora composition at the end of the Mesozoic era killed those reptiles not adapted to the new alkaloids produced by the newcomers.

2.2   Insects and plants in the past

It seems that the oldest terrestrial arthropods ever found were two centipedes and an arachnid 414 million years ago, i.e., during the Silurian. It also seems that complex, multicellular creatures were not restricted to the sea during the Ordovician. There are fossil traces of animals who moved into lakes and rivers and who were able to survive out of water. Theses traces date from 450 million years ago, but no fossils of these creatures have been found. They might have been woodlice or myriapods. (See Figure 2.2.)

It was during the Silurian, that the vascular plants first appeared. Scorpions and millipedes already existed, and it is very possible that during the early Silurian (Ordovician), the first examples of life outside water started, including arthropods feeding around epicontinental seas on mats of green

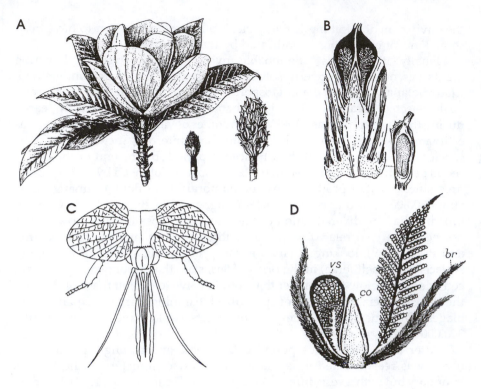

Figure 2.2 Pollination and insect feeding in the past. **(A)** *Magnolia grandiflora* L., a cantharophilous and primitive flower (original magnification 0.5×). On the right side are the gynaecium (megasporophyll) and infrutescence. (Adapted from Emberger, 1960.) **(B)** A floral bud of *Cycadeoidea dakotensis* (Bennettitales). Longitudinal section through mature cone shows the mass of sterile tissue surrounded by a complex folded microsporophyll and by basal leaves. On the right side is a seed showing the embryo with two cotyledons. Mostly boring insects infested the cone. (Adapted from Moret, 1943.) **(C)** Head, with piercing mouthparts and winged prothorax, of *Stenodictya lobata* (original magnification 4×; Palaeodictyoptera) from the Upper Carboniferous of France. (Adapted from Scott and Taylor, 1983.) **(D)** *Cycadeoidea* sp. (Benettitales). Bisexual flower opened with apical cone (co), covered with macrosporophylls and microsporophylls (vs), folded on the left, and opened on the right; br = hairy bracts. Normally, it seems, the flower does not open. (Adapted from Moret, 1943.) **(E)** *Williamsoniella coronata* (Bennettitale) pollinated via various insects during the Mesozoic. On the right, a "flower" with a megasporophyll surrounded by several simple microsporophylls; this flower probably opened. (Adapted from Moret, 1943.) **(F)** An Upper Carboniferous marshy tropical forest reconstruction, with Lepidodendraceae and Sigillariaceae on the left, and Equisetales, Ginkgoales, and Lycopodales on the right; arborescent Pteridosperms are in the background. Palaeodictyoptera lived in this forest, along with some modern insects. (Adapted from Jeannel, 1979.) **(G)** A cassid (Coleoptera: Chrysomelidae), ventral side, from a diatom quarry (France; Upper Miocene). The climate then was warmer than now, and this beetle could be closer to species of the Asian *Aspidomorpha* living on Convolvulaceae along the seashore than to the European species of *Cassida*. (Adapted from Balazuc, 1989.)

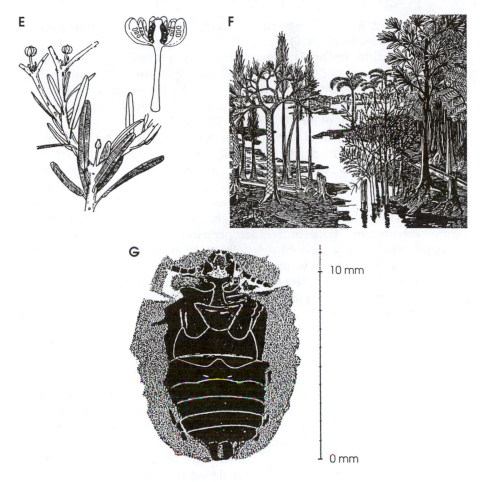

Figure 2.2 (continued)

and red algae, Charophyceae, and on cyanophytes abundant even before the Cambrian. It is evident that the scorpions, aquatic or not, were always carnivorous, as well as the primitive arachnids and centipedes, and the millipedes and certain arachnids were scavengers feeding on the ground on the vegetal mats. They had abundant detritus, spores, seeds, and (mostly) decaying vegetation.

It seems that the fossil remains of the earliest known insects were found in North America. Early insects did not fossilize well because they had soft bodies and no wings. These fossil remains date to about 390 million years ago, i.e., during the Silurian, and belong to the jumping bristletails (Archaeognatha). Entomologists believe that insects can be even older and could have evolved from the Ordovician, i.e., 400 to 440 million years ago, from a centipede or a millipede which fed on land plants or algal mats. From the beginning there were carnivorous forms feeding on phytophagous insects (Labandeira et al., 1988).

Trigonotarbid arachnids were probably spore-feeders that fed on the sporangia of species of *Aglaophyton* found during the lower Devonian (Rolfe, 1950). Evidence of early land arthropods in the Devonian is well known (Shear et al., 1984). Early land animals including primitive insects are abundant in the Devonian of North America (Shear et al., 1984).

Acari (mites) appeared during the Devonian and were probably mostly plant eaters (wood, fungi, litter), even if some were feeding on spores or pollen. According to Shear and Kukalova-Peck, nothing suggests that the ecological role of millipedes has changed during the 400 million years of their evolution. They were mostly detritus feeders in the past — for example, species of *Arthropleura*, a wholly extinct group that lived from the Devonian to the Carboniferous. Some individuals probably were up to 1.80 m long and perhaps longer. They resembled trilobites. Waterlot (1949) surmised that they were living in fresh water in or around wet plants such as ferns, as certain Isopoda now do. However, paleontologists know that they were terrestrial animals, feeding in humid surroundings on plant litter. Wood of fallen lycopod trees and tracheids of these plants have been found in their guts. Also, they were dusted with pollen from seed ferns, probably a part of their environment.

The Upper Carboniferous entomofauna was rich and diverse. Many species evolved feeding mechanisms for sucking sap or nectar or a combination of sucking and chewing. Palaeodictyopteroids used their beaks to suck up spores from the cones of lycopods and to drink juices from the ovules of Pteridospermae. We are reduced to hypothesis in certain cases, but the relationships with plants are unquestionable.

During the Permian period, the flora and the fauna changed slightly, and gymnosperms and cycadoids slowly replaced tall lycopods and tree ferns. Many insects, probably more than 50% of them, fed on plants. Butterflies and moths appeared in the early Cretaceous period and participated in the pollination of flowering plants as they evolved.

Probably, as mentioned by Shear and Kukalova-Peck (1990), the Paleozoic Paleoptera and Neoptera had cutting ovipositors similar to modern insect ovipositors used to deposit eggs in slits cut in the plant stem. Many modern insects, among them several Coleoptera, behave in the same way. Probably Palaeodictyoptera also dug holes in the plants to feed on megaspores. Their mouth parts probably tore apart the cones of tree lycopods and cordaites to reach spores and pollen. The gut of some fossil species show spores inside. Other species with narrow beaks might have fed through the micropyles. Plant sap, a sugary substance, also was taken by highly specialized species. Also, there were predators among those carboniferous Permian period insects, probably some bloodsucking species that fed on amphibians, but Protodonata were the dominant predators of plant-feeding insects.

Herbivory was largely distributed among Paleozoic insects even if it is difficult from damaged leaves, cones, or ovules to deduct insect interference. Spores, though, are present in the fecal pellets, and the primitive Coleoptera of the Permian period may have been detriticolous, fungivorous, or even pollen-spore eaters. To Carpenter (1971), there is no doubt that the beaks of the Palaeodictyoptera and related orders were used for piercing and sucking liquid nutrients, presumably of plant origin and probably from some of the succulent lycopods of that period.

During the Mesozoic era, there were also many phytophagous insects feeding on seeds, spores, and pollen. Some beetles, such as the Protoscelinae, were probably frequenting the pseudoflowers of some Benettitales. A large proportion of these beetles were confined to conifer trunks (Cupedidae) and the ovules and strobili of cycadoid gymnosperms (Arnol'di et al., 1992). Benettitales bore bisporangiate strobili, with the megasporophylls and the microsporophylls in a spiral inside a cluster of sterile leaves. That structure resembles today's angiosperm flower. Recent analysis make these earlier plants ancestors of the modern flowering plants. Probably those "flowers" were differently colored and scented and perhaps produced nectar, at least the opened ones, which attracted butterflies and other pollinators. Some palaeobotanists believe that the cones of the Cycadeoidea did not open at maturity, but rather disintegrated internally and the ovules were self-polli-nated (Taylor and Taylor, 1993). Many cones, however, show evidence of insect borings in the receptacle and in the area of the pollen sacs, implying the presence of beetles (Crepet, 1979). Crepet et al. (1974) suggested that this pollination system contributed to the decline of the Cycadeoidea during the Cretaceous period and a shift of their insect pollinators to the angiosperms rapidly evolved during that period. In contrast, the cones of the William-soniaceae opened at maturity and probably bore colored flowers. Benettitales, as well as Cycadales, had xerophytic trunks with inflated tis-sues stocking water. Those trunks were attacked by several borers, among them beetles.

It is possible that beetles feeding inside the Benettitales cones did the same kind of boring as do the present-day larvae of *Palophagoides vargasorum*, discovered recently by Kuschel and May (1990) in Chile. The larvae of that primitive Chrysomelid (Palophaginae) begin their development in the male strobili of *Araucaria* spp. well before they burst open with ripe pollen. They feed mainly on the pollen.

There is not any contradiction between the ideas of Taylor and Taylor (1993), who write that the algae Charophyceae gave rise to land plants which evolved to become arborescent, and those ideas of Takhtajan (1991), who believe that the herbaceous state for the angiosperms is an example of neo-teny. It is evident that after a long evolution towards the arborescent form, as a means of protection against the insect attacks, that some angiosperms reverted to an herbaceous state. It is possible that the first flowers emerged during the Triassic in swamps (Cornet, 1993), but they diversified much later during the middle of the Cretaceous period (100 million years ago).

During the Mesozoic and Cenozoic eras, the interrelationships between insects and plants narrowed to pollination, mines, galls, lodging, etc. Since the ants evolved during the Jurassic and the Cretaceous period, they rapidly became social insects, and their relations with plants became narrower and narrower. Species of *Oecophylla*, at the Eocene, probably already were social. Spiders during middle Devonian (380 million years ago) were already spinning webs (Shear et al., 1989). By then they forcibly captured various arthropods, including plant-feeding species. Guilds, probably as complex as those today, already existed. Evidently during the end of the Paleozoic the complexity increased with the diversity of the arthropod fauna.

If the aftermath of the famous K/T boundary seems to have had a serious effect on plant life (Collinson, 1986; Tschudy et al., 1984), it did not affect various factors causing the diversity of insects (Whalley, 1987; Labandeira and Sepkoski, 1993). It seems, however, that the rate of insect evolution slowed down after the appearance of flowering plants during the Cretaceous period (125 million years ago). According to Labandeira and Sepkoski (1993), insects then shifted from pollinating cycads to pollinating flowering plants. Finally, at the end of the Permian period (245 million years ago), there occurred the worst mass extinction of insects, with 65 insect families suddenly disappearing. An enigma remains: Did the diversification of flowering plants during the Tertiary lead to diversification of the insects? Many have evoked a sort of coevolution between flowers and insects — let us say, a diffuse coevolution. It seems that there is a mutual influence between the two kingdoms, plant and animal, and that the insects greatly multiplied to reach today's diversity. There in not much difference between Baltic amber fauna and today's fauna, except for the northern invasion of tropical insects.

2.3   Life in amber

The history of amber has become familiar through many books dealing mostly with the Baltic amber fauna (Larsson, 1978; Krzeminska et al., 1992; Poinar, 1992; Poinar and Poinar, 1994) and other amber, including that from other regions. The increased study of DNA sequencing from insects 25 to 40 million years old and particularly the study of DNA from weevils (Nemonychidae) 120 to 135 million years old (i.e., from the Cretaceous period) has become a useful tool for evolutionary studies. This was the dream of the novel and movie *Jurassic Park*, but it has not been done using insects that fed on the blood of dinosaurs, but more modestly from weevils feeding on the pollen of the male cones of *Araucaria* trees from Lebanon, or at least this seems the most likely food source. DNA studies on fossil insects have been strongly critized recently.

Most of the insects from Baltic amber are related or identical to extant ones, and their trophic relationships with the plants, if any, are easy to find out. In old and recent amber, pollinating bees with pollen-collecting apparatus are common. Many plant-feeding chrysomelids and cerambycids have been trapped in Baltic amber, probably during flight. Many sapsucking Hemiptera and Homoptera are also abundant there. Poinar (1992) affirms that a dense, humid rainforest covered at least a great part of the Baltic area. At that time, part of Northern Europe was subtropical. Then the climate cooled, and tropical insects disappeared.

2.4 Fossil evidence of herbivory

There is considerable evidence of plant feeding among the fossil records. An excellent review by Taylor and Taylor (1993) cites data on this from many recent publications. Other reviews of the same topic were made by Scott and Taylor (1983), Scott et al. (1985), and Takhtajan (1991). They have studied the evolutionary trends in pollination tactics.

2.4.1 Coprolites

The first coprolites described from rocks of the Ordovician were probably from semiaquatic arthropods. Coprolites from all ages have been described in relation to stems, leaves, roots. Generally, they are linked with wound tissue, namely in fossils of the Carboniferous period. These show deep tunnels in cordaitean roots with coprolites inside, along with much other evidence of wood-boring activities in the past, particularly on Jurassic conifers.

2.4.2 Defoliation, mines, wound tissues

As we have seen previously, Flanders suggested the hypothesis of caterpillars defoliating the conifers and other trees during the K/T catastrophic events and pushing the dinosaurs to their death by starvation. This does not seem to be the case, as the ammonites also became extinct in the seas of the world at the same time. However, the traces of damaged leaves and flowers are numerous during Mesozoic and Cenozoic times. According to Taylor (1987), there are no reports of ring irregularities caused by defoliation, but it is very difficult to interpret correctly the fossil data available to us.

Evidence of insect damage to pinnules of the Carboniferous ferns by neuropterons have been described by Scott and Taylor (1983). Brooks (1955) mentions an Eocene angiosperm leaf with semicircular marginal incisions caused by a species of *Megachile*, a bee.

There are reports of fossil mines in leaves from the Carboniferous to the Tertiary. Several are attributed to Lepidoptera. The case of a flea beetle egg deposit on *Alnus parvifolia* leaves from the middle Eocene is well known (Lewis and Carroll, 1991). The eggs are attributed to a species of *Altica*. Since most of the species of this genus have free larvae, mines are not common, and some species are root feeders. Therefore, Alticinae mines must be found (as they are for Diptera, Hymenoptera, or various Coleoptera) primarily on Tertiary angiosperm fossil leaves to prove the case. The difficulty is to find the real host of the leaf mines. Fossil evidence of lepidopterous leaf miners on *Quercus* spp. starts from the middle Miocene, but they may have existed much earlier.

Recently, Labandeira et al. (1994), found well-preserved leaf damage of the middle Cretaceous Dakota flora (97 million years ago); three distinctive insect feeding traces are present. These were assigned to two extant genera of two families of Lepidoptera: *Stigmella* and *Ectoedemia* (Nepticulidae) and Phyllocnistinae (Gracillariidae). That discovery shows that during the middle of angiosperm radiation the leaves of woody dicots were already exploited by leaf miners. The authors conclude that "the major lepidopterous lineages probably occurred during the late Jurassic on a still gymnosperm dominant flora."

Necroses sites of insect wounds either by piercing (Palaeodictyoptera and Megasecoptera) or chewing mouth parts are common in plants during the Devonian and the Permo-Carboniferous. Sapsucking was already spread among the insects from the lower Devonian, primitive species of *Rhynia* having necroses tissue probably caused by insects.

2.5   Gut contents

Lycophyte tracheids have been found in the gut of the primitive myriapod *Arthropleura*. This is not surprising, as arthropods lived in the humid forest litter in the Permo-Carboniferous forest (Scott and Taylor, 1983). Fossil seeds and pollen grains are known from the stomach of several insects.

The herbivorous dinosaurs probably fed on cycadophytes and ferns during the Triassic, then on conifers during the Jurassic and probably on angiosperms during the Cretaceous period. There are people who say that the angiosperms killed them. Insects also fed on the same plants and were progressively adapted to the morphological and chemical changes which occurred during the Mesozoic era, mainly the diversity of the alkaloids and related substances.

2.6   Pollination

Pteridophyte spores on coprolites indicate the prevalence of spore-eating arthropods in the Paleozoic (Scott et al., 1985; Takhtajan, 1991). Self-pollination has evidently arisen as a secondary phenomenon, and self-pollinating species are usually the ends of evolutionary lines. Several people attribute the decline of the Benettitales to self-pollination. Floral pigments and floral odors, attractants for pollination (physical, chemical, or mechanical), evolved in the geological past, but they were surely present before the advent of the angiosperms. Birds and butterflies can see red, and other colors seem attractive to many insects, although many of them see these colors differently than we do.

Beetles were probably among the earliest pollinators of the flowering plants, but many other insects pollinated primitive plants during the end of the Paleozoic and the Mesozoic eras. The development of floral tubes during the Cretaceous period gave rise to highly specialized pollinators such as Hymenoptera, Diptera, Lepidoptera, birds, and bats. Angiosperm evolution led to anemophily from entomophily, probably during the Albian (Lower Cretaceous) period (Takhtajan, 1991), due probably to a scarcity of pollinators but also, as in Cyperaceae, to the return to entomophily, as in *Rhynchospora* species. Even if anemophily existed among the Pteridospermae, insect pollination already was in practice during the Carboniferous. It is doubtful that giant (or small) species of *Arthropleura* pollinated pteridosperms during the Carboniferous, and pollen found on their body or on their legs perhaps was there by accident (Rolfe and Ingham, 1967). Crepet and Friis (1987) think that the Benettitales were pollinated by chewing beetles, but pollination of the flowering plants was done by insects during the Miocene by beetles, butterflies, moths, and Hymenoptera. According to Takhtajan (1991), the color of the flowers evolved from green to yellow (by loss of chlorophyll), then to white, orange, pink, blue, violet, and red. The evolutionary sequence that took place during the Cretaceous period involved xanthophyll, carotenoids, and anthocyanins. The development of the floral tube during the Cretaceous period favored insects able to reach their nectar, leading to specialization among these animals. Zygomorphic and actinomorphic flowers developed and provided an opportunity for concealment of nectar and pollen for privileged pollinators. It even seems that there were extrafloral nectaries between the seeds of species of the Paleozoic cycads of the genus *Phasmatocycas* to serve as attractants for some pollinators or to evict others. It is difficult to know (Mamay, 1976), but extrafloral nectaries must have evolved with flowering plants to keep out the ants and to feed the parasitoids (see Figure 2.2).

2.7 Galls and domatia

Larew (1992) gives a good review of fossil galls. Present-day galls are the same as the old ones and are a physical record of plant interactions with other organisms. Galls have been defined as an abnormal growth caused by the attack of a pest. The interaction is very old; galls are known from species of the Pteridospermae genus *Odontopteris,* living during the Permian period (280 million years ago). These are pinnule deformations. Insect galls are also known from the Triassic and the Cretaceous period. However, it is from the Tertiary that the most valuable fossil galls are known, due to the many different insects: cecidomyiids, cynipids, aphids, and also mites. Galls from *Populus* spp. are known from the Miocene and from *Quercus* spp. during the Pliocene. Galls are tri-dimensional and are more difficult to interpret as fossils on flattened leaves.

Fossil myrmecophytes, plants bearing myrmecodomatia, are still unrecorded. Acarodomatia, or mite houses, are well known from the Australian

Eocene (O'Dowd et al., 1991). Recent mite plants, acarophytes, are described later in this book.

Grass accumulations dating to the early Miocene have been found in Nebraska (Thomasson, 1982). These seem to have been constructed by harvester ants, but this remains hypothethical.

2.8   Paleomimetism, homochromy, and aposematism

Jeannel (1979) analyzed Theobald's book on fossil insects and criticized the confusion between the relationship of a fern leaf and an insect. This is just paleomimetism, and these examples were already common during the Upper Carboniferous. For instance, Scott and Taylor (1983) used a suggested case of mimicry to show the similarity between the pinnules of *Odontopteris callosa* and the cockroach *Phylomylacris villeti*. The *Neuropteris* pinnules very much resemble insect wings, which does not seem to be accidental. (See Figure 2.3.) We cannot know much about the homochromy, coloration, or behavior of insects of the past. We can see only the analogies between insect wings and the venation patterns of many Carboniferous ferns or Permian period *Glossopteris* species.

More problematic is the supposed mimicry between the merostomata (horse crab) *Euproops* species and the leaves of an arborescent lycopod (*Lepidodendron* spp.). We must remain prudent when making these speculations, mostly because horse crabs were probably aquatic even in the Carboniferous.

Carpenter (1971) showed pictures of many Paleozoic insects with bright contrasting wing markings. Some have eyespots already on the forewings, such as those of *Protodiamphipnoa* from the upper Carboniferous, as are the extinct mantids. Such spots are considered as deterrents to frighten the predators or at least to divert their attention. The insects of that period were certainly as colorful as the present ones, although, as Carpenter writes, we have no idea what those colors might have been. Even the facet-like organs on the wings of certain extant holometabolous insects (Martynov, 1924, 1925; Jolivet, 1955; Forbes, 1924) are present on the wings of many Carboniferous and Permian insects, but these apparently are not all holometabolous types. These stigmata are much larger and they can be slightly different in origin and structure (Carpenter, 1956). They may be degenerated wax glands, but their function is not very clear. Those organs have nothing to do with eyespots or wing markings, they are not brightly colored, and they are hardly visible on the wing. They are not involved in homochromy.

2.9   Conclusions

All plant-arthropod relationships that exist today probably already existed in the geological past: leaf- and spore-feeders, sapsuckers, and pollen-feeders. In litter, millipedes and cockroaches were feeding on rotten plant material and spores. Centipedes were predators of small creatures, including insects.

Beetles, mites, and millipedes inhabited rotten wood. Pollination was done by different insects during the Paleozoic and the Mesozoic eras, but by similar ones during the Cenozoic.

Camouflage existed even during the Paleozoic, and plants must have developed early morphological and chemical defenses against herbivory. Not only were insects feeding on plants, but dinosaurs also fed on them during the Mesozoic era, and later on mammals joined the others.

It is clear that arthropods must have played an important role in plant-animal relationships in the past, as they lived in the forest canopy, forest floor, litter layers, and soil. Their survival in a moist tropical climate where decomposition is quick is a kind of miracle, but this miracle occurred several times, even if many species disappeared for ever.

One hundred million years or more of plant-herbivore association means several times 10 millions generations occurred mostly in the tropics. Spencer (1988) has asked how coevolution and chemical diversification depend on length of generation and length of association. No one really knows for sure, but insect-plant associations have been present from the very beginning, and insects diversified along with the plants. Selection pressure was necessary to give birth to an insect able to detoxify a new secondary chemical — is that not coevolution? A lot of questions remain and will be discussed later in Chapter 13.

chapter three

An early twentieth century classification of food plant selection among phytophagous insects

Since the earlier classification of Hering (1950) of food habits among the phytophagous insects, quantities of papers have been published in the U.S. and England and by the Wageningen school in Holland. These have modified the original concepts. The classification was re-interpreted by myself (Jolivet, 1959, 1992) and, even if theoretical, the new interpretation provides a better understanding of those types of selection than do the classical divisions of monophagous, oligophagous, and polyphagous. Let's examine first the botanical instinct that was so dear to Jean Henri Fabre, the French entomologist (1825–1889). Schoonhoven (1990, 1991) completely reviewed the subject using his wide experience with host plant-insect relationships during his stay at Wageningen Agricultural University.

3.1   The botanical sense or instinct; Maulik's principle

I discussed in my book, *Insects and Plants* (1992), the topic of plant-insect relationships. I selected several examples of moths, butterflies, beetles, and even parasitic fungi. To Jean Henri Fabre goes credit for the idea that there exists a sort of infallible instinct with which insects are able to recognize species of plants that are taxonomically related, even if they are morphologically quite different. There are cases where some South American moth species of the genus *Thyridia* recognized that species of *Brunfelsia* do not belong to Scrophulariaceae but to Solanaceae. The plant has some strange-looking flowers, and those flowers change color after fertilization from white to yellow or from purple to white. A species of *Brunfelsia* bush has multicolored flowers and is fragrant at night. All the species of *Thyridia* feed on

Solanaceae, the only "exception" being the one found on *Brunfelsia*. In that case, an insect corrected the botanists.

As reviewed by Schoonhoven (1991), the botanical instinct took on another aspect. Insects not only detect the attractiveness of a plant, but they also sense their "distaste" through the means of chemical repellents. We now know how these are analyzed by an insect. Of course, electrophysiology has permitted narrow analyses of the insect reaction, but codification and monitoring of the brain still remain a mystery. Schoonhoven (1990) wrote: "A multitude of ecological studies has revealed that insects and plants do not simply live together, but rather they interact with each other. They suffer from each other and they adapt to each other, because they need each other. They both are as they are because of their coexistence. The natural world would change its appearance rapidly and drastically if all insects were to be removed." Is that coevolution? Imagine a world without bees, butterflies and moths, ants, beetles, and dragonflies. Of course, something esthetic would be missing, but many plants would disappear and eventually so, too, would man and other mammals.

Hering (1950) uses the botanical sense to explain the plant choice by leaf miners. This works rather well, meaning that the chemical affinities of the plants are detectable by mining insects and plant parasitic fungi. According to Hering's classification, this works only for the second- and third-degree monophagy and simple oligophagy. It is just a question of good sensitivity. What was named Maulik's principle is just the corollary of the botanical instinct.

The principle formulated by Maulik (1947) was mostly designated for leaf beetles, but it works also for any specialized leaf-feeder, including parasitic fungi. Maulik said, "If a phytophagous insect accepts as food a whole group of plants belonging to one or more genera, another insect that accepts one of those plants will accept all the plants of the group."

Really, this does not work well in the case of certain insects such as species of *Leptinotarsa*, which are usually feeders on *Solanum* spp., but they do not accept the same food plants even from the same genus. Maulik's principle proposes that if an oligophagous insect feeds on one plant, it finds on it attractive chemicals and the plant is devoid of repellents or antifeedants. Then if it feeds on another species of plant, other insects which feed on the first plant probably will accept the second plant. However, we have seen that the reception of the repulsive chemicals can be very subtle, and Maulik's principle, if true on a rough scale, is not 100% sure. It has a statistical value, however, and is a way to predict host plants with 50 to 60% accuracy.

I refer here to my small book (Jolivet, 1992) for details about the botanical instinct and several examples quoted there. It is a fact that "Maulik's law" gives some means of predicting the food choice of the species of certain genera of Chrysomelidae such as *Crioceris, Lilioceris, Lema*, etc. and many others. Species of *Chalcomera, sensu lato* (Chrysomelinae) in Australia (Daccordi

1994), in New Zealand (*Allocharis*), and in Southeast Asia (*Phola*) feed on Scrophulariaceae and Verbenaceae (Tubiflorae) and closely related families. Otherwise, it is extremely difficult to predict the food choice of any Entomoscelini (Chrysomelinae). New observations find different host plants from different families, which may prove that a group is essentially artificial. *Iscadida* spp. feed on species of *Rhoicissus* (Vitaceae), and it seems that all the species observed feed on the same plants. Polyphyletism is not enough to explain the diversity of host-plant feeding. Long-time isolation can also be invoked. For instance, the genus *Timarcha* has been split since early Tertiary between the western U.S. stock (*Americanotimarcha*) and the Eastern stock (*Timarcha*, *Timarchostoma*, and *Metallotimarcha*). The long time that elapsed during the separation was enough to change completely the food habits of Rubiaceae and Plantaginaceae and relatives in the East and Rosaceae in the West. Probably the *Timarcha* species during the Mesozoic era were modestly polyphagous and they evolved differently after separation mostly because they probably disappeared from the western U.S. with the recent glaciation and persisted, fully isolated, in protected places in the western Pacific states. Let's also point out that the botanical relationships between Rubiaceae and Scrophulariaceae on one side and the Plantaginaceae on the other side recorded in the old flora were rejected in new plant classifications. *Timarcha* species seem to confirm the views of the old botanists, and the two (or three) families seem very closely related. Rubiales, Plantaginales, and Scrophulariales are closely placed in Cronquist's recent book (1988) (see Figure 3.2).

It is chiefly in the subfamily Chrysomelinae that the species are restricted to several related plants, although species of Alticinae show some very narrow specialization leading sometimes to pure monophagy. As we will see later on, the leaf beetles are really exceptionally polyphagous (see species of *Systena* and *Diabrotica*), but hidden behind their apparent polyphagy is a large oligophagy habit. Basically, *Diabrotica* spp. are Cucurbitaceae feeders. Normally leaf beetles are oligophagous, meaning that they feed on plants chemically and botanically related.

The sort of "sixth sense" that enables the insect to recognize the taxonomic affinities of plants is of a chemical nature and it is logical that plants of the same genus or the same family produce the same chemicals. As Schoonhoven (1990) wrote, "Taxonomic relationships in many cases are synonymous with chemical relationships and insects do not search for plants that are classified within a certain taxon but rather for all of those plants which fit into a certain chemical profile." The profile in the case of Brassicaceae, for instance, can be extremely narrow, the "mustard oils" being slightly different from plant genus to plant genus, even from plant species to plant species. In the case of certain insects, such as papilionid butterflies, an olfactory sense is able to discriminate acutely between the essential oils. Taste and sight are also involved. Butterflies are capable of learning, and their choice is partly instinct, for all that means, and partly previous experience. The *Heliconius* butterfly species, for instance, learn to avoid *Passiflora* spp. leaves with real or false eggs. That principle will be discussed later on. When swallowtail butterflies in Israel, for example, alternatively choose plant species of Apiaceae and Rutaceae, it is because the essential oils present in the leaves are fairly similar. These insects do not show a good "botanical instinct" since the two families, even though in the Choripetales, are very distant from each other. It shows the importance of the presence of common chemicals in the food choice of phytophagous insects.

Secondary plant compounds have a protective function (Fraenkel, 1959, 1969). These are one of the reasons for the insect's choice. These secondary plant substances are not connected with the normal metabolic processes of the plant, at least as far as we can understand. Also, these occur in animals, but most of the known natural products are of plant origin. Many insects sequester toxins from their vegetable food. According to Harborne (1977), the chemical structure of about 3000 plant secondary compounds, mostly alkaloids, are now known, and many more are isolated every year. Most of these chemicals were isolated from higher plants. There is a wide chemical diversity among these secondary compounds. Hegarty (1992) lists among those that can have defensive functions terpenoids, alkaloids, cyanoglucosides, polyphenols, flavonols, saponines, glucosinates, and amino acids.

These compounds are specific feeding stimulants, but Jermy (1958; 1976a,b; 1983) added the concept of feeding deterrent and insisted that "host-plant specificity in phytophagous insects is determined mainly by the botanical distribution of plant substances inhibiting feeding."

This last point became a focus of attention recently (Schoonhoven, 1991; Bernays and Chapman, 1987). So, what is the main reason for food selection: attractants or deterrents? I believe there is a balance between both chemicals, phagostimulants being as important as feeding deterrents.

Electroantennography recently permitted entomologists to prove not only the existence of specific phagostimulant cells among caterpillars and Colorado beetle larvae (Schoonhoven, 1967), but also of feeding-deterrent receptors. These receptors are integrating complex stimuli, such as the ones linked with sinigrin (mustard oils) and permit very complex reactions according to the species. Stimulus quantity and stimulus quality have been analyzed in the deutocerebrum of the Colorado potato beetle, but we are still far away from deciphering the complex code of botanical instinct (Schoonhoven, 1990).

Food-preference induction shows the adaptability of certain phytophagous insects to certain foods and the possibility of modifying the food preferences. A paragraph will be devoted later on to the so-much discredited Hopkins Host Selection Principle, but this is not exactly the topic I am discussing here. Sometimes, a synthetic diet makes a new food perfectly acceptable. The caterpillar of the tobacco hornworm, *Manduca sexta*, a sphingid, accepts, after being grown on an artificial diet, cabbage (Brassicaceae) or dandelion leaves (Asteraceae), plants which are totally different from the normal food spectrum (Schoonhoven, 1967). This insect is very adaptable and even feeds on *Atropa belladonna* (Solanaceae), accumulating atropine in its tissues.

Saxena and Schoonhoven (1978) have shown that the same *Manduca sexta* caterpillar normally repelled by the smell of citrate present in citrus fruits can be conditioned to this substance and even attracted to it. Raising species of *Drosophila* flies on a medium to which mint has been added produced adults attracted by the mint odor, normally a strong repellent. Experience can modify the "instinctive" food preferences, and probably, in nature, what has been called allotrophy or xenophoby has no other origin. However, very often what we naively perceive as allotrophy — such as in the case of the galerucine beetle, *Agelastica alni*, abandoning Betulaceae and feeding suddenly in Western Europe on Rosaceae — can be inscribed into their genes. The oriental species of *Agelastica* feed normally on Rosaceae, and that brutal food change in Europe can be called a form of "atavism".

So, it seems that the botanical instinct could be reprogrammed and that the chemical receptors are more adaptable than previously believed. Also, a plant is not a fixed entity; one species can change its chemistry, an ability that is being used to develop resistant plant species.

Behavioral differences in food plant selection are known within the silkworm and for several other moths reared in the laboratory, such as *Yponomeuta padellus* (Yponomeutidae), for which experiments show several biological races feeding on species of *Sorbus, Crataegus,* or *Malus.* It seems that in some experiments, caterpillars of the race living on hawthorn (*Crataegus* spp.), when fed experimentally on apple trees, still produced female moths that preferred to lay eggs on hawthorn. In that case, the food change has not modified the ancestral food habits.

Since Fabre's days, the notion of botanical instinct has evolved. It presupposes a peripheral chemoreceptor system, sensitive to phagostimulants and deterrents, and a central nervous system that "computerizes" the patterns and makes the decision as to what to eat and not to eat.

It is evident that food generalists have more potentialities than do specialists. They can detoxify more toxic chemicals and enzymes and even store them for a defensive purpose. Botanical instinct applies only to oligophagous species with limited detoxifying abilities. Myriapods, for instance, feeding on decaying vegetable matter, are specially adapted to that diet and do not show any food selection abilities.

However, despite all the laboratory work, the "botanical instinct" of insects remains an enigma, an "abominable mystery" as Darwin would have said. The knowledge gained in 100 years has only served to alter the questions being asked (Schoonhoven, 1990). We do not know what determines the almost infinite variation of chemoreceptors defining their specificity.

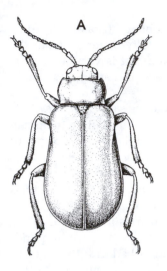

Figure 3.1 Mono-, oligo-, and polyphagy among some leaf beetles (Chrysomelidae). **(A)** *Coelomera cajennensis* (Fabricius) (Galerucinae), which is monophagus on *Cecropia* spp. in Venezuela. (Adapted from Jolivet, 1995.) **(B)** *Leptinotarsa decemlineata* (Say) (Chrysomelidae), the Colorado potato beetle, which is oligophagous on some species of *Solanum* and several other genera of Solanaceae in the U.S. (Adapted from Jolivet, 1995.) **(C)** *Cneorane ephippiata* Laboissière (Galerucinae) from China, which is oligophagus on various species of Fabaceae. **(D)** *Mimastra pectoralis* Laboissière (Galerucinae) from Vietnam, which is polyphagous on 11 families of plants. **(E)** *Galeruca* (*Emarhope*) *rufa* Germar (Galerucinae), which was captured in France on *Convolvulus* sp. (Convolvulaceae) but is probably polyphagous. **(F)** *Oides laticlava* Fairmaire (Galerucinae) from China, which is polyphagous on many plant families. The Galerucinae are often oligophagous or polyphagous, while the Chrysomelinae are generally oligophagous. (Parts C, D, E, and F adapted from Laboissière, 1934.)

3.2   *Botanical selection; bridge species*

I must simplify herewith the complex classification of Hering (1950) and Jolivet (1954). Although holding true for most cases, this classification is rather artificial. I refer to my book (Jolivet, 1992) for more details. I distinguish herein monophagy, oligophagy, and polyphagy, which can be only partial or total. The last state is extremely rare (termed pantophagy). Monophagy and oligophagy are, however, relative, depending on the availability of potential host plants (Thompson, 1988). (See Figure 3.1.)

3.2.1   *Monophagy*

The larvae and adults of insects that feed only on one species of plant are called, specifically, monophagous. This is not a rare thing, and a few cases of monophagy are known for leaf beetles, primarily some Alticinae (species of

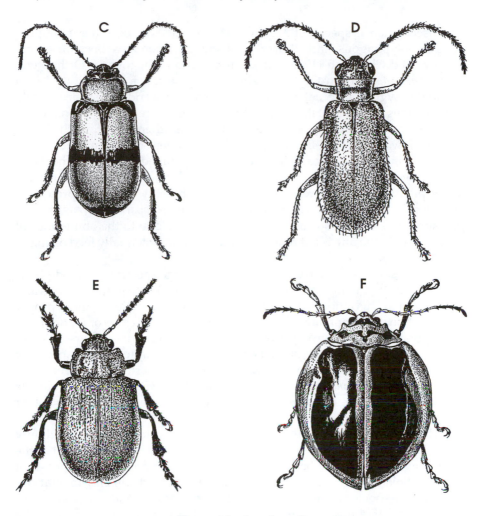

Figure 3.1 **(continued)**

Longitarsus), some Chrysomelinae (species of *Chrysolina*), and even more in the primitive subfamilies (Palophaginae on *Araucaria* spp.). One *Papilio* butterfly, from New Guinea, the beautiful *Papilio ulysses*, seems to be strictly limited to a Rutaceous plant, *Evodia accedens*. The plant grows only in the plains, but the adult butterfly has migrated through the mountain chains into the interior. Some entomologists have seen the butterfly laying eggs in the mountain gardens on citrus trees, where the larvae hatch and die, being ultra-specialized to feed on only one plant. Other species of *Evodia* feed species of *P. ulysses* elsewhere and even species of citrus in the Moluccas. There are 45 species of *Evodia* in the Old World tropics, but despite that big choice the caterpillar remains very specialized.

Insects such as species of Acrididae (the locusts) can be either monophagous or polyphagous, and all degrees of specialization have been observed.
Bernays and Chapman (1994) cite as strictly monophagous an Orthoptera,
Botettix argentatis, from the North American desert. It feeds only on *Larrea
tridentata* (Zygophyllaceae). Normally, grasshoppers and locusts are largely
polyphagous. Some locusts in the Ethiopian desert are often seen on a poisonous shrub, *Calotropis procera*, probably because there is not much choice.
Only yellow aphids and Danaid caterpillars frequent the plant. The locusts
expel, as a means of defense, air mixed with toxic blood (hemaphorrhea) and
are thus well protected. The locusts are potentially polyphagous and can eat
many plants. Their apparent monophagy is probably linked to protection
and a lack of competition. The brown planthopper, *Nilaparvata lugens*, is a rice
parasite, but its being so common on the cereal is due to the abundance of
rice. In reality, the bug is not monophagous and can potentially feed on many
other Poaceae.

Insects such as the agromyzid flies have a large spectrum of food plants
(Spencer 1990) ranging from moss, horsetails, and ferns to gymnosperms and
angiosperms. This is not coevolution but rather a colonization without proper
modification of either plants or insects. However, many cases of monophagy
exist among mining Agromyzidae and even among other miners (Diptera
and moths). This monophagy can be very strict and limited to one species of
plant only. The moth *Tyria jacobaea* (Arctiidae) is perfectly faithful to its
unique host plant, *Senecio jacobaea*, an Asteraceae. *Longitarsus jacobaea* (alticine
leaf beetles) feed on the same plant, but in experiments will accept other
species of the same genus.

The species of the genus *Heliconius*, a New World butterfly, feed only
at the larval stage on vines of *Passiflora* species (Passifloraceae), but at the
adult stage the butterfly has the tendency to collect pollen from species of
Anguria (Solanaceae) and several other flowers (Gilbert, 1972). The relationships of the species of *Heliconius* with the species of *Passiflora* can be very
narrow (only one species of the genus) or rather wide (several species). We
will explore that topic later on, but the selection can be visual, the butterfly
being perfectly able to recognize the shape of the leaf of the *Passiflora*
species selected, even though the shapes of *Passiflora* leaves are extremely
diverse.

As we have seen previously, some *Heliconius* butterflies or *Longitarsus*
flea beetles can select one species of a botanical genus but others select some
or all the species of this genus. Some genera of plants are monotypic (one
species only), but most of them are polytypic, that is, made of several or
many species. Evidently, phytophagous insects restricted to a monotypic
genus are strictly monophagous (first-degree), but others feeding on several
or all the species of a botanical genus are also monophagous to a lesser
degree. They can be considered as belonging to second-degree or third-
degree monophagy

The fly, *Liriomyza cannabis* (Liriomyzidae), which mines leaves of *Cannabis sativa*, the Indian hemp and a psychotropic drug, feeds on a monotypic genus, the genus *Cannabis*, and resort to first-degree monophagy. A caterpillar, *Eloria noyen*, feeds exclusively on the leaves of the coca plant, from which cocaine is harvested (*Erythroxylum coca*). Even though the genus *Erythroxylum* contains 250 tropical species, the moth seems monophagous on only one species, a trait that is used specifically for destroying plantations when the caterpillars are dropped over coca fields in Peru from an helicopter.

Examples found in Jolivet (1992) for second-degree monophagy include the leaf miner, the fly *Phytomyza abdominalis*, and the hymenopteran *Pseudodineura mentiens*, which restrict themselves to species of the genus *Anemone* of the botanical group (or subgenus) *Hepatica*. Other species of the two genera feed only on species of the *Anemone* group *Pulsatilla*. Even some Uredinea (rust fungi) species of the genus *Puccinia* show more or less the same selectivity. Similar behavior can be found among the beetles, miners or not.

The generic monophagy (or third-degree monophagy) is more common. The fly *Agromyza nigrescens* and the wasp *Fenella voigti* live on all species of the genus *Geranium*. The same phenomenon is well known among the caterpillars of butterflies and moths and many leaf beetles. Many Chrysomelidae are restricted to *Euphorbia, Rubus, Galium, Solanum, Rumex, Polygonum, Mentha*, etc. and have the particularity to be able to feed on all the species of the genus, though they often prefer some of them.

3.2.2   Oligophagy

The problem is more complex here, as the insects must feed on many species or genera of related families of plants. Some cases which refer to oligophagy can hardly be explained, except perhaps in chemical terms (disjunct oligophagy or temporary oligophagy). In the case of Donaciinae, for instance, aquatic at the larval stage, the adult plant choice is mostly guided towards plants of the same environment. In that case, polyphagy is the best term since these families are unrelated. Earlier I coined the term "ecological selection" for that case.

3.2.2.1   Systematic oligophagy

These are phytophagous insects which feed on related plants. In first-degree oligophagy, the insects feed on related plants belonging to several genera of the same family. This choice is the one made by many phytophagous Coleoptera, Hymenoptera, Diptera, and Lepidoptera (see Figure 3.2). Sometimes the choice is limited to one or several subgenera. For instance, the New World species of *Timarcha* feed on species of *Rubus, Fragaria*, and sometimes *Rosa* (Rosaceae), while the Old World species feed on different species of genera of related plant families.

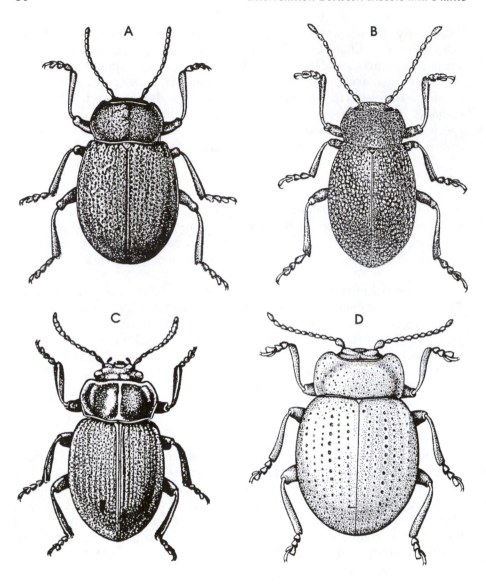

Figure 3.2 Apterous Phytophaga (Coleoptera: Chrysomelidae), with fused elytra, feed on herbaceous plants or shrubs. (A) and (B) feed on species of *Rosa* and *Rubus* (Rosaceae); (C) feeds on *Espeletia* (Asteraceae); (D) feeds on *Rhoicissus* (Vitaceae); (E) feeds on species of *Adenostemma* (Asteraceae), and (F) feeds on an unidentified species of Asteraceae. **(A)** Male *Timarcha* (*Americanotimarcha*) *cerdo* Stål, 1850, from Cannon Beach, OR, at sea level. **(B)** Female *Timarcha* (*Americanotimarcha*) *intricata* Haldeman, 1853, from Corvallis, OR, at 300 m. **(C)** Male *Elytrosphaera* (*Elytromena*) *nivalis* Kirsch, 1883, from Tungurahua, Ecuador, at 3700 m. **(D)** Female *Iscadida ståli* (Vogel, 1871), from eastern Transvaal, at 500 m. **(E)** Male *Elytrosphaera* (*s. str.*) *xanthopyga* Stål, 1858, from Viçosa, Minas Gerais, Brazil, at 700 m. **(F)** Female *Elytrosphaera* (*s. str.*) *melas* Jolivet, 1950, from the eastern Andes, Bolivia, at 2000 m. (Adapted from Jolivet, 1995.)

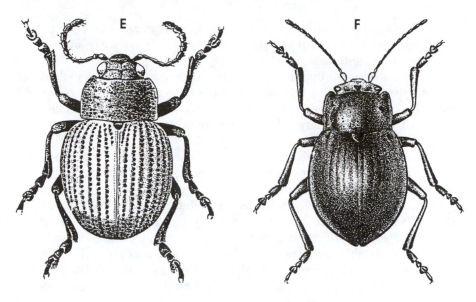

Figure 3.2   (continued)

The Colorado potato beetle is potentially oligophagous, even if at the beginning, in the wild, it is monophagous, feeding on only one species of *Solanum* (*S. rostratum*). All the species of the genus *Leptinotarsa* are oligophagous on Solanaceae, Asteraceae, and Zygophyllaceae (Hsiao, 1986, 1988; Jacques, 1988). In the laboratory, *L. decemlineata* has shown a wider potential on Asteraceae (*Lactuca sativa*), Brassicaceae (*Capsella bursapastoris*), and Asclepiadaceae (two species of *Asclepias*). Finally, the Colorado potato beetle can complete its development in nature only on 14 species of Solanaceae and the plants cited above, but it never feeds willingly on species of *Capsella*, *Lactuca*, and *Asclepias*. In nature, the beetle can feed on *Hyoscyamus niger*, *Lycopersicum esculentum* (the tomato), and 12 species of *Solanum*. It is a typical oligophagous beetle with hidden potentialities in the laboratory, such as for the silkworm.

The fly *Dizigomyza morio*, just as for several North European species of *Timarcha*, feeds on herbaceous Rubiaceae but rejects similar exotic species or genera. Hering (1950) named that rejection "xenophoby". Xenophoby is easily understood as having some differences in secondary compounds, either attractants or repellents, even if the fundamental chemistry of Rubiaceae is present in all plants.

Similar cases of first-degree oligophagy can be observed among parasitic fungi, such as *Peronospora parasitica* (Pers.), which lives on more than 100 species of Brassicaceae and among Acari or mites (Erigophyidae), such as species of the genus *Vasates* or *Aceria*. Those two genera are adapted to various species of Solanaceae belonging to the genera *Lycopersicum*, *Solanum*,

Datura, Physalis, and *Petunia.* Never have any *Leptinotarsa* species shown such plasticity by being restricted to only three genera of Solanaceae and mostly to species of the genus *Solanum.*

In second-degree oligophagy, the larvae or adults feed on plants of various genera belonging to related plants of the same order. Examples I cited in 1992 are the weevil *Ceutorrhynchus contractus* and the fly *Scaptomyzella flava,* which live on Cruciferae, Resedaceae, and Capparidaceae, all Resedales or Papaverales.

In third-degree oligophagy, the insects live on plants belonging to a variety of genera in different orders, but which are still related groups. For instance, species of Lepidoptera of the genus *Elachista* live on Gramineae, Cyperaceae (Glumiflorae), and Juncaceae (Liliiflorae). The fly *Chylizosoma vittatum* lives on Liliaceae and Orchidaceae (Liliiflorae and Microspermae). All those botanical groups are related and share some chemicals that make the species palatable. We can also state that most species of *Timarcha* in southwestern France, Spain, and North Africa feed on Plantaginaceae, Dipsacaceae, and Rubiaceae, but, very exceptionally, certain species feed on Scrophulariaceae, Brassicaceae, and Asteraceae. This can also be considered to be a third-degree oligophagy; relationships between Plantaginaceae and Scrophulariaceae are very common among various leaf beetles (*Chrysolina* spp.), but Asteraceae and Brassicaceae are not very closely related.

Many other examples of oligophagy are reported by Hering (1951) among leaf miners: Poaceae (Gramineae) and Cyperaceae; Cyperaceae and Juncaceae; Caprifoliaceae and Oleaceae; Chenopodiaceae and Polygonaceae; and Primulaceae, Myricaceae, and Amentaceae, etc. The Amentaceae are an heterogeneous assemblage of catkin-bearing families all more or less related: Salicaceae, Juglandaceae, Betulaceae, Fagaceae, etc.

Now consider the theory of "bridge species", which are found in both the phytophagous insects and parasitic fungi. According to this theory, attributed to Hering, two species that live on two different family groups of plants, not closely related, are able to transfer from one to another group through other plant species, the bridge species. These bridge species do not need to be related botanically to one of the families or groups of families involved. There exists, however, the need of some common chemical substances, sometimes known to be essential for the two insects (or fungi) species. This is the "common denominator" proposed by Hering (1949, 1951).

There must be in this bridge species a common chemical ground which is palatable to the insects. That applies wonderfully well to leaf miners but also to some leaf beetles or caterpillars, external feeders. I must say that species of *Plantago* are bridge species to many *Chrysolina* species when feeding on Lamiaceae and Asteraceae, and Scrophulariaceae and Asteraceae, and for *Timarcha* species sharing Scrophulariaceae and Rubiaceae. Curiously enough, the older plant classifications, as said before, closely associated Rubiaceae and Plantaginaceae, and also some recent books (such as the book by Swain, 1963) does the same based on chemical relationships of the plants involved. However, even if Cronquist (1981, 1988) placed the Campanulales in between these two, it does not mean he is separating them. His classifications are mainly based on plant structure.

From that common ground, each animal species or plant parasite can theoretically migrate beyond its normal plant family group towards another species group. Among the bridge species, the most common is the garden nasturtium *Tropaeolum majus* (Tropaeolaceae), a native of the Peruvian highlands but widely cultivated in all parts of the world. The bridge effect works well for both the insect species and the parasitic fungi.

The example given by Hering (1951) of *Tropaeolum majus* being a bridge species for the two European species of *Scaptomyzella*, a fly, is characteristic. One species, *S. incana*, mines Chenopodiaceae and Caryophyllaceae (Centrospermae) plus species of *Tropaeolum, Trigonella,* and *Anthyllis* (Fabaceae). The other species, *S. flava,* mines Brassicaceae, Papaveraceae, and Resedaceae; *Tropaeolum majus;* and then the *Trigonella* sp. and *Pisum* sp. (Fabaceae). Both species meet on the bridge species *Tropaeolum majus.* There are 86 species of *Tropaeolum* from southern Mexico to Patagonia, but only a few are cultivated, and none more extensively than *T. majus.*

The rust *Puccinia subnitens* (Uredinales) is found on the leaves of Chenopodiaceae, Polygonaceae, Amaranthaceae, and Brassicaceae. *Tropaeolum majus* also serves here as a bridge species between those two groups of plants. Several moths or weevils that mine Brassicaceae also live on *Tropaeolum majus.* They find the same chemicals (mustard oil-like substance) among those plants, i.e., myrosin. Will they also go to Centrospermae? We are not sure, because these latter are lacking some of the chemicals, such as the thioglucosides, which are shared with the Capparidaceae, Resedaceae, Brassicaceae, Limnanthaceae, and Tropaeolaceae.

Thus, according to the bridge species theory, if a species migrates to a food plant species which provides common and agreeable chemical substances, it will accept as its host plant the other species by gradually becoming accustomed to the chemicals of the bridge species. Natural selection can easily explain that change in behavior. However, such colonization is the fruit of slow evolution, often unfinished in many insect families.

3.2.2.2   Combined oligophagy

This group includes phytophagous insects which may live on all or practically all species of the genera of a single plant family, plus one species of plant in another family unrelated to the others. For example, the flea beetle *Altica oleracea* lives on Onagrariae and Lythraceae, but also on *Polygonum aviculare* (Polygonaceae), an unrelated plant. Actually, those previous plants are the normal choice of the species, but numerous others have been mentioned in the literature. There may be some confusion with other species or biotypes or with a case of allotrophy that is common among the Alticinae as a conservative gesture and an act of survival. Our knowledge of the biology of the Alticinae is not very advanced. What we consider as the same species on different plants can be a group of biotypes having each other as a different host plant. That was already suspected in 1925 by Peyerhimhoff in his studies on the Algerian Alticinae.

The previous case also could be a case of disjunctive oligophagy. Some species of *Timarcha* in Spain feed on two different species of Brassicaceae, which are totally unrelated to their other food plants.

The selection of *Altica oleracea* could also be explained as a case of ecological selection, all of the plants eaten by them being found in humid areas, as are the Donaciinae. These insects, being relatively polyphagous, are forced to accept the plants available in its ecological niche. For instance, the Australian Paropsini (Chrysomelinae) feed on Myrtaceae and Mimosaceae. This is not really polyphagy but is not far from it. In Western Europe and Western Siberia, the chrysomeline leaf beetle species of the genus *Oreina* feeds on Asteraceae and Apiaceae, both plants sharing the same mountainous biotypes.

3.2.2.3   Disjunctive oligophagy

Insects live only on a limited number of unrelated plants of different orders of the plant kingdom. This case is relatively infrequent among the insects, miners or not.

Liriomyza eupatorii (Diptera) feeds on the *Eupatorium* sp. (Asteraceae) and *Galeopsis* spp. (Lamiaceae). Also, the moth *Lyonetia ledi* lives on species of *Ledum* (Ericaceae) and *Myrica* (Myricaceae). The two plants belong to the same community (ecological selection?) but they are certainly unrelated. Also, those cases sometimes can be related to the ecological selection or selection of plants of the same ecological habitats, but what is the meaning of ecological selection? Nothing is clear about that matter. Many other explanations are found: biological races or biotypes, mutation, allotrophy. It remains chemically, physiologically, and taxonomically unexplained.

3.2.2.4   Temporary oligophagy

Oligophagy is said to be temporary if, in its different stages of development, an insect species constantly feeds in succession on different and unrelated plants; this means at specific stages of its life cycle. This is seen in the case of those insects with complicated life cycles such as aphids or fungi such as rusts (Uredinales) and others. Among the leaf miners, the best examples are several species of Lepidoptera of the genus *Coleophora* of the *lixella* group. During their first larval stage, these insects live exclusively inside the calyx or in the seeds of Labiatae, but later in the autumn or at the beginning of the following spring, they move to various species of grasses (Poaceae) where they make large mines. It is difficult to explain the transfer from one family to another so different taxonomically (dicots during their early stages and then an monocots), a transfer probably made necessary by insufficient food during the long period of development.

Examples of temporary oligophagy among the Coleoptera are difficult to find. These exist, however, and when better known the biology of floricolous and pollenivorous beetles, such as species of *Orsodacne* (Orsodacninae), several Alticinae, species of Oedemeridae, and in general miners and gallicolous species will probably offer more examples. Some species of Oedemeridae are well known to be oligophagic, but others are confined to the pollen of a single species (e.g., *Oxacis subfusca*). The Papilionidae (Lepidoptera) in the Middle East and elsewhere will feed successively on Apiaceae (Umbelliferae) and Rutaceae (*Citrus*) according to the season. This may be explained by the presence of essential oils in both families, the food source pressure, but not by botanical relationships. It is a matter of survival for the butterfly.

Many rusts produce two different types of spores during the duration of their life cycle, each on different, unrelated plant families. There are similarities between the elaborate cycle of aphids and the rusts. To explain this strange behavior, two hypotheses are equally acceptable: (1) original polyphagy with later specialized monophagy (the rusts); or (2) original monophagy which evolved later on to polyphagy on very different food plants (*Coleophora* spp.).

3.2.3   *Polyphagy*

Polyphagous insects feed on a great number of plants belonging to unrelated genera in different orders. Here also we can distinguish two degrees, but the distinction is very subtle. This distinction was introduced by Hering (1950).

3.2.3.2   *First-degree polyphagy*

This is found in many species which feed almost at random on a wide variety of botanical orders, but of the same class. Species of Lepidoptera of the genus *Cnephasia* live on various dicots and seem to be attracted by herbaceous plants. Diptera of the genus *Phytomyza* (*P. atricornis*) and *Liriomyza* (*L. strigata*) do the same thing. Meloid beetles and several leaf beetles (Alticinae: *Systena* spp.; Galerucinae: *Galeruca* spp.) are equally polyphagous, the first on cultivated crops, the second on environmental wild plants. It is very difficult to classify that kind of polyphagy.

3.2.3.2   *Second-degree polyphagy*

This is the utilization by insects of plants of several classes, both monocots and dicots. The weevil *Orthochaetes insignis* and the moth *Cnephasiella incertana* show this type of food selection, if we can speak of their food selection. They live in humid areas surrounding freshwater ponds or along river banks (ecological selection).

The caterpillar *Amsacta meloneyi* (Arctiidae) appears in enormous numbers at the beginning of September in certain parts of the presahelian zone in western Africa, mostly in irrigated areas. These hyperactive caterpillars, as did the soldiers of Attila, cover hundreds of meters each day, devouring everything along their way, including both cultivated and wild plants, monocots, dicots, and even certain cryptogams. However, it rejects spicy plants such as red peppers (imported in Africa) and the native latex-bearing plants with a milky sap, such as Asclepiadaceae and Euphorbiaceae. *Calotropis procera* has its own predators but is rejected by the caterpillar of this moth, which eats most voraciously another local species of Asclepiadaceae, *Leptadenia hastata,* which has translucent, not bitter, latex. This is extremely interesting, because species of *Calotropis* and *Leptadenia* have similar attractants, and in the Old World they are selected by eumolpine beetles. The eumolpines feeding on *Calotropis* spp. in Asia cut the veins of their leaves. It is probable that the caterpillar of the *Amsacta* spp. has not yet learned to do this.

Food selection in polyphagous insects, such as species of *Amsacta*, is controlled by the lack of repellent chemicals. The species must have a great capacity to detoxify a large quantity of alkaloids and other secondary substances. Species of *Amsacta* feed on the Madagascaran periwinkle *Catharanthus roseus* (Apocynaceae), a native of southern Africa and devoid of latex. These caterpillars absolutely reject the native *Adenium obesum* (Apocynaceae), a very toxic, thick-stemmed, fleshy-leaved xerophyte with beautiful pink flowers. This plant produces cardiac glucosides and has few herbivorous enemies.

A more polyphagous species, the aposematic cricket (*Zonocerus* sp.) and others eat and thoroughly digest the latex of the most toxic Asclepiadaceae. They use these chemicals as a means of defense, but yellow aphids are always present on their leaves.

Weevils also are largely polyphagous. A good example is the New Guinea species complex of *Oribius*, and in the Mascarenes islands (Reunion, Mauritius, Rodriguez) and in Madagascar the species of the genus *Cratopus*. Myriapod Diplopoda, which are linked to hypogaeic habitats such as in forest litter or under logs, bark, or stones, are also widely polyphagous. *Spinotarsus caboverdus,* introduced into Cape Verde islands in the Atlantic, devours any plant but mostly the cultivated species. These extant species of Diplopoda have reduced or absent eyes, rigid exoskeletons, and few spines, while their carboniferous ancestors had larger eyes, bifurcated spines, and soft, flexible exoskeletons. There is much evidence that the fossil forms were external feeders, living openly on leaves of lycopods in the Paleozoic forest. The diplopods sequester toxins in their blood. Several tropical species (New Guinea, Haiti, Sumatra) expel quinones at a great distance as a means of defense (e.g., *Polyconoceras* sp. in New Guinea). It is impossible to know now if the giant species of *Arthropleura* from the Carboniferous period also had defensive glands, but this seems most probable. It must have been a repulsive beast but it fed only on lycopods.

3.2.4 Pantophagy

Pantophagous insects feed on almost all green plants and even on decaying vegetable matter. As the most extreme examples we can include here the

Clytrinae (leaf beetle) larvae (Chrysomelidae) which feed on vegetable detritus in ant nests, on ant eggs, on larvae, and on ant excreta. The mining Tendipedinae (Diptera: Chironomidae) not only feed on living matter from the mined plant but also on plankton organisms of plant and animal origin and on decayed detritus.

Normally, pantophagous insects accept coniferous plants, angiosperms, ferns, mosses, algae, and lichens as food. Fly species of the genus *Sciara* feed on Asteraceae and the cryptogams of the genus *Marchantia* (Hepaticae). This behavior could also be called third-degree polyphagy. The fly larva feeds on highly evolved Asteraceae and very primitive liverworts, a very large choice indeed.

A last word on aphagy, which is when the adult feeds on neither animal nor vegetal matter. Several chironomid flies do not feed, and neither do the adults of several moths such as the Saturniidae and many other groups. Food accumulated during the larval stage is sufficient for the duration of the adult stage. Females of species of *Anopheles, Culex,* and *Aedes* mosquitoes feed on blood and nectar but males feed only on nectar. Real aphagy is rare even among the Chironomidae; most of the species feed on flower nectar as adults but even some South African species feed on human and animal blood.

3.2.5 Xenophagy or allotrophy

Xenophagy results as another consequence of the lack of food deterrents in the host plant, which is thus a palatable plant outside the normal food spectrum of the insect species. This is a drastic change in the food habits of an animal, which is sometimes brought about by extreme habitat stress. This often happens to leaf beetles. For certain Alticinae, though, the change concerns only the adults, and the larvae continue to feed on their normal host plants.

This should not be confused with temporary oligophagy, such as that found in certain hydrophilid beetles. These water beetles are carnivorous in early stages but later on they become herbivorous. One of Kammerer's students early this century claimed that he exchanged the heads between herbivorous hydrophilid and carnivorous dytiscids by grafting. According to the scientist the water beetles changed their habits: the dytiscid became herbivorous and the hydrophilid became carnivorous. This feat was probably a hoax, as the Kammerer school was known for performing such crazy experiments: After being found guilty of having artificially modified the legs of a toad (*Alytes obstetricans*) by injecting ink in the jaws, Kammerer committed suicide. Such was the Kammerer school in Vienna!

Xenophagy is also a complete and sudden change in diet, the herbivore becoming a carnivore or vice versa, predators becoming necrophagous, coprophagous becoming necrophagous or carnivorous, and so on. However, we do not deal here with that kind of change, but only with the abrupt change in host plant among phytophagous insects. Xenophagy has its parallel among parasitic fungi (xenoparasitism).

Butterflies of the genus *Papilio* and most of the Papilionidae may be divided into several groups according to their food choices. Three groups (among others) can be found commonly feeding on species of *Aristolochia*, on Rutaceae (citrus and allies), or on Umbelliferae (Apiaceae). The transition from Rutaceae to Umbelliferae (and vice versa, which rarely takes place) can be explained by chemical analogy (there are no taxonomic relationships between the two families) and bridge species that contain ethanol such as the rutaceous plant species of *Dictamnus*, perennial herbs with aromatic leaves, or *Pelea*, an aromatic tree. In countries such as Israel, the transition is direct, without bridge species. Xenophagy in parasitic fungi occurs when the "abnormal" host plant is damaged or weakened, which in turn helps the fungus to colonize a new host.

Numerous Diptera and Lepidoptera leaf miners exhibit xenophagy. Many "abnormal" mines come from the urgent need of the female to oviposite, and in the absence of the right chemical from the normal host plant, it lays eggs on a new host plant. Sometimes the larvae are able to develop in that new situation, but generally they cannot complete their cycle. Caterpillars of *Coleophora fuscedinella* have been observed mining flowers of species of *Caltha* (Ranunculaceae). These larvae had evidently fallen from an overhanging tree (*Alnus* spp.) which was their ordinary and normal host. The same species was also found mining a species of *Lysimachia* (Primulaceae) below a birch tree (*Betula* sp.). Similarly, the caterpillars of *Coleophora flavipennella* have been found mining *Helianthemum* sp. (Cistinae) below an oak (*Quercus* spp., its normal host plant). Many other examples of xenophagy among Lepidoptera and Diptera miners are known and would take too long to cite here. We find cases of switchs from Scrophulariaceae to Loganiaceae, from Guttiferae (*Hypericum* spp.) and Asteraceae (Compositae) to Lamiaceae, from Oleaceae to Saxifragaceae, from *Malus* sp. (Rosaceae) to Vitaceae, from Rosaceae and Amentiferae to *Humulus* spp. (Cannabidaceae), from *Betula* spp. (Betulaceae) to *Ribes* (Grossulariaceae), etc. How are these abrupt changes in the normal and basic food plant to an alien species or genus explained?

There are many proposed explanations and no one of them is fully satisfactory; for instance, an abnormal plant odor (repellent) can be masked from the egg-laying female by the odor of the normal food plant, a strongly fragrant species such as a species of Lauraceae, Verbenaceae, Apiaceae, Rutaceae, or Lamiaceae. Sometimes that can be the explanation when the two plants, the usual and the abnormal host, coexist in the same habitat.

Also, this switch can be the consequence of an abrupt mutation. The attractiveness of a chemical can be modified and result in the birth of a new biotype, a new strain of a species, thus, an emerging species. The deterrent effects of the new plants can also not have any effect on the new biotype. Sometimes this innovation remains without further development.

May I repeat here that for a plant to be acceptable as a new host, the insect must find in it the attractive substance (olfactive and gustative) — or at least no specific deterrent — and the insect must not be exposed to any toxic substance within the new host plant. All these conditions are rarely met. The Colorado potato beetle eventually accepts and lays eggs on a species of *Solanum* which slowly kills the larvae. Exactly this same thing happens when a species of *Papilio* lays eggs on a Rutaceous plant toxic to its larvae.

Many other factors seem responsible for the xenophagy of leaf miners. Greater succulence or juiciness (and the nontoxicity or the absence of anti-feedants) of the leaves of many cultivated species and higher temperatures of greenhouses or of walls exposed to the sun can slightly modify the behavior of an insect. All these factors bring about biological disturbances which disrupt the normal biological equilibrium.

It must also be stressed that, besides the real cases of xenophagy, some cases may be distinguished only after a deeper knowledge of natural chemical and taxonomic relationships of the botanical families (Amentiferae and Rosaceae, for instance) is learned.

The urgent need of the female to deposit its eggs on any plant sometimes can be responsible for the phenomenon of xenophagy. Often the normal development of the larva is not completed on the new host. However, sometimes it happens that the cycle is completed. If we believe in the capability of Lepidoptera to learn, then adults born from those larvae later can show an attraction to the new plant, and by selection a new biotype is born. We are dealing here with a quasi-heretic Hopkins principle and shades of Lamarck. The idea is only speculative, and for an orthodox Darwinist it has to be rejected. It has never been proved that the food plant change persists beyond the adult.

Also, the destruction of certain plants over wide areas by herbicides and cultures plays a role in the aberrant behavior of the phytophagous insects. In Western Europe, the biodiversity reservoir, the former hedge rows have practically disappeared and with them the fauna of apterous insects (see Figure 3.2) and parasitoids, which has had a marked effect on host plant selection. Failing to find its normal host, an insect will lay eggs on a novel plant, often leading to the death of its offspring; mining larvae very often have limited power of flight and no great chances of finding their "normal" host plant. When the strictly oligophagous *Lyonetia clerkella* (Lepidoptera: Lyonetiidae) — which normally lays its eggs on Rosaceae — lays its eggs on *Salix* spp., it is pure heresy for a botanist.

As Schoonhoven (1991) said, discrimination of acceptable plants from unpalatable ones is based on chemical characteristics of which the plant kingdom shows an almost endless diversity.

3.2.6   The old clichés

In Jolivet (1992), I mentioned several problems, which the plant chemists are slowly elucidating following the discovery of the chemical structures of secondary plant substances. These substances — dear to Dethier (1941 through 1993) and Fraenkel (1959, 1969) — the lack of deterrents (Jermy, 1958; 1976a,b;

1983; 1984; 1987; 1988), and the presence of attractants and phagostimulants explain most of mysteries of host plant selection. The role of experience (learning) also plays a significant role in the food choice of the herbivorous insects.

The list of the "old clichés" is as follows: xenophoby and xenophily, Maulik's principle, the role of hybridization and chimeras, and butterfly memory. We also will consider later resistant plants in which agriculture places great hopes to help eliminate insecticide pollution. The botanical instinct of insects has been dealt with at the beginning of this chapter and will not be discussed again. For the insect, it is just a matter of detecting the peculiar chemistry of acceptable host plants. Insects use chemoreceptors to detect feeding deterrents, and phagostimulants are just the ones making the difference in selection after the computerization of those stimuli in the insect brain. Xenophagy is the total rejection by many insects, mono- or oligophagous, of a foreign plant, even closely related to its usual host. There are degrees in that rejection, and in certain cases the insect which accepts a foreign plant is unable to complete its life cycle if it feeds on it.

Female butterflies sometimes oviposite on plants which are toxic to their larvae. The example of *Papilio ulysses* is not unique. *Troides priamus* in Queensland (Straatman, 1962) will oviposite on *Aristolochia elegans* introduced from Brazil. The butterfly finds attractive stimuli but the caterpillars die rapidly. Such mistakes have been found in Queensland among many butterflies: *Eurycus crassida*, *Papilio demoleus*, *Euploea eichhorni*, etc., and this must be common everywhere. This means that introduced plants are well protected against foreign predators and that it takes time before new insects are adapted to new plants.

On the contrary, some insects are pre-adapted and willingly accept imported plants close to their original host selection. This has been named xenophily, but the phenomenon is more rare than xenophoby. A good example of xenophily is the colonization of two *Nothofagus* species by Lepidoptera in Southern Britain as reported by Welch and Greatorex-Davies (1993).

Many examples of xenophoby are found among leaf beetles (e.g., Colorado potato beetle) and leaf miners (Lepidoptera, Diptera, and Coleoptera). This also occurs among rusts (Uredinales). It seems that the geographical isolation of plant species between the Nearctic and the Palearctic regions since the Eocene has brought with it a divergence of proteins and some heterosides. No only is the insect generally repelled by the plants but the larvae usually die when eating these leaves.

3.2.6.1 Maulik's principle

Maulik published his theory in 1947, intending it to be a means to deduce, by analogy with related species, the food choice of another insect. It is only the application of common sense, but it works generally very well in a group of species when based on simple biochemical selection. It works with oligophagous species, but among leaf miners there are many exceptions. For instance, if one species of *Lilioceris* feeds on a member of the Lily family, there is a good chance that another species of the genus *Lilioceris* also will feed on a member of Liliaceae. That deduction also holds at the tribe or subfamily level but in a rather loose way. The matter was discussed in more detail earlier in this chapter.

3.2.6.2 *Plant hybridization and chimeras*

Feeding an insect on a synthesized diet can modify completely its food choices. It is the same as when a larva feeds on a plant hybrid or a chimera. A chimera is a mosaic of tissues from different species or genera of plants. It is made by grafting. Hering (1950) has shown that leaf miners such as *Nepticula nanivora* (Lepidoptera) and others can adapt to species of different genus of plant through hybrids and chimeras. The same applies to *Parornix* and *Lithocollitis* spp. (*Phylonorycter* spp.; Lepidoptera: Gracillariidae and Lithocolletinae).

3.2.6.3 *Animal hybrids*

Hybridization between two animals, both of which are phytophagous species or subspecies, can for some unknown reason so deeply modify the hybrid metabolism that the offspring can accept a host plant rejected by the parents. Many examples are known among such Lepidoptera, such as the hawk moth *Celerio euphorbiae s. str.* crossed with the subspecies *C. euphorbiae mauritanica*, both of which feed only on species of *Euphorbia* (Euphorbiaceae) and not on species of *Salix* (Salicaceae). The hybrid formed from male *C. mauritanica* × female *C. euphorbiae* = × *wagneri* and the reverse cross can be fed entirely and without any problem on species of willow (*Salix*). Salicales have no botanical relationship to Euphorbiaceae (Geraniales), and so far as known the biochemistry of the two families is very different.

Similar behavioral disturbance is known among hybrids of the hawk moths, *Pergesa elpenor* and *Celerio hippophaes*. These disturbances can be compared to the effect of plant hybrids or artificial diets on the behavior of the insect; otherwise, such drastic changes can hardly be explained.

3.2.6.4 *Butterfly learning or butterfly memory*

It is a well-known fact that species of *Heliconius* butterflies as adults collect mostly pollen and nectar on certain flowers, namely from species of Cucurbitaceae. When ready to lay eggs, the female become sensitized to the green color and the shape of *Passiflora* leaves and will lay its eggs on the species on which the larvae feed. We know that the shape of the*Passiflora* leaves differs from each other according to the species. The selective behavior of the butterfly is what Fabre referred to as botanical instinct, but a butterfly and even a beetle are capable of learning, and the previous experience of a larva can modify the behavior of the adult. To Benson et al. (1975), the striking adaptation to passion vines by the species of *Heliconius* is the result of parallel evolution in the two groups. Whether it is true or not, the genus is a living laboratory and could be used to study the effect of modifications of the caterpillar food plant.

All butterflies and moths, when egg laying, become sensitized to the color green and the shape and the odor of the host plant. Previously, when collecting nectar and pollen, only bright colors such as red (among insects only the butterflies are sensitive to that color), orange, and blue interested them. However, the olfactory sense is the primary selective factor for the mature female (and the only factor for moths blind to colors). Flowers pollinated by moths during the night are fragrant and mostly white.

Yponomeuta padellus (Lepidoptera: Yponomeutidae) consists of two biological races, one on apple trees (species of *Malus*) and another on hawthorn (species of *Crataegus*), which are closely related plants (both Rosaceae). The race living on hawthorn, even if fed experimentally on apple trees, produces female moths that still prefer to lay eggs on hawthorn and vice versa. In this case, the food change does not seem to have modified the ancestral food habit of the two biotypes. Fraenkel (1969) acknowledges that some butterflies become conditioned to the secondary substances of the initial host when transferred to another host (*Papilio aegeus, Pieris brassicae*).

The experience with *Yponomeuta padellus* does not contradict the famous and much-discussed Hopkins Host Selection Principle. According to this principle, the adult insect of oligophagous and polyphagous species has a tendency to lay eggs on the host plant species that was its food during the larval stage. This is not in contradiction with the example shown by species of *Yponomeuta* which deals with biological races and not with food selection within the same species. The "principle" has been verified many times with various caterpillars, but it is also frequently contested.

Futuyma et al. (1993) reared the galerucine *Ophraella notulata* on *Iva frutescens* and *Ambrosia artemissifolia* (Asteraceae) and obtained the progeny of four possible crosses. This study provided no evidence of effects of the maternal host plant on offspring feeding and oviposition, which also contradicts the Hopkins principle.

Alternative host plants chosen by an oligophagous insect generally contain the same attractive substances. However, the mustard oils in the case of the Brassicaceae, Resedaceae, Tropaeolaceae, and Capparidaceae (alkyl isothiocyanates) can be slightly different in species of *Brassica, Reseda, Cochlearia, Sinapis, Rorippa*, etc. This explains the strict selection of several monophagous or oligophagous species.

The olfactory reaction of species of *Drosophila* suggests associative learning (Atkins, 1980). For the adult fly, peppermint oil is normally a repellent, the insect being attracted only by scents of decaying fruit which are slightly acid and yeasty. If the fly larvae are reared in the presence of mint solution mixed with the normal rearing medium, the resulting adults are attracted to it. This could be a key factor in the Hopkin's principle.

Newly hatched larvae of the tobacco hornworm, *Manduca sexta*, a sphingid, feed on an artificial diet, whether or not it contains citral. Last instar larvae exhibit clear feeding preferences and orientation towards the diet on which they have been cultured (Saxena and Schoonhoven, 1978). This is more proof of induced preferences for certain foods.

Dethier and Yost (1979) have clearly shown in their experiments that the caterpillars of the same tobacco moth, *Manduca sexta*, do not learn to reject a plant which previously made them sick. They have been fed on *Atropa belladonna*, insecticide-treated *Nerium oleander,* or a species of *Petunia.* The larvae, after recuperation, accept without hesitation the previous plants and do not learn from the experiment. This seems a totally different reaction from that shown by slugs or vertebrates such as rats (the Garcia effect). Food-aversion learning is well known among rodents, which makes them very difficult to control. Slugs seem more clever than insects, which is rather surprising. However, Dethier (1980) has performed another experiment with completely different results; he reared two species of woolly bear caterpil-lars, *Diacrisia virginica* and *Estigmene congrua,* on *Petunia hybrida,* a toxic solanaceous plant. The caterpillars prefer that plant to five of nine common food plants. Ingestion of petunias caused acute illness from which the cater-pillars recovered. After recovery, the larvae no longer preferred petunias, showing that they have the capacity for food-aversion learning.

Food-aversion learning seems to have been demonstrated in the polypha-gous grasshopper, *Schistocerca americana* (Bernays and Graham, 1988). The locust learned to avoid a food associated with an artificial aversive stimulus, nicotine tartrate.

All this does not confirm or refute the hypothesis of the induction of food habits among the invertebrates, at least among oligophagous species. More experiments are needed because the results sometimes have been contradictory.

As mentioned previously, there is a loss of specificity for several insects after being reared on an artificial diet (Schoonhoven, 1967). The sphingid, *Manduca sexta,* normally a solanaceous feeder after feeding on an artificial diet, will accept species of *Brassica, Taraxacum,* etc. A neutral medium brings a gustatory independence from the original host plant and gives one more proof of the possible extension of the food spectrum of an insect when secondary substances do not play their selective role.

It is also evident that an induced change in the food choice of the adult of a butterfly or a moth after a change of its larval food is not inherited and transmitted to the offspring. This would be Lamarckism; we are eons from proving that.

3.3   Plant resistance to insects

Many books have been written on the topic of plant resistance to insects, some very recent (Pandah and Khush, 1995). Molecular biology is a modern tool used to insert a resistant gene into a plant to make it toxic to the insect predator. Several attempts have been made to add the bioinsecticidal properties of *Bacillus thurigensis* to various plants, such as tomatoes, tobacco, potato, cotton, etc. Transgenic plants are still a pioneering subject to be explored in the future. Tissue culture and molecular biology also have opened new fields in host plant resistance studies. Protoplast fusion techniques can produce otherwise unobtainable hybrids and transfer genes for pest resistance.

From the early times of agriculture, during the post-glacial period, humans in the Middle East, the high plateaus of South America, Mexico, and Thailand began slowly to improve cereals and legumes, consciously or not, by selecting the biggest seeds (corn, wheat, barley, rice, sorghum, oats, and rye), the best tubers (potatoes), the best fruits, and the plants most resistant to diseases and pests. Continued selection from the dawn of agriculture to now has produced better seeds and better fruits or legumes and has improved the resistance to diseases or pests. Research to produce insect- and disease-resistant plants has long been a primary objective of agronomists and geneticists. Clones among poplars (*Populus* spp.) that are resistant to leaf beetles have been selected. Mangoes resistant to scale insects and rice resistant to pests and diseases have developed slowly, but they already have improved production. IRI in the Philippines and various institutions in Peru, Columbia, and Mexico are working on producing the best rice, potato, and corn plants by selecting and crossing the best varieties. Unfortunately, the best varieties are not always the resistant ones, and sometimes the wild species are more tolerant to pests and diseases than the cultivated ones.

Some antifeedants have been isolated from various plants, and there is a possibility of using them as biological insecticides. For instance, antifeeding diterpenes active against the caterpillar of *Spodoptera litura,* a serious pest of crops in Southeast Asia, have been isolated from 13 species of Verbenaceae (Hosozawa et al., 1974). They are common among species of *Clerodendron* and *Vitex,* but what is a strong repellent to many insects remains attractive to several others.

Certain plants, such as the gymnosperm *Ginkgo biloba* (the maidenhair tree, which originated in China), *Azadirachta indica* (the neem tree, a native of Burma), and many others, including *Tanacetum cinerariifolium* (pyrethrum), *Melia azedarach, Lonchocarpus nicou, Derris trifoliata* (with *Lonchocarpus,* a source of rotenone), etc., produce antifeedants or insecticides. Unfortunately, when they are imported, they are not immune to local pests and their use can be quickly discouraged by their lack of resistance and fragility to sunlight. Eucalyptus trees do not host any leaf miner attacks outside the Australo-Papuan area, but they are seriously eaten by termites in the Neotropical world. Hevea trees remain practically immune to pests in Africa, but not so in Asia.

Kursar and Coley (1992) state that many tropical plants produce entire canopies of red, white, yellow, or light green young leaves. Sometimes this is a means of attracting pollinators (species of *Poinsettia, Mussaenda, Rhynchospora*), but the authors believe that delayed greening involves a delay in the input of chlorophyll, nitrogen, and energy, thus causing herbivores to feed on these new colored leaves. This matter is not very clear, since, for instance, I have never seen the red young leaves of cecropia trees in South America being accepted by herbivores which normally feed on the tough green leaves (*Coelomera* spp. and other galerucines). Nichols-Orians (1992) suggests that the acceptability of young and mature leaves varies with the amount of light in the environment. Young leaves being devoid of epiphylls affects the acceptability of leaves to leafcutter ants. The leaves also contain less tannins. The chemistry of the leaves varies with the time of the day and light intensity, but if they are devoid of repellents and toxic compounds, they are always accepted by "their" herbivores.

Asclepiadaceae and Euphorbiaceae (latex plants) have in their country of origin many insect pests adapted to their toxins. Either these insects detoxify the chemicals (e.g., aphids) or they cut the leaf veins, stopping the flow of latex to their feeding area. Many techniques are used by leaf beetles, caterpillars, and other insects to stop or reduce the flow of sap to leaves. For instance, the volatile resin exuded from the stem bark of *Commiphora rostrata* (Burseraceae), a Kenyan plant, plays a role in plant defense (McDowell et al., 1988). It is well known that the larvae of several highly toxic flea beetles are used as arrow poison by natives; *Diamphidia* spp. feeding on *Commiphora* spp. is one example. Another flea beetle (*Polyclada* sp.) feeds on Anacardiaceae (Jolivet and Hawkeswood, 1995; Jolivet, 1967). These beetles do not stop the sap flow in the plants but instead sequester the toxins for their own protection. A number of other strategies are used by other beetles and caterpillars.

The squash beetle, *Epilachna borealis* (Coccinellidae), as do many other beetles, chew a circular trench through all but the lower epidermal leaf tissue of the plant. That way it isolates the area on which it is about to feed and avoids the cucurbitacine defense substances (Tallamy, 1985). Vein cutting blocks latex flow to intended feeding sites (Dussourd and Eisner, 1987). *Labidomera clivicollis*, a leaf beetle, cuts the veins of milkweed leaves (*Asclepias syriaca*), a behavior found in many Eumolpinae in Asia (e.g., *Platycorinus undatus*, *Corynodes* spp., etc.) and elsewhere. However, *Chrysochus auratus* in the U.S. does not seem to proceed this same way on *Asclepias* and *Apocynum*, but that remains to be verified.

Several insect larvae found feeding on species of *Eucalyptus*, such as *Perga dorsalis* (a sawfly), remove the remains of leaves on which they are feeding by chewing through the leaf petiole (Weinstein, 1990). This can be interpreted as removal of chemical cues used by parasitoids in host searching or as a means to prevent the transmission of induced defense materials from leaf to leaf. Experiments to prove this, however, have not been convincing.

Many caterpillars sabotage the defensive devices found on their host plants (Euphorbiaceae, Moraceae, or Asteraceae; see Compton, 1987, 1989). In Costa Rica, the sphingid *Erinnyis ello* feeds on *Cnidoscolus urens*, a very well-protected Euphorbiaceae. The plant possess urticating hairs and sticky latex. Before attempting to feed on a leaf, a caterpillar grazes the hairs from the petiole, then constricts the petiole, effectively stopping the latex flow into the leaf (Dillon et al., 1983). Several leaf beetles (cassids and chlamisids) also shave the leaves before eating them. Some Brazilian species of *Platyphora* do the same.

A species of *Bursera* in Mexico produces an array of monoterpenes distributed in a network of resin canals (Becerra and Vanable, 1990). When an entire leaf of *Bursera chlechtendalii* is plucked, it triggers a fine spray of terpenes (the squirt-gun response). A leaf beetle has found a way to block the terpene bath response by biting the midrib at the blade base.

Thus, the more the topic is investigated, the more we find new ways used by the specialists to stop the latex or resin flow. We can expect in the future many additional surprises. Trunks of the big cactiform *Euphorbia* spp. in East Africa (*E. tirucalli*, etc.) spray latex when cut. If you get it in your eyes, you must rinse them out immediately and be very careful; it can blind you. Few insects feed on the *Euphorbia* spp., but those that do either sequester the poisons or find a way to protect themselves from the flowing latex.

Toxic Asclepiadaceae, such as species of *Calotropis*, which are natives of Africa and Asia, have their parasites (danaid butterflies, eumolpine leaf beetles, and aphids), but they are few and the plants remain well protected. In Venezuela and Central America, where *Calotropis* spp. have been imported and became established, the bushes seem to be enemy free. The flowers in the Old World are visited by carpenter (*Xylocopa* spp.) bees, ants, and some butterflies, although this remains unobserved in America. It will take a very long time before these newly established plants will develop a new coevolutionary relationship. Even so, the species of *Xylocopa* are specially adapted for the fertilization of the flowers of this plant.

One can also believe that certain plants can repel insect attacks not only with toxins but also with hormone mimics such as the growth inhibitors found in coniferous trees and ferns.

Hybridization, which can extend, as we have seen before, monophagy and oligophagy, can also produce the converse, i.e., plant hybrids resistant to insect and disease attacks. Unfortunately, very often hybrids lose the edible features of the parents and can add the bad features of both partners, thus turning out to be failures. Also, research has been oriented towards clonal varieties by systematically testing resistance and selection of spontaneous or induced mutations.

In the fight against the Colorado potato beetle, 80 years of selection have not been able to find a really resistant variety of *Solanum tuberosum*, the potato plant. In fact, the beetle is now resistant to all insecticides, from organochlorines to pyrethroids and even to *Bacillus thurigensis*. However, even if the damages due to the beetles have been somewhat overestimated (particularly in Europe, perhaps), the development of a resistant potato plant would be welcome.

The species of *Solanum* have been classified into four sections according to their susceptibility to the Colorado potato beetle: (1) highly susceptible species; (2) moderately susceptible species; (3) moderately resistant species, and (4) resistant species. The cultivated potato belongs to the second category, as it is not the most favored plant by the beetle. It was not the original host plant in the Mexican area of its original home. It is also evident that the mechanical obstacles (hairs, sticky trichomes, lack of a smooth surface for attachment of the larvae) are often as repulsive to the insect pest as are the internal chemistry, feeding deterrents, or lack of phagostimulants and attractants. Other factors such as trichome density and the angle of insertion must also be considered (Pillemer and Tingey, 1976). It is certain that the glandular and nonglandular plant hairs (trichomes) are a factor in insect resistance of plants. Leafhoppers sometimes can be physically impaled on leaf hairs. Ants are frequently seen glued onto these hairs (species of *Lavoisiera*, *Salvia*, etc.). Leaf pubescence can be simply a deterrent to some beetles. Females of the criocerine (*Oulema melanopus*, for instance) are reluctant to lay eggs on densely setose leaves of wheat. These hairs cause a heavy mortality to first instar larvae (Schillinger and Gallon, 1968). Also, leaf pubescence plays an important role in resistance of selected clones of *Fragaria chiloensis* (the beach strawberry) to *Otiorhynchus* weevils (Doss et al., 1988). The same happens in many other plants (Schoener, 1987, 1988). As described by Kennedy (1986), some insects have evolved ways of "tiptoeing through the trichomes". Carnivorous plants, for instance, harbor an associated fauna immune to trapping setae in their watery traps (e.g., mosquito larvae).

As already mentioned, caterpillars (Hulley, 1968), leaf beetles, cassids, chlamisids, and several Chrysomelid larvae (Vasconcellos-Neto and Jolivet, 1994) sometimes attack the plant's defense mechanism by mowing trichomes before feeding. To Potter and Kimmerer (1988, 1989), the thick glabrous cuticle and tough leaf margin of *Ilex opaca* (holly) are more important than the spinose lateral teeth in deterring edge feeding caterpillars. Also, the high levels of saponines in young leaves provide a good protection against many herbivores. On the other hand, leaf fall in the water oak (*Quercus nigra*) seems to be the result of physical damage by leaf miners and not an induced defense of the plant (Stiling and Simberloff, 1989).

There is no relationship between plant attractiveness for egg laying and larval feeding. Several species of *Solanum* neglected by the Colorado beetle adult are sometimes favorable for larval development and vice versa. Sometimes these beetles accept a plant species which kills or sterilizes them.

The goal of the struggle against the Colorado potato beetle was to obtain, by a clever crossing of two species of *Solanum*, a hybrid that retains the food qualities of the potato tuber but also possesses the noxious properties of other species such as *Solanum demissum*. It is impossible to maintain a permanent colony of the Colorado potato beetle on *S. demissum* or on one of its hybrids with the tuber potato *S. tuberosum*. The reason seems complex. *S. demissum* does not allow the larvae to feed properly and also it produces a strong repellent, the alkaloid demissin. Most of the hybrids obtained between the potato and *S. demissum* and other American species show variable resistance to *Leptinotarsa decemlineata*, because the hybrid qualities sometimes are recessive. On the other hand, not all of these hybrids produce tubers similar to the ones produced by the cultivated potato, but this is only part of the story. Fifty years after initiating research on this subject, the problem is still far from being solved.

A classical example of natural resistance is the case of the African varieties of cassava, *Manihot utilissima*, originally a tropical American tree (Euphorbiaceae). The bitter varieties with high starch content are protected from predators by the presence of cyanogenetic glucosides.

Examples of plants resistant to insect pests include the classical resistance of the American grape vines to phylloxera, a dangerous pest of grapes in Europe; excellent results also have been obtained by the discovery of resistant alfalfa, barley, beans, maize, sorghum, rice, sugar cane, and corn. Of course, even though this resistance is to several insect pests, several fungi, and viruses, it is not to all pests. The discovery of resistant varieties of rice in the Philippines, India, and Thailand has improved the quality and the quantity of rice grown there, and in neighboring countries such as Vietnam. This has enabled them to emerge from a huge financial deficit to self sufficiency. They even export their surplus to China. Research is going on to find resistant cotton, pumpkins, onions, peanuts, sweet peppers, soybeans, and tobacco with some success, and it is hoped that producing transgenic plants in molecular biology laboratories will be accomplished one day.

Several timber trees are naturally resistant to termites. Kogan (1977) distinguished in plants an ecological and genetic resistance which is essentially morphological or chemical. The influence of environmental factors on the expression of that resistance should not be neglected.

Recently (Sagers, 1992), it was found, through experiments done in Panama, that caterpillars receive a variety of benefits from living within a leaf roll. These were believed to protect the insects from the hazards of weather, from parasitoids, and from predators; however, the main effect seems be a reduction of the chemical and physical defenses of the leaf (tannins) without altering the nutritional status of the host tissues. In the case of the Neotropical shrub, *Psychotria horizontalis*, shading by the roll of leaves decreased the toughness and leaf tannin concentration by 15%. Herbivores prefer the leaves that have grown within a leaf roll over flat and normal leaves.

chapter four

Relationships between insects and fungi

Numerous fungi, either parasitic or saprophytic, infect insects, not to mention the harmless external Laboulbeniales and internal Eccrinales also associated with them (see Figure 4.1). I consider the fungi as plants in this book, used in the vernacular, even if they are today often treated as a separate kingdom of living organisms. True, the fungi are devoid of chlorophyll, making them distinct from the green plants, and some, such as the Myxomycetes, are amoeboid and move slowly on old logs and across soil. I know that studies made during the past 35 to 40 years show, according to their ribosomal RNA, that fungi are closer genetically to animals than to plants. As a result of these and other researches, living things are now classified into five kingdoms, with fungi being one of them. The five kingdoms are Plantae, Fungi, Animalia, Protista, and Monera (the bacteria and cyanobacteria). Other more recent classifications are pushing the fungi even closer to being considered animals. Daily changes in the higher classification of organisms lead me to keep the old eighteenth-century classification. I am not the only one to keep the fungi within the division of plants, albeit as aberrant plants, if you wish. Kew kept them in this territory along with cyanobacteria and lichens (which, of course, are not a group by themselves, being composed of two distinct organisms). It is interesting to note that a designation in Latin is still required by the International Codes for fungi, as it is for the green plants.

Fungi and insects have coevolved, and the fungus gardens of termites and ants are a living example of a very old and successful association for both of these insects (see Figures 4.1 and 4.2).

4.1   Parasitic and saprophytic fungi

Many kinds of fungi are capable of growing on insects and, as a result of this parasitism, killing insects. Most of them are really facultative parasites, as they are not confined to living substrates. These fungi kill the insects even though their parasitism is not obligate. Some species have yet to be cultivated

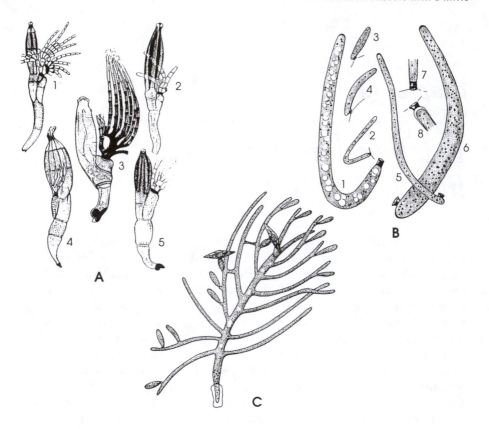

Figure 4.1   Fungi and insects. **(A)** Some *Laboulbenia* spp. parasitic on Chrysomelidae. (1) *L. bertiae* Balazuc on *Ceralces natalensis* Bates from Congo; (2) *L. grayi* Balazuc on *Pseudadorium oberthuri* Bachyne from Madagascar; (3) *L. dorstii* Balazuc on *Phenrica aemula* Weise from Amazonia; (4) *L. motasii* Balazuc on *Podagrica goudoti* Harold from Madagascar; (5) *L. skirgielloae* on *Podagrica goudoti* Harold from Madagascar. All are epizoic fungi. (Adapted from Balazuc, 1988.) **(B)** Trichomycetes, with their suckers, are parasitic inside the gut of insects. (1 to 2) *Paramoebidium procloeoni* Manier parasitic inside Ephemeroptera nymphs, in France; (3 to 8) *Paramoebidium chattoni* Dubosc, Leper, and Tuzet are parasitic inside *Simulium* spp. larvae, in France. (Adapted from Manier, 1950.) **(C)** Trichomycetes. *Stipella vigilans* Leger and Gauthier, which are parasitic inside the gut of *Simulium ornatum* Meigen, in France; thallus and zygospores. (Adapted from Manier, 1950.) **(D)** *Termitomyces fuliginosus*, mycelium hyphae, around the fungus mass built by the termite *Acanthotermes acanthothorax*, in West Africa. (Adapted from Heim, 1977.) **(E)** Leafcutter ant carrying a piece of leaf back to the nest. (Adapted from Dumpert, 1988.) **(F)** Bromatia bodies of a leafcutter ant fungus (*Atta* spp.). (Adapted from Wheeler, 1925.) **(G)** *Microtermes subhyalinus*, in Africa. Fragment of a fungus mass, enlarged. Mt = mycotete; V = mycelum velvet; M = mycostele. (Adapted from Grassé, 1982). **(H)** *Xylaria termitum* (Ascomycetes) growing over a termite mound abandoned by its inhabitants; *Microtermes* sp. from Madagascar. (Adapted from Heim, 1977.) **(I)** *Termitomyces microcarpus* carpophore over a termite hill and fungus debris. (Adapted from Grassé, 1982).

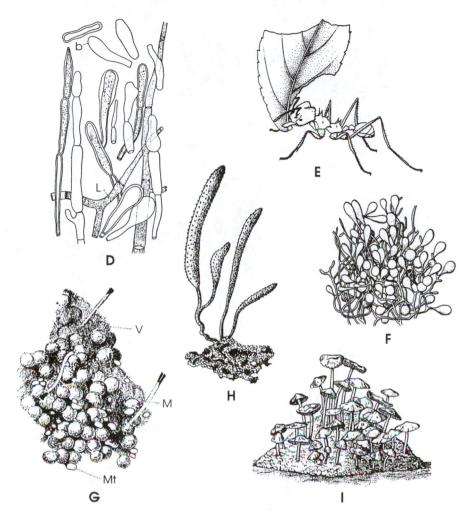

Figure 4.1   (continued)

apart from insects, which may indicate true parasitism. For example, some species of Entomophtoreae are found only on insects, but *Aspergillus flavus* and *Mucor hiemalis* have been found elsewhere. Many kinds of fungi are normally saprophytes. Those on insects infect the host by penetrating through the outer integument or through the gut walls. The continuity of the outer integument is interrupted in several ways: by stigmata, the openings of the tracheae, through pores of glands, the point of insertion of scales, and the digestive tract, which is lined by chitin only on the fore and hind gut. The midgut is then less protected from infection. Fungi species of the genera *Aspergillus, Beauveria, Cordyceps, Metarrhizium, Entomophthora,* and others infect their victims through the integument usually when the epicuticular wax is,

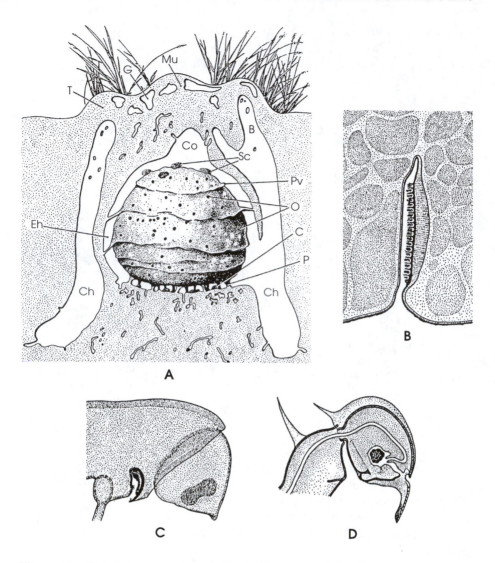

Figure 4.2 Fungi, insects, and evolution. **(A)** Young nest of *Bellicositermes natalensis*. In the middle is the habitation with the fungus; C = stones, B and Ch = aeration tunnels, Co = dome, Eh = free space, G = galleries, Mu = wall, P = pillars, Pv = folds, O = opening, Sc = sawdust, T = turrets. (Adapted from Grassé, 1949.) **(B)** Mycangia or receptacles used to carry symbiotic fungi; fungus-filled organs of a *Sirex* spp. larva. (Adapted from Batra and Batra, 1967.) **(C)** A scolytid beetle in longitudinal section carrying fungus spores in pockets situated at the base of the anterior legs. (Adapted from Batra and Batra, 1967.) **(D)** Section through the head of a fungus-eating ant showing the fungus pocket. (Adapted from Batra and Batra, 1967). **(E)** Evolution within the plant kingdom showing that when the gymnosperms declined the angiosperms became abundant, probably through the acquisition of chemical defenses by the angiosperms. (Adapted from Ehrlich and Raven, 1964.)

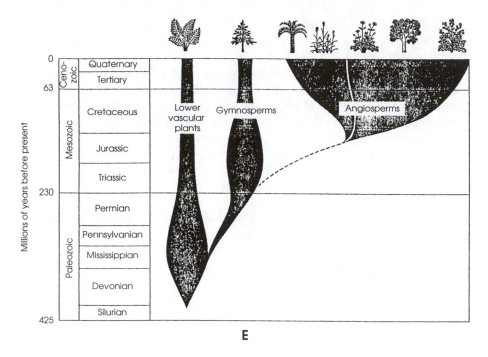

Figure 4.2   (continued)

for one reason or another, removed or abraded. The fungus *Empusa* sp. is also common in flies and certain other insects. It is also true that special microclimatic conditions are necessary for some infections to take place. For instance, the heavy parasitism of orthopteran locusts by fungi in Thailand always happens at the end of the rains when microclimatic conditions are favorable. These fungi develop in the blood, fatty tissue, or muscle fibers of the insect, according to the nature of the infection and of the host. Infections by *M. anisopliae* of the Colorado potato beetle is an example. In this case, it is only after the death of the host that the muscles, nervous system, and fat body are completely consumed by the fungus.

Symptoms of a fungal disease infection include behavioral symptoms, such as ascending plant stems and foliage, which seem well suited to efficient dispersal of the fungal spores. Ants parasitized by *Cordyceps* sp. are often seen in the tropics at the top of a grass stem, dead and covered by the fungus masses. Ants parasitized by trematodes worms (flukes) behave in the same manner. It this case, the grass and the ants both may be swallowed by sheep, which are in turn, infected with the parasite. Restlessness is a common symptom for these insects when parasitized, and this causes them to wander away before climbing and dying. Paralysis is also a common symptom, along with color changes in the infected insect (to creamy, black, brown, etc.).

When the dying insect climbs up a grass stem, it dies attached to its support by its paralyzed legs or tied by the fungus hyphae growing from its body (e.g., *Cordyceps* spp.).

Some parasitic fungi produce chitinases used to dissolve chitin. These fungi show different degrees of host specificity, while the host species also presents varying degrees of resistance to these infections. The rarity of mycoses in cerambycid beetles is believed to be related to the isolation of the larval stages within the wood. Only certain families of beetles are susceptible to infection by species of Laboulbeniales, and the distribution of these parasitic infections among the possible beetle hosts sometimes seems illogical.

The use of fungus diseases for pest control has never been entirely successful, even in a very humid climate. A recent review of the fungal pathogens of the leaf beetles has been done by Humber (1996), and the effectiveness of these pathogens are thoroughly discussed.

4.2   Ectoparasitic fungi: Laboulbeniales

Several monographs on the Laboulbeniales have been published. In addition to the classical works of Thaxter (1901–1926) and the monographs of Tavares (1985) and Balazuc (1988, 1989), a recent review of the associations of these fungi with the Chrysomelidae has been made by (Weir and Beakes, 1996). The Trichomycetes (in the hind gut of insects) and the Laboulbeniales (on the exoskeleton), both Ascomycetes, are rarely pathogenic to their host. Laboulbeniales are merely commensals and are fixed externally on various parts of the insect or arthropod body. They produce ascospores sexually, and transmission is often achieved during copulation. They parasitize Diplopoda, mites, cockroaches, beetles, earwigs, flies, bugs, ants, termites, Mallophaga, Orthoptera, and Thysanoptera. Coleoptera are the most common hosts (30% of the beetle families). Their specificity is relative but sometimes narrowly restricted. Nutrition of the fungus is obtained by means of a very small haustorium which penetrates the host integument. Some species produce rhizoids that penetrate into the body cavity. The host species of the Trichomycetes belong to families that have wet biotypes, live in forest litter, or are truly aquatic.

No Laboulbeniale has ever been observed on a weevil, while the Chrysomelidae, similar in life habit, are widely parasitized. This fungus lives on beetles, including leaf beetles, on the posterior part of the elytra, the pronotum, the underside of the body, the antennae, and the legs.

On several species of New Guinean weevils and other beetles (colydid for example), is found a miniature forest growing on the elytra and sometimes on the pronotum. This forest is composed of several different kinds of plants, among them mosses, liverworts, fern prothaslli, algae, cyanobacteria, and various fungi, producing what Gressitt et al. (1965) refer to as epizoic symbiosis. The fungal spores germinate into the depressions with mucus of the integument of the beetles' backs. There are no Laboulbeniales among that miniature suspended garden, a phenomenon that will be discussed in more details later on in Chapter 8. It is not uncommon to find weevils in the mountains of Borneo or Panama (*Geobyrsa* spp.) with green or blue algae on their elytra. All those beetles live in higher altitudes in the moss forest. However, as of now, the New Guinean epizoic symbiosis is unique.

The big difference between the Laboulbeniales parasitism and epizoic symbiosis by cryptogams is evident considering that the Laboulbeniales thales fungi penetrate the chitin of the back with an haustorium, but the epizoic cryptogram grow entirely externally on the surface, even if fed by the mucus of the elytral glands.

4.3   Endoparasitic fungi: Trichomycetes or Eccrinids, and endosymbionts

The Trichomycetes or Eccrinids are fungal associates of arthropods living attached in the hind gut. It is probably a polyphyletic grouping. The validity of some species described this century has been questioned by Lichtwardt (1986) for various reasons, mostly of only taxonomic concern.

Trichomycetes are found among myriapods, crustacean, and Insecta (Collembola, Ephemeroptera larvae, Odonata larvae, Plecoptera larvae, Trichoptera larvae, Coleoptera: Hydrophilidae, Passalidae, Scarabaeidae, and Diptera larvae). They probably exist elsewhere, but they are never been found as parasites of weevils or leaf beetles. They may exist in the larvae of aquatic species such as the Donaciinae, but this has not been reported. No depigmented cyanobacteria such as the species of *Oscillospira* in rodents has been found in the gut of insects or even in species of Myriapoda.

Trichomycetes are either unbranched or branched fungal thalli attached to the lining of the hind gut of arthropods, living there and obtaining their nutrients from the ingested material of the host. However, as an exception, one trichomycete, *Amoebidium parasiticum*, grows externally on the exoskeleton of various aquatic arthropods (Lichtwardt, 1986).

Most of the arthropods harboring Trichomycetes live in fresh- or saltwater habitats, but millipede and beetle hosts live in humid terrestrial habitats, and these hosts are scavengers or detritivores. "Clean" beetles, such as the Chrysomelidae, Bruchidae, and Cerambycidae, do not harbor any of these fungi. Attachment in various parts of the gut varies with the parasite species, but generally they live on the chitinous lining of the hind gut.

Trichomycetes are a "lower" or more primitive fungi. Their host specificity is relative, depending on the species. With rare exceptions, some are innocuous commensals.

Here must be mentioned that some yeast-like internal fungi, as is the case for some bacteria, are symbiotic and provide insect hosts with vitamins: these are symbionts. They are situated in various parts of the insect alimentary tract in specialized cells or in the Malpighian tubules. These symbionts help the insects to digest cellulose (wood feeders) or blood (bloodsuckers), but they are erratically distributed among the plant- or blood-feeding insects. Among the leaf beetles, for instance, only species of some subfamilies have mycetocytes or bacteriocytes for this purpose. Roughly speaking, the endo-symbionts provide various nutrients that cannot be synthesized by the insect or are in short supply in the diet (Becker, 1994; Crowson, 1981). Transmission of the symbionts is done by egg smearing during egg laying.

4.4   Special insect-fungi relationships; Attini, termites, and others; insects feeding on or in fungi

An excellent revision of insect-fungi relationships was provided by Batra and Batra (1967), and several symposia have been held on this topic. Grassé (1982) summarized the relationship of termites, while many publications exist about the relationships of ants (Attini), particularly the work of Cherrett (1972, 1986, 1988, 1989). (See Figure 4.2.)

Many insects, including many flies and beetles, feed on fungi. They sometimes disseminate fungal spores. Among the Diptera, Mycetophilidae larvae are well-known mycophagous specialists. They burrow in their food and live in tunnels in a slimy secretion. Many beetles (Staphylinidae, cisids, mycetophagids, histerids, endomychids, some scarabeids) feed on fungi or on molds. Quite a few insects have a special taste for fungi and some cave insects have no other food.

Paulian (1988) cited mycophagous species among the beetles that attack inferior fungi (Fungi Imperfecti) and those mycetophages that feed on superior fungi (Ascomycetes and Basidiomycetes). Six families of beetles are feeders on the Myxomycetae, those gelatinous and slowly moving fungi. Other beetles are fly predators within the decaying fungi. Many Diptera feed on fungi, including underground species such as *Tuber melanosporum* and others. Some species of the genus *Suillia* (Heleomyzidae) seem to be partial to truffles; for instance, the females of *Suillia tuberiperda* lay their eggs in the ground above the truffles. The young larvae dig into the soil and go directly to the fungus. Looking for flies is a good way to find truffles, although not as efficient as using a dog or a pig as the hunter.

There are leaf beetles (Alticinae) feeding on algae, lichens, mosses, and various cellular cryptogams, but none before Morelet's observations (Jolivet, 1996) had been found feeding on fungi. *Crepidodera* (=*Chalcoides*) *aurea* (Geoffroy) feeds willingly on the conidial spores ("lawns") of the fungus *Venturia* spp. (Ascomycetes) found on the leaves of its normal host plant, *Populus tremula* (Salicaceae). This behavior contributes to the dissemination of a disease on the aspen trees, because the beetle fly very well from host to host. When the leaf beetle, *Gastrophysa viridula*, feeds on *Rumex* spp. leaves

infested by *Uromyces rumicis* (a rust fungus belonging to the Uredinales), it also grazes part of the fungus (Hatcher et al., 1994).

Cecidomyid or Itonidid larvae are coprophagous, saproxylophagous, or zoophagous, or they live around the mosses. Others live in fungi growing on coniferous resin or in decaying vegetable matter. Some are gallicolous, either as producers or as inquilines. These galls contain relatively host-specific fungi which grow in the plants' tissue and form a thick layer inside the wall of the gall. Apparently, the female fly introduces spores of fungus when she lays her eggs. It seems that the larvae feed mostly on the digestive remains of the internal tissue of the gall induced by the fungus and not on the fungus itself. Several coccids of the genus *Aspidiotus* (scale insects) live inside the fungal tissue of species of the genus *Septobasidium* which is found on trees. The scale insect feeds on the sap of the trees and is protected from predators even if the fungus feeds on some individual scale insect in the colony. It is an association which is advantageous to both. The fungus does not exist without the insect (Batra and Batra, 1977).

4.4.1   Hymenoptera; ants

Several ants, namely *Philidris (=Iridomyrmex) myrmecodiae,* and related species probably feed on the various mushrooms growing on the inside walls of their myrmecophilous plants (species of *Myrmecodia, Hydnophytum,* and related species), epiphytic Rubiaceae of southeastern Asia. This will be discussed in detail in Chapter 7. Several other examples have been found among the myrmecophytes.

Species of wasps of the genus *Sirex* (Hymenoptera: Symphyta, or sawflies) and related genera are associated inside their wooden galleries with several fungal species of the genera *Stereum* and *Daedalea.* Probably the fungus, transmitted by the female with her ovipositor, helps the larvae to digest wood. The wasp's ovipositor is extremely strong and used for boring and drilling deep into wood. The future female larvae carry an inoculum of fungus in a waxy cavity behind the thorax, between the first and second abdominal segments.

Some tropical American ants (Attini) have developed the most advanced stage of fungus gardening. Some species are found in the U.S., but most species inhabit the Neotropics. Fungus-growing ants, as well as termites, occupy a similar ecological niche, but the fungus-growing ants live exclusively in the Western hemisphere, while the fungus-growing termites are confined to the Old World tropics.

Attini ants cut and collect fresh leaves on which to grow their fungus gardens. They feed on the sap of the leaves, but they also rear their larvae on the fungus hyphae and staphylae, the swollen hyphal tips produced by the symbiotic fungi. However, the hyphae provide a small source of food for the workers of *Atta sexdens*, and the fungus may provide up to 9% of the respiratory energy requirements, the remainder presumably being provided by

plant sap (Bass and Cherrett, 1995). The workers prune their fungus and stimulate its growth. In either case, the result is fungal growth (Bass and Cherrett, 1994).

The species of several genera of leafcutter ants, such as *Atta*, *Cyphomyrmex*, *Acromyrmex*, use a sort of manure (compost) made chiefly of cut leaves, but they also use petals and other plant material on which they plant the fungus (for more details, see Jolivet, 1986). The fungus is eaten by and provides food mostly for the larvae. The 400 known fungus-growing ant species belong to only one tribe: the Attini.

It was Belt (1874) in Nicaragua who discovered ants using the leaves for making compost and growing fungus. Before then everyone thought that the ants were feeding on the leaves. The ants carry the leaves over their backs (hence, parasol ants). They cut so many leaves and there are so many workers, they can rapidly defoliate an entire tree. However, they are very conservative, and in a native forest they quickly change their collection site. The leaves regrow rapidly. It is only in cultivated gardens and in desert regions that the ants can become pests. Geoffroy St. Hilaire once wrote, "Either Brazil must get rid of the leafcutter ants or the leafcutter ants will get rid of Brazil." Of course, that is a bit exaggerated, and the ant problem is an ecological problem of which man is, as usual, directly responsible.

After a very large atta ant queen is fertilized, she drops her wings and looks for a place to nest in the ground. She digs a cavity into the soil and remains in it. She has carried into the soil the fungus material collected on her body and stored in an infrabuccal pocket. This pocket takes the form of a spheroidal sac and opens into the mouth cavity by means of a short narrow canal. The material from the pocket is planted in the soil cavity, and starts the fungus garden used to feed the future colony. Among this material are hyphae or threads from the previous meal of the future queen, and these hyphae will act as seeds to start the garden. For many more days, the queen will fertilize these developing fungi with her own excreta, utilizing even some of her own crushed eggs both as manure and for food. She constantly cultivates the garden with her antennae and legs. Some eggs which are laid become workers, fed in the larval stage with the eggs of the queen. Then, when they are adults, they feed on the fungus they are growing by themselves after having collected new leaves. They also drink at the sap exuded from the leaves they cut. Excreta of caterpillars, plant debris, and anthers and petals from flowers are also collected by some ants to improve the quality of the compost or replace the leaves. They very much like the leaves of the roses and the flowers of *Hibiscus* spp., which are invaders in the ants' native country but now naturalized inhabitants. Generally small colonies collect a great variety of compost materials; however, in the laboratories of Europe, atta ants generally are fed on bramble leaves.

As noted previously, the leaves are cut and carried by the workers to their nest in the ground, where the leaves are mixed with saliva and covered with excreta. On this compost the ants place some mycelium as seed for their

subterranian garden. After a certain time, the fungal mass develops and resembles a sponge, rather similar in appearance to a termite garden. In the laboratory, with the aid of plastic boxes and transparent tubes, it is easy to see the building of a fungus garden and the separation of the task between the castes. Recently Bass and Cherrett (1996) found that leafcutting ants (*Atta sexdens*) prune their fungus to increase and manage their productivity. That is extra proof that the ants do their gardening very carefully.

Atta ant species are not the only leafcutters. Species of *Megachile* solitary bees, commonally called leafcutter bees, construct cells in rotten wood or in soil made out of cut fragments of green leaves. *Chalicodoma* bee species also build nests of chewed paper and leaves. These bees have dentate mandibles fit for cutting leaves. Similarly cut leaves have been found as fossils. However, species of *Megachile* bees do not cultivate fungi. These cut leaves serve only as protective plugs within the underground nests.

Under the constant care of the worker ants, the garden takes shape. The fungi produce special terminal bulbs on hyphae. These are variously called bromatia, gongylidia, staphylae, or kohlrabies. Only ant gardens produce this nonreproductive element as food for the ants. Bromatia do not appear spontaneously in culture media. Ant larvae are placed in these gardens by the workers. It is here that all the members of the colony feed, but most of the food is used by the larvae.

The nests may become so large that sometimes a bulldozer is necessary to break them up before proposed construction can begin. Underground atta ant nests can reach a considerable size; the craters of the nest and the naked soil around it resemble the outside of the nests of other ants, particularly the harvester ants. A nest can cover 100 m^2 and possess hundreds of entryways. Atta ant workers move in columns onto trees and bushes to cut the leaves, and their enormous numbers can be very destructive when complete defoliation of a tree occurs. As protection from the voracious atta ants, some trees, especially species of *Acacia*, have either developed toxic substances in their tissue or have myrmecophilous associations with species of *Pseudomyrmex* ants. In the South American tropics, a species of *Cecropia* (the parasol trees), an outgrowth of secondary forests, is very rarely the prey of atta ants because the trees are rather well protected by species of *Azteca* ants inhabiting the tree.

A large nest may contain a hundred gardens, some of which are suspended from roots in the ceiling of the chamber. The temperature and humidity of the nest are very important, and the ants keep these constant by opening and closing nest entries. Moist, dew-covered leaves are not collected; therefore, much of the harvesting is done at night when the leaves are dry.

As it is frequently the case among fungus eaters, almost every species of ant cultivates its own species of fungus. It is difficult to make them change fungus species artificially. These fungi are specialized for the ant species. Thus, the ants probably have attained the most evolved state of gardening among insects. When ants do not eat bromatia, they lick them for their secretions. These bromatia look much the same as the white spherules (mycotetes) of termite gardens. There the nymphs feed exclusively on them.

As long as the ants cultivate their garden, fungus does not sporulate or produce fruiting bodies. However, the species of several genera of those cryptogams have been cultivated or found in open air after the withdrawal of ants from their former nest. The fungus does sporulate when unattended by ants and by these spores can be identified. How are the bromatia produced? Probably by constant pruning of the mycelia and by the antibiotic effect of ant saliva and excreta. These antibiotics not only influence the growth of the symbiotic fungi, but also prevent germination of other species of cryptogams.

An atta ant colony is polymorphic. Minute workers, the "minimes", almost exclusively cultivate the fungus and feed the larvae; the "medium" workers collect leaves. A third form, large ants, are the last to develop and have mainly a defensive function similar to the soldiers of other species.

The colonies of fungus ants vary considerably in their organization. Some species make only one garden (*Acromyrmex* spp.); others have only one type of worker, and in still others even the queen collects leaves. Some fungus gardens are covered by an envelope, a part of the fungus itself.

Epiphyll growth organisms (i.e., liverworts, crustose lichens, bacteria, algae, and fungi) on the surface of leaves repel the fungus-growing leafcutter ant, *Atta cephalotes*, from harvesting leaves on tropical plants (Muller and Wolf-Muller, 1991). So what is often avoided by the phytophagous mites living in acarodomatia sometimes can also be protection against species of Attini.

4.4.2   Coleoptera

The wood-boring beetles (Scolytidae, Platypodidae, and Lymexylidae), most of which are called "ambrosia beetles", penetrate wood where their larvae are nourished by velvety fungi which develop on the walls of the burrows. They cannot survive without their specialized fungi (Ascomycetes and Fungi Imperfecti), which are slowly being identified. The beetles carry fungal spores in different ways (crop of the female beetle or in excreta, the brushes of setae found on the head of the female, thoracic cavities, prothoracic pores), but in the species of many genera the beetles carry the fungal spores in pockets (the mycangia) situated at the base of the anterior legs of adults. The spores rapidly germinate inside tunnels carved into usually freshly cut wood that is rich in sap. The beetles, both larvae and adults, feed on fungi rather than on the wood itself. The characteristic smell of yeast attracts adults, as may be seen when females and sometimes males fly around in the evening near empty beer cans. The galleries dug by these insects contain adults of both sexes, larvae, and eggs. The galleries have a characteristic shape and color according to species. They are stained a dark brown or black by the mycelium owing to the action of the fungus on the wood. It seems also that there is a close association between a beetle species and "its" fungus. It is the formation of these yeasts which are prevalent inside the tunnels, and these are affected and maintained by the insect (see Figure 4.2C).

4.4.3   Termites

As stated above, the leafcutter ants, the species of *Atta*, cultivate their fungi only in the New World, while the fungus-growing termites grow theirs only in the Old World tropics. The fungus-cultivating termites all belong to a single family, the Macrotermitidae. Species of *Macrotermes* and *Odontotermes* make their nests inside high, sturdy mounds. Other species, such as those of *Microtermes*, are subterranean and the nest is only slightly visible. The biology of those termites is well known. It has been thoroughly studied by Grassé (1982) and his school (see Figure 4.2).

Organic matter, mostly collected from wood chewed and partly digested by the workers, makes up the termite fungus gardens. It consists of pellets agglutinated together and covered by the velvety-appearing fungus. Sometimes there is a single garden which can weigh up to 30 kg, but generally there are numerous gardens dispersed throughout the nest. The inner cavity of the nest is lined with a mixture of saliva and dust.

The cavity is often ventilated by a complex system of vertical conducts reaching the outside. The gardens have a spongy appearance and are grayish or brownish in color. They are kept moderately humid. Some gardens are firm, while others are soft, brittle, and fragile. On the surface of the gardens are seen small, white, shiny, pearl-shaped spherules, the "mycotetes", bundles of threads emanating from the surface of the mycelium strands of the garden. These whitish spherules are collected by workers and sometimes eaten or generally offered to certain nymphs. It does not seem that the king, queen, and young nymphs eat the fungus spherules, but the alate and reproductive forms take this fungal material with them in order to seed new fungus gardens.

In contrast to most termites of the temperate and tropical areas which harbor intestine protozoa for digesting cellulose, no protozoa seem to be living inside the intestine of fungus-growing termites. From this observation, it has been deduced that cultivated fungi predigest cellulose and provide vitamins. These termites cannot live on wood only and continuously graze on the garden material. They use their own fecal matter to fertilize the fungus, which is then eaten and thus recycled.

Species of several genera of fungi grow inside termite nests, but species of *Termitomyces* are the most common and cannot be found elsewhere. The garden is not a pure mycelium culture, since species of *Xylaria* (Ascomycetes) and *Termitomyces* (Basidiomycetes) grow together. However, the velvet covering is exclusively due to the second fungus, *Termitomyces* spp. Species of *Xylaria* can be also found outside the termite nests.

Fruiting bodies do not generally appear when the termites are inhabiting the nest. However, under certain favorable conditions, *Termitomyces* species reproduce sexually. The spherules grow, reproduce, and produce carpophores with strong caps. When they mature and grow out of the nest, they disseminate these spores. Heim (1977) has identified these species of *Termitomyces* as being closely related to *Agaricus*.

In Africa and India, termites sometimes remove the upper layers of their gardens and spread them outside the nest during the rainy season. Soon mushrooms, with their sporophores (caps) appear. In this manner the spores of various nests are mixed together. The termites then collect the mycelia produced from this mixture of spores. It seems that in doing this the termites bring about a cross-fertilization of their fungi.

4.5   Conclusions

As we can see that the relationships of insects with fungi are diverse, and many more associations remain to be discovered (see Figure 4.2). We have seen that fungi can kill insects, bringing under some microclimatic conditions serious epizooties. Also, the fungi associated with species of *Attini* ants and those associated with Macrotermitidae are necessary for their survival. The ants actually grow, prune, fertilize, and render their fungi aseptic. This is a perfect example of successful agricultural practice. The termites are also apt gardeners. They fertilize and build the substratum but seem less careful about maintaining the garden and do not really clean it. Nevertheless, it is successful.

There is a close association between ants and termites and their respective fungi. The fungi produce food for their gardeners from their substratum, and the insects eat either fungus bodies or the substratum predigested by the fungus, or both. Although fungus is prevented from producing reproductive structures while in the host's nest, it receives from its insect host a safe ecological niche and often a means of dispersal.

The situation is reversed for certain coccids. In this instance, the insects are sheltered by species of the genus *Septobasidium,* a fungus. The scale insects feed on plant sap and provide food to the fungus from their own bodies. Several people see a similar association (Joel, 1988) between carnivorous plants and insects. The plants feed the insects with an abundance of nectar and extrafloral nectar at the expense of several of them being sacrificed on the altar of mutualism. It is a point of view that we are not obliged to share.

Fungi (and algae), as discussed further on, can grow on the elytra of certain weevils in New Guinea and elsewhere. As for the Laboulbeniales that we see fixed on the body of several arthropods, these fungi are generally harmless to the carrier. The transmission of these species of Laboulbeniales from one insect to another is still unclear (De Kesel, 1995). It seems that infection is either from the contaminated soil or by direct contact between individuals of the same host species. The spores ooze out of the perithecium and form adherent threadlike structures which favor direct transmission. These spores have a short life span, which is, perhaps, the reason for the rarity of the infections. The gut-frequenting Trichomycetes are also pure commensals similar to the protozoan gregarines and do not do any harm to the host.

In the case of the fungus gardens, there is a more or less perfect symbiotic relationship. First, a dissemination system for the fungus is provided. This may be simple, or sometimes, elaborate. Second, it is certain that due to antibiotics secreted by these insects or by the fungus itself, only fungi useful to the host are allowed to grow. All others are totally inhibited. Such substances are also partially responsible for the fungus metamorphosis, which without its gardeners, does not produce bromatia or spherules, but rather returns to its archaic form of reproductive body. Some new findings show otherwise; the saliva and excreta of ants and termites are presumed partially responsible for these changes. In the case of the woodboring beetles, however, there are no spherules, or bromatia, but the cryptogams of the gardens are replaced by yeasts in these galleries, a more assimilative medium. It also seems that temperature conditions, quantity of carbon dioxide in the air, acidity, and so on produce similar modifications of the fungus and that this phenomenon is not only produced by insects themselves. Laboratory trials are not absolutely conclusive, however, and much remains to be found and to be proven in this fascinating field. No one has produced spherules or bromatia in the laboratory, nor has it been possible to produced an artificial gall. Insect "savoir faire" is necessary.

Fungi, such as bacteria, can be endosymbiotic in special organs, the mycetomes, in certain plants, or in blood-feeding insects. These help in digesting the food and providing the necessary vitamins. Fungi can be "pollinated" by a fly, and many kinds of mushrooms provide even sweet and glutinous spores. The ergot on cereals has many ways of attracting flies and provides them with a reward. An heterothallic Ascomycete, *Epichloe typhina*, is fertilized by a fly, *Phorbia phrenione*, which is an important transporter of their conidia. Not only does the ergot *Claviceps purpurea* provide sugary rewards, but many fungi are also associated with particular flies, and we know that most species of Mycetophilidae are associated with mushrooms in their larval stage.

chapter five

Physiology of food selection

During the past hundred years, at least since the Verschaffelt article (1910), a great deal of research has concentrated on the physiology and chemistry of food selection. Many papers and many books were published, some concerned more with the genetics of food selection (Futuyma and Slatkin, 1983), and others more on physiology and chemistry (Bernays and Chapman, 1994). Many symposia have been held, some in Holland (Wageningen) under de Wilde and Schoonhoven and their school, others in Hungary with Jermy, and even in France in Pau. So many are the papers on the topic, and so fast the advances in chemistry and physiology, that it is very difficult to keep up to date.

Electrophysiology has helped in the investigation of the reactions of insects to odors and flavors through the use of electroantennograms and electropalpograms. A working compensator (in Wageningen) also helped to interpret correctly the reactions to colors and odors of the Colorado beetle in a wind tunnel. A lot of information has been similarly gathered throughout the world and published during the last 15 years. Our knowledge has improved considerably since Fabre's day (starting in 1886), but as Schoonhoven (1990) said, "The problem of botanical instinct remains in essence unsolved." The more we know about the chemical structure of repellents and attractants, the more problems are raised.

It is not my goal here to review the physiology of the phytophagous insects. Others have done it better than I. I just want, in a simple way, to review part of the problem.

According to Fraenkel (1959), the basic food requirements of all insects seem to be very similar and much the same as those of higher animals. Such requirements include the amino acids of the vitamin-B group, a sterol, and the basic necessary minerals. The "primary" substances which occur in all living cells are not responsible for the choice of the food plant by an insect. On the contrary, these choices are made by the selection of "secondary" plant substances, such as glucosides, saponines, tannins, alkaloids, essential oils, and organic acids, nearly 30,000 of which have been described for plants. These are the gustatory and olfactory substances (see Figure 10.1B).

To Fraenkel (1969), and many others, food specificity of an insect is based on the presence or absence of these odd compounds found in plants where they serve either as repellents or attractants. These also, eventually, induce feeding. A plant may be attractive and at the same time poisonous, but through selection such poison events are eliminated. The phenomenon occurs rarely in nature and has been observed mostly in the laboratory. Essential oils can be the reason for a common attractiveness of certain species of *Papilio* to Apiaceae (Umbelliferae) and Rutaceae. In this case there are no antifeedants or repellent.

Host plant selection is also related to the chemoreceptors and the olfactory cells. More and more in recent years, the localization of these cells has been explained. There are olfactive receptors on the antennae of caterpillars and gustatory receptors in various positions in all phytophagous insects. The interactions of the chemicals with the receptors bring information to the central nervous system of insects, and the result is food plant recognition (see Figure 5.1C).

To Jermy (1984), host plant selection is governed primarily by chemoreception. The emergence of new trophic profiles results from the changes in the insect chemosensory systems. Adaptation to the nutritional quality of the new host plant, and detoxification of its eventual poisons must be a secondary process.

As far as we can judge, insects have the five main senses the same as we do, and sometimes these are more highly developed. Many of these insects, such as bees, see ultraviolet light and, as a result, their sight is rather different than ours, but they do see colors, although with some restrictions. They hear well, and their gustatory and olfactory senses are quite well developed. While human beings distinguish four gustative sensations — bitter, salty, sweet and acid — the insects seem to distinguish more varieties in the range of alkaloids and other chemicals. The sense of touch is not negligible and particularly comes into play to distinguish between smooth or hairy surfaces. They have special organs and cells adapted for the perception of the environmental world, especially and most importantly their food environment.

Olfactory receptors are often localized on the antennae and those for taste on the palpi. Often, especially among flies, beetles, and butterflies, taste is also perceived via ventral pads on the tarsi. In Lepidoptera, the tarsal receptors inform the insect about the exact place of food on the substrate after their smell and sight have brought them from a distance. The antennae of lepidopterous larvae have tactile and olfactory organs (sensilla trichodea and basiconica).

So, food selection depends on sight, smell, and taste. The sense of touch sometimes plays a role equal to the other senses and also can be a reason for the local selection of certain parts of the host plant: smooth vs. hairy, rough vs. tender. Secondary substances seem responsible for selection or repulsion but sometimes can be responsible for insect poisoning or at least slowing down the development of the insect. When an insect successfully selects a new host plant, it must be pre-adapted to it; otherwise, it is the result of a long selection trial period.

Here we will distinguish types of trophic selection as physical or chemical. Certainly this may be considered to be old-fashioned entomology, but these observations are the result of good sense and they are not in contradiction with the most recent discoveries.

5.1   Mechanical or physical selection

Selection involves primarily the sense of touch, but an obstacle can be purely mechanical, and thorns or spines can prevent an insect from feeding on a given plant. Holly leaves (*Ilex aquifolium*, Aquifoliaceae) are so smooth that many insect larvae cannot grasp them. Those leaves are also surrounded by thorns. If you cut into their dentate edges, several caterpillars, such as *Lasiocampa quercus*, will feed voraciously on what otherwise is an inaccessible food.

A monophagous (or oligophagous) insect can refuse a plant that has the same chemicals and secondary substances that exist in its normal host plant due to the toughness of the leaves or the presence of hairs too dense to enable normal chewing. In this case, no deterrents are present; the plant get its own protection from its mechanical unpalability. Insects that eat species of *Galium* (Rubiaceae) such as the European species of *Timarcha* and the galerucine *Sermylassa halensis* reject *Galium aparine*, the goosegrass, a rambler with reflexed hooks on stems, and they also reject *Rubia peregrina*. Both plants, when the leaves mature, are too hard and too hairy to be eaten by these beetles. However, they feed readily in the spring on young and tender shoots of the same plants. In the Balearic islands, *Timarcha balearica* accepts *Rubia peregrina*, its mouth parts probably being well adapted to chewing that rough food. The American species of *Timarcha* on the West coast of the U.S. feeds on tender leaves of strawberry and bramble bushes. They always bite the tender tips and the new leaves.

Starting to feed on a leaf is sometimes a problem for an insect. Gregarious beetles are more successful in opening a leaf, and all the larvae immediately follow the one that has opened a hole at the first feeding site.

Aphids sometimes fight for a feeding site. Foster (1996) reports that in India the horned aphid *Astegopteryx minuta*, a very small insect, fights with its brothers on bamboo leaves. The winner is the larger aphid, and it inserts its stylets into the precise site on the leaf from which the defender has withdrawn its stylets. Fighting between males for females or fighting against potential predators is well known among insects, but fighting for food rarely has been mentioned. Using a pre-existing feeding site, gained either by fighting or by using an abandoned site, confers significant time benefits (see Figure 5.1A).

It is a common observation that, in Europe, Asia, or the U.S., the species of *Chrysomela* choose the tender leaves of the poplar trees, mostly the new

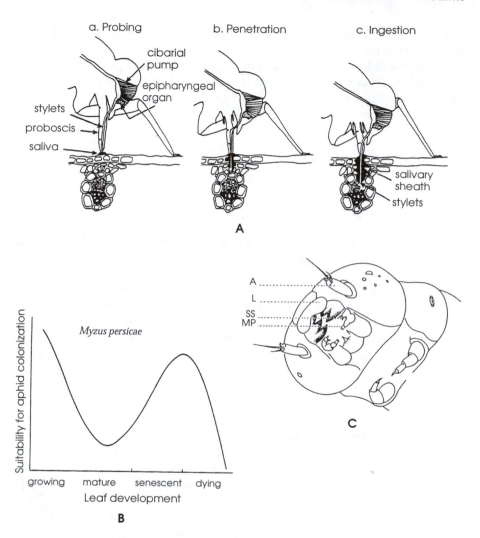

Figure 5.1 Feeding, sensilla, electropalpograms, and pollination by ants. **(A)** Diagram of the probing, penetration, and ingestion phases of host selection by an aphid. (Adapted from Dixon, 1985.) **(B)** Relative suitability for colonization by *Myzus persicae* of leaves of the successive stages of development. (Adapted from Kennedy et al., 1950.) **(C)** Ventral view of caterpillar head; A = antenna, L = labrum, SS = maxillary sensilla styloconica, MP = maxillary palpi. (Adapted from Schoonhoven, 1972b.) **(D)** Galea of a caterpillar bearing two sensilla styloconica. **(E)** Impulse frequencies in maxillary sensilla styloconica of *Manduca* sp. sexes (Lepidoptera: Sphingidae) when tested with various chemicals. NaCl = salt, SAL = salicin, SOL = solanin, and INO = inositol. (Adapted from Schoonhoven, 1969b.) **(F)** Ant pollination and floral structure. *Myrmecia urens*, the male ant pollinator of *Leporella fimbriata*, attempting to copulate with the labellum of the orchid; bar = 5 mm. (Adapted from Peakall, 1994.)

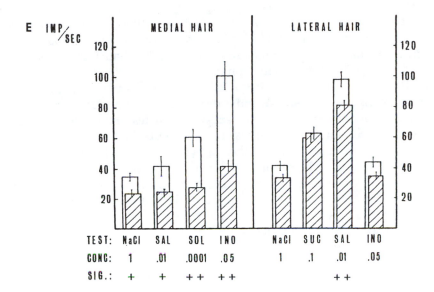

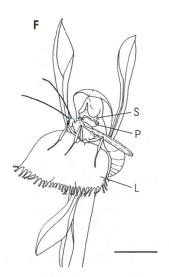

Figure 5.1   (continued)

growth or the new shoots. *Paropsis* spp. in Australia and New Guinea also often prefer the tender leaves of eucalyptus, but sometimes full-grown trees can also be defoliated.

The little cabbage diamond moth, *Plutella xylostella*, lays eggs more easily on a rough, unequal surface than on a smooth one. Mustard oil (sinigrin) is one of the selective factors for moths, butterflies, and flea beetles feeding on cruciferous plants, but factors other than secondary compounds are also involved.

The bean jassid, *Empoasca fabae*, is disturbed by the density of hooked hairs (trichomes) of certain varieties of beans. *Aphis craccivora* reacts in exactly the same manner. Epidermal hairs, hooks, and stellate hairs are sometimes responsible for host-specific aphid response to the pubescence in selecting their host plants (Dixon, 1985). The exudate of some glandular hairs of *Solanum berthaultii* traps aphids and is responsible for the plant's resistance to the aphids. Resins in coniferous trees and silicates or oxalates of other plants provide certain plants with resistance to pests. Not all, though, since some flea beetles normally feed on the silicaceous leaves of some species of *Equisetum*. Secreting trichomes in many plants glue the ants to the leaf or stem, thus preventing them from reaching their flower nectar. In Brazil, in the Serra do Cipo and various mountains of the Serra do Diamantina, trapped ants very often are seen on the stem of the species of *Gaylussacia* (Ericaceae), *Barbacenia*, and *Vellozia* (Velloziaceae). They were trapped in sticky hairs while looking for extrafloral nectar.

Many leaf beetles (Eumolpinae) have succeeded in feeding on latex-bearing plants: Asclepiadaceae, Apocynaceae, Euphorbiaceae, and others. On *Calotropis gigantea* are found various species of *Platycorinus*, cutting the veins and feeding with impunity on the leaves. Latex-bearing plants are a real obstacle for most insects, even if several aphids have succeeded in detoxifying the poison and others have found a way to stop the latex flow. Species of Eumolpinae are specially attracted by asclepia plants and feed openly on some species devoid of latex, such as the species of *Euryope* in Senegal, which are abundant on the various species of *Leptadenia*, including *L. hastata*. This species is eaten by caterpillars which normally avoid latex plants.

It is evident that the chemical composition of a leaf varies with the age of the plant, shady vs. sunny conditions, tropical vs. not, time of the day, and so on. No two plants are chemically identical. Young leaves of many tropical plants are red when young, as in the species of *Cecropia* trees. The flovonoids cause the red color of the new leaves of tropical plants, and it is well known that petaloid leaves surrounding the flowers can be yellow, white, red, or orange in color. The galerucine beetles, *Coelomera* spp., generally avoid these red leaves, but several insects will feed on them. The chemical composition of these young leaves is different from the older green ones (Kursar and Coley, 1992), and the lack of chlorophyll reduces the energy cost to the plant.

Younger leaves are richer in protein and older leaves in carbohydrates. *Manduca sexta*, a sphingid used in laboratory tests, feed on different aged leaves, probably to obtain a balance of food of different composition (Bernays and Chapman, 1994).

Two insects that feed on the same plant — one during the day, the other during the night — may be considered to have, in a sense, two ecologically different diets. Sugar concentration in the leaf, for instance, varies considerably, depending on the time of day and, hence, the amount of light reaching the leaf.

Caterpillars of *Acronicta abscondita* (Noctuidae) prefer stunted individual plants of species of *Rumex* and *Euphorbia*. Such cases are not isolated. Caterpillars of species of the genus *Acidalia* (*Scopula*) (Geometridae) eat only wilted leaves, and some *Pterophorus* spp. (Pterophoridae) larvae cut certain veins of the leaf, causing the tissue they are going to eat to wilt. Males only of the flea beetle, *Gabonia gabriela*, are strongly attracted by the wilted leaves of species of *Heliotropium* (Boraginaceae). This could be the expression of a kind of "lek behavior" of the males, but in the fall many other leaf beetles show a strong attraction towards wilted leaves of trees (Boppre and Scherer, 1981). This probably has a link with the sequestration of pyrridizin alkaloids. However, the large polyphagous caterpillar of the African arctiid *Amsacta meloneyi*, which eats practically everything that does not bear latex while on its short migratory course, prefers fresh leaves to wilted ones. The caterpillar is in a hurry to feed during the short rainy season of northern Senegal. Some Lepidoptera caterpillars and some leaf beetles are so well equipped with strong chewing mouthparts that they can eat hard tissues and the most thorny or hairy leaves. These are, however, exceptions.

Chemical factors in the composition of the leaf also are important once the mechanical obstacles have been passed successfully. Insects living in water are less touchy about their selections of plants, as long as these are always soft and easy to eat. Some Brazilian species of Dynastinae, such as *Chalepides fuliginosus*, feed on the bottom of some mountain rock pools on Cyperaceae (*Eleocharis* spp.) but probably also on various herbs, mosses, and algae. Species of *Donacia* are well-known polyphagous eaters in both the larval and the adult stages.

I will come back later to the problem of damaged trees and the "messages" they are supposed to send to healthy ones. It is well known that herbivore damage increases the level of alkaloids and modifies the chemistry of the whole plant. That also has an influence on the attacks by polyphagous insects.

5.2   Selection of feeding sites

We have seen previously that some aphids fight over a feeding site on a bamboo leaf. The feeding site is sometimes limited to a specific part of the leaf by leaf miners, but aphids also display area selection when sucking plant juices (see Figure 5.1B). Leaf-mining larvae choose, depending on the species, the cells of the two epidermis layers of a leaf — either the palisade tissue or the palisade and parenchyma tissues. The selection often depends on the differences between the chemical compositions of the different layers. If the egg is laid on the wrong side of the leaf, the larva often refuses to feed and perishes. Attacks by the leaf miners can also modify the metabolism of the leaf and, consequently, the nature of the products ingested. The most striking example is that change caused by the caterpillar of the nepticulid moth, *Stigmella hemargyrella* (= *Nepticula basalella*). The leaves mined by these caterpillars remain green in the feeding area a long time after the departure of the pest and even after the leaves fall to the ground.

Among the sapsucking insects (leafhoppers, aphids, reduviids, penta-tomids, and others) and root eaters (eumolpines, elaterids, alticines, etc.), the choice of the plant tissue is very limited. That choice seems to be based on the sense of taste or perhaps, more precisely, on the ability to distinguish various pH gradients. Very little is known about the way insects with ovipositors select their host's tissue (see Figure 5.1A,B).

In gallicolous species, for instance, this selection is very specific. It is possible that chemoreceptors provide the insect with all required information. When laying eggs over a vegetal support, the female leaf beetle disposes of sensory receptors at the top of its ovipositor.

Selection is less restrictive for caterpillars, but it does exist. Large caterpillars eat all the lateral part of the leaf to the stalk, but the small-sized ones often start at the apex of the leaf where the veins are less developed. Other caterpillars skeletonize the leaf, leaving the veins, while others eat only the upper or the lower surface. This kind of selection is more mechanical than chemical and is in accordance with the feeding abilities of the insect.

The examples of food selection could be multiplied. The method of selection is eminently variable in the same insect and on different plants. Some local selection is more mechanical (as in *Ilex* spp.), while other selection seems to be more chemical or a combination of both of these.

When insects are adapted to a part of the plant (root, stems, flowers, fruits, seeds, or leaves), then by knowing that their selection site seems constant, we can experimentally alter the selection of the part the insect will use. However, the insect generally makes this change very reluctantly. The Colorado potato beetle will feed on potato tubers and stems in addition to the leaves which are its normal diet, but only if forced to do so. In Russia, they eventually feed on rotten tubers in the fall in order to survive. In that way they swallow the nematodes abundant in the plant tissue.

Some insects, such as the caterpillars of the polyphagous moth *Heliothis* spp. sometimes become stem borers — for instance, in the internodes of the *Cecropia* tree. It will accept purely synthetic food only if forced, because the normal attractants, such as the green color, the smell, and the tissue texture are absent. Very often in the so-called synthetic food, leaf extracts are necessary at least as appetizers.

The caterpillars of *Eupoecilia (=Phalonia) ambiguella* (Cochylidae), which lives during the spring on grape flowers and other plants, and during its second generation on the fruit of the same plants, shows some adaptation to the host's development. The moth *Larentia incultaria* (Geometridae) does the same with leaves and capsules of *Primula* spp.

Among the gallicolous cynipids, such as *Biorrhiza pallida*, there are alternating generations on an oak (*Quercus* sp.), one generation of the cynipid being on the aerial parts of the tree, and the other only on the roots. Root-feeder larvae are generally much less selective than the adults. Sometimes, the larvae of some flea beetles, pests of cultivated plants, are selective on some plants at the adult stage and only feed on the roots of the same plants at the larval stage. However, the matter is completely different with the eumolpines, which can be selective as adults and polyphagous, especially on Poaceae, as larvae. That is true for most of the species of Eumolpinae.

As mentioned by Dethier (1954), the chemical composition of the green plants varies with the time of the day, the season, the stage of development of the plant, the tissues fed upon, the climate of their environment, the nature of the soil, and also the insecticides and fertilizers used to cultivate the plant. There were even some agronomists in France suggesting that insecticides increased the fitness of the weeds.

The chemistry of the plant is not uniform and permanent for all individuals of the species and its varieties. Consequently, this explains in part at least the local or seasonal behavior of insects and also the ability of certain plants to resist these insect species. Schoonhoven (1972) writes that many insects can evaluate the nutritive composition of their food and choose accordingly. These insects, even if theoretically polyphagous, are in fact selective. The apparent random selection of trees by leafcutter ants in their native forest is in reality directed by subtle differences in toughness or palpability of the leaves and the result is conservation. At time, ants are more clever than humans, who indiscriminately burn or cut trees without any planning.

5.3   Visual selection

It is a well-known fact that butterflies, during oviposition (or the egg-laying process), cease being interested in brightly colored flowers and instead become attracted only to the green leaves of their larval host plant. During this period, they rest on green or blue-green surfaces and actively move their anterior legs, probably to get olfactive and gustative perceptions of the substratum via the tarsal receptors. Species of *Pieris* butterflies oviposite readily on paper wetted with sinigrin solution if the paper is yellow, green, blue, or white, but not if it is violet, red, or black (Traynier, 1986). This experiment could mean the preeminence of view over smell and taste. The females are incapable of associating the last group of colors with sinigrin.

Some moths such as *Manduca sexta,* a sphingid, cannot find their host plant if their antennae are cut; that action, therefore, totally prevents oviposition. This detection is first visual, but then olfactory. A third factor, taste, may intervene. The females of the cabbage butterfly *Pieris brassicae* and the diamond moth (cruciferous feeders exclusively) will lay eggs on any surface, provided the anterior legs stay in contact with a cabbage leaf or even some mustard oil on filter paper. However, we have seen previously that the color of the support is also important. Some aphids are attracted by the yellow color of the aged leaves. Among the saturniid and several other nocturnal species of Lepidoptera, the perception of the host plant (oak, birch, etc.) seems to be chiefly olfactive and localized on the basiconic sensillae of the female antenna. The rhododendrons fertilized by moths in New Guinea have white flowers and are scented, while the red-flowering ones are odorless and pollinated during the day by birds. It seems that several leaf beetles, such as *Agelastica alni* on alder in Europe and *Chrysomela populi* on poplars and willows are partly attracted by the green color of the leaves and the smell of the tree.

It is wrong, however, to say that all caterpillars make no choice for themselves and that the adult chooses the plant for them. Certain butterflies and moths deposit their eggs at random when flying near the host plant without alighting on it. In this case, the caterpillars must make a food choice. It is the same with several leaf beetles such as *Iscadida* sp. (Chrysomelinae) which feed in South Africa on Vitaceae *(Rhoicissus* spp.). The larvae have to climb onto the host immediately after hatching. In other cases, such as species of leaf beetles of the genus *Syneta* in the Northern Hemisphere (the weevil *Otiorhynchus* sp.), the beetles drop their eggs on the ground, but the larvae dig into the ground and feed on roots. For all root feeders, only a positive geotropism is needed to dig into the soil, and often some attractants from the roots themselves is necessary. Those attractants have been isolated from some roots.

The sight of a caterpillar or a larva is much reduced by the ocellae compared to that of the adult, but it seems that they can sometimes perceive the color and the shape of things. Many caterpillars, such as those of *Heliothis armigera,* a major pest, are evidently attracted by green, and an observer crossing a tomato or cotton field in the tropics can find them on their clothing if the cloth is green. Visual selection, therefore, even if not always dominant, plays a real role in food choice.

Aphids appear to respond mainly to visual cues showing a marked orientation to yellow (Dixon, 1985). That is why water pans painted yellow are so attractive to them. According to Dixon, color is a good indicator of the nutritive status of the plant. Young and senescent foliage is generally yellow in temperate climates, while young shoots tend to be red in the tropics. This is, however, a poor indicator of a plant's taxonomic status. Aphids are more attracted by color than by odor, even if there is also an odor-induced orientation in aphids. Statistically, aphids lay eggs mostly on yellow-green or orange surfaces, provided the colors are not saturated. The colors chosen are often in relation to the specific host characteristics. Aphids do not follow generalized behavior patterns, so one species can be very different from another.

Many adult chrysomelid beetles, often brightly colored themselves, distinguish colors rather well. Species of *Chrysolina* can separate violet from green, blue from yellow, and orange from violet. Certain leaf beetles are even more selective. They are capable of distinguishing various tones of pale green, but not dark green (Schlegtendal, 1934). These insects probably use this ability in their host plant choice and in mating. Only adult butterflies see red, but this trait is useful for finding nectar-bearing flowers. Mating between beetles is generally done without mistakes, but some leaf beetles do end up with a mate of a different species or genus. Sometimes, wrong mating between *Chrysolina violacea* and *Timarcha tenebricosa* can be explained by the similarity of color between the two species. *Timarcha* species feed on *Galium* species and *C. violacea* on *Glechoma hederacea* (Lamiaceae). We cannot invoke the influence of the same host plant. Sometimes, the interspecific matings between leaf beetles can be explained by their common hosts (species of *Salix, Mentha*, etc.), but sometimes there is no satisfactory explanation.

Recent experiments with the watercress leaf beetle *Phaedon cochleariae* have shown that at least one phytophagous species has a specific reaction to yellow and green wavelengths, which is clearly associated with host plant selection. However, color itself is not the only criterion for selection, and the role of tint and light intensity must not be overlooked.

The conclusions drawn from the research on the watercress leaf beetle have shown clearly that this insect is attracted to its cruciferous host plant by the yellow and green colors, but also by the odor of mustard oil. Mustard oil odor provokes feeding, and this act continues to be induced by the plant glucosides. Leaf texture also affects its feeding, but experiments show that sight and smell, at least at the beginning, are the chief elements of food selection.

It is well known that some caterpillars, with their very limited sight, distinguish three main colors: blue, green, and white. Color plays a significant role in the choice of flowers by their adults, especially for honey-gathering activities, as well as during the choice of the host plant where it will deposit its egg. For instance, species of *Macroglossum*, a day-flying hawk moth, primarily visit blue flowers when young, and later, when nearly ready to oviposite, they prefer yellow-green plants and flowers. Yellow-green is the color of *Galium* spp., a Rubiaceous plant, which feeds the hawk moth caterpillars.

These reactions are not isolated. Even night-flying moths are sensitive to color, but in a different way. Flowers during the night must add some kind of fragrance to be visited. They are often white. It seems that brightly colored flowers are invisible to nocturnal visitors, but the moths can distinguish pale colors. Moths are generally blind to red, but day-flying moths such as species of Uraniidae, Arctiidae, and others are sensitive to colors the same as butterflies. Bees are also able to detect various colors (both violet and ultraviolet), but in a different part of the color spectrum, a phenomenon that will not be discussed in detail here.

Phytophagous beetles, such as Chrysomelidae and some Coccinellidae, as well as carnivorous ladybird beetles, will follow a dark band painted on white paper and turn when the band changes its direction. This tendency seems to be linked with the habit of climbing up stems and branches to reach leaves (ladybird beetles visit leaves to find their prey). Caterpillars show a crude perception of shapes through their ocelli. This perception, even though imperfect, helps them to find their host plant when they have dropped to the ground, but it is also associated with chemical senses which distinguish the shady forms they are perceiving. From past research, it appears that the nature of the stem is not enough to provoke the caterpillar into climbing; it is also necessary that at the top there be a screen which casts shade on the ground. Of course, because of its small ocelli (a very primitive sight organ), the caterpillar cannot discern the shape of leaves as the adult butterflies do.

Wandering solitary caterpillars, such as those of *Amsacta meloneyi*, avoid latex-bearing plants but devour everything else in their way. They must have a more acute sight mechanism than sedentary caterpillars. The larvae of certain beetles also are attracted by vertical ferns.

A good example of the influence of the shapes of the host is cited by Kogan (1977) for the American moth, *Autographa jota* (Noctuidae) which generally does not lay its eggs on the dandelion, *Taraxacum officinale*, the favorite food of its larva. It seem that this is due to the low habitus of the plant. The adult prefers the soybean, with a tall and vertical habitus, which means that the larvae must migrate to the proper host. Species of Neotropical butterflies of the genus *Heliconius* are very interesting because their caterpillars feed on leaves of several specific passionflower plants, *Passiflora* spp. (Passifloraceae). These butterflies have excellent vision and associate the shape of the leaf with food, even though these leaves vary enormously among the 350 species of the genus *Passiflora*. Strangely enough, the butterfly is associated at the adult stage with the pollen of several Cucurbitaceae (species of *Anguria* and *Gurania*) but not exclusively.

We know that butterflies are reluctant to oviposite on leaves already with eggs or caterpillars. Certain species of passiflora vines have modified extrafloral nectaries which copy the eggs of *Heliconius* spp. This is thus used as a repellent against adult butterflies. The female of the pipevine swallowtail *Battus philenor* (Feeny et al., 1983), in southeastern Texas, searchs for its larval food plants, *Aristolochia reticulata* and *A. serpentaria*, and forms search flights looking for the shapes of the host leaves. Finally, the female alights on the required *Aristolochia* species recognized by contact chemoreception. No doubt the birdwing butterflies of southeastern Asia (species of *Ornithoptera* and allies) do the same, as they are quite specific among many different species of *Aristolochia*. *Battus philenor* visually discriminate between the different species of *Aristolochia* — ones with broader leaves and others with narrower leaves (Papaj, 1986).

It is rather rare to find similar host plant specificity among adult butterflies, and *Heliconius* spp. are artists in that field (Gilbert, 1975). It was Gilbert who discovered the repulsive effect of fake eggs or egg mimics (modified extrafloral nectaries) among the species of *Heliconius* butterflies. Among most of the butterflies, the presence of real eggs on a host plant has a strong dissuasive effect on the female.

With the visual senses are associated various tactile receptors: photo-, geo-, anemo-, and hydrotropisms are involved. These factors, together with the chemical factors, play important roles in the choice of food and oviposition. However, the main determinants in food selection are the chemical factors, which are further described in the following paragraphs.

5.4   Chemical selection (odor and taste)

When aphids land on plants, they are in the middle of the vegetation and they respond to olfactory and visual cues (Dixon, 1985). After settling, an aphid recognizes the structure and chemistry of the plant surface, the nature of its internal tissues, and the quality of the phloem sap. Sensilla of the tip of the proboscis are tactile receptors and chemoreceptors. Probing, penetration, and ingestion of sap obey very complex mechanisms in which perception of the leaf surface, olfaction, and gustation play a role. The chemical selection of the insect is so guided by odor and taste. The very delicate nerves of the chemoreceptors guide the insects toward its favorite feeding spots and help them to distinguish the internal chemistry of the plant. Aphid antennae have many sensilla, and with their mouthparts they contribute to plant and site selection.

Nicotine, nornicotine, and anabasine produced by the glandular hairs of *Nicotiana* (tobacco) and the petunia are a physical defense of the plant against aphids. The presence or lack of sinigrin is a selective element for *Brevicoryne brassicae*. Secondary plant substances act as attractants or deterrents for the aphids as they do for other insects.

Aphids have (as do many Hemiptera, such as pyrochroid bugs) basiconic sensillae, chemoreceptors that are responsible for food choice. Simply stated, for a plant to be acceptable food, it is necessary that: (1) the plant produce an attractive smell, (2) its taste is agreeable, and (3) it does not contain any chemical that is repulsive, inhibiting, sterilizing, or poisonous to larval or pupal development. Of course, there must not be present any antifeedant or repellent to prohibit acceptability by the insect, and the other critera must be efficient for a long-term selection. That is the result of evolution, but when new plants are introduced, an insect pest, such as the Colorado potato beetle, may accept a plant that can sterilize or poison it. For example, a species of *Solanum* may be attractive to the beetle, but feeding on this plant slows down growth or eventually simply poisons it. This is similar to cows in New

Guinea being killed by eating cycad flowers in an environment not known by the animal in its native Europe.

Chemical selection is less evident among aquatic insects where taste and smell are combined, as all those stimuli are mixed together in a water solution. In any event, aquatic plants have protective chemicals which act as repellents from which the soft tissues of herbivores offer little protection. Ostrofsky and Zettler (1986) examined 15 species of aquatic plants for alkaloids. They found two to nine alkaloids per plant, which are potential deterrents for herbivory. Alkaloid arrays showed little similarity among plant species, even among species of the same genus, such as species of *Potamogeton*. From this we must conclude that aquatic insects also have ways to detect attractants and deterrents among these aquatic plants.

Thorsteinson (1960) summarized the feeding processes of an insect as follows: $F = -I - D + Esn(Ep)$, which means that an optimal feeding response (F) exists when the substratum is devoid of inhibitors (I) and repellents (D) and contains the necessary chemotactic stimuli (E) in order to encourage feeding. Stimuli (E) are sapid products (Esn) and sometimes special kinds of "pungent" stimuli (Ep) which are facultative. The secondary substances belong more to Ep, however. This formula requires taste, but does not mention "smell" (except if it can be put among the Ep stimuli), which is often equally important for attraction and the continuation of feeding.

The Cape Verde islands are situated in the Atlantic 500 km from Dakar, Senegal. There food is scarce, as it normally rains only one day a year. Even this is better than places along the Peruvian coast, where it rains once every hundred years on the average. In Cape Verde, xerophytic plants such as the native species of *Dracaena*, which are rich in resins, and several imported species of *Nicotiana* are extremely toxic to mammals and insects, but these are sometimes eaten when nothing else is available. The locust *Pyrgomorpha cognata* sometimes has been observed to feed on *Datura innoxia*, an imported plant growing almost spontaneously on the islands. The plant is known for the high level of poisonous alkaloids within its leaves. Dead or dying locusts and their abnormal egg laying have been seen on the leaves caused by the plant's reproductive inhibitors. Cattle have been fed on species of *Dracaena* at the peak of the dry season and have become sick and died. The same occurs when cattle fed on Cycadales in Australia, New Guinea, and South Africa. Their injury seems irreversible.

Similarly, the widely polyphagous caterpillars of *Heliothis armigera* (Noctuidae), which normally feed on tomatoes, cotton, beans, maize, and other crops, accept in Cape Verde cabbage leaves that are repellent but not toxic, because of the scarcity of available food.

The birdwing butterfly *Troides priamus richmondensis* normally feeds on *Aristolochia praevenesa* but also lays eggs on *A. elegans*, an introduced plant in Australia, on which the larvae perish. Oviposition is stimulated by the presence of specific attractants, and because the imported plant and the butterfly did not coevolve, selection is nonexistent (Straatman, 1962). In the same way, the

Colorado potato beetle oviposite on *Solanum nigrum* and *Datura nutiloides*; however, the larvae do not develop on these foreign plants. Selection can be very strict. For instance, *Lema trivittata* (Say) feeds primarily on *Datura* spp., and *L. trilineata* feeds mostly on species of *Solanum* and *Physalis* (White and Day, 1979). It is well known that only some biotypes of the Colorado potato beetle feed on *Lycopersicum esculentum*. Detoxification of the poisons from some species of plants is under genetic control of the beetle, and that is probably the way new species arise from biotypes of an insect species.

Rejection of foreign plants as food due to some differences in the secondary substances or alkaloids is rather common among insects. It is what Hering (1950) used to call xenophoby. Attraction, followed by death, is due to the presence of the attractants and the lack of repellents in the new plant. Obviously, the insect is not adapted to these chemicals and cannot detoxify the poisons. Rarely, exotic species are accepted and not destructive to the insects (xenophily). Separate evolutionary processes produce separate results. A good example of xenophily is the successful acceptance of *Nothofagus* spp. by some moths introduced into England. General acceptance or rejection is more or less in relation with the distance between the place of introduction and the country of origin of the plant. The greater the distance, the better are the chances of rejection (Hering, 1950).

Generally, butterfly adults discriminate their host plants better than their caterpillars do. However, the silk moth *Bombyx mori*, which has lost through domestication not only its flight abilities but also some selective sense, lays eggs anywhere. Even the caterpillars of several strains of these moths accept many unrelated plants as food.

Sinigrin constitutes a chemical barrier to the U.S. swallowtail butterfly *Papilio polyxenes* (Erickson and Feeny, 1974). Normally restricted to Umbelliferae (Apiaceae), the caterpillars, at concentrations of 0.1% of sinigrin, all die. *P. polyxenes* is bounded by chemical barriers of both a toxicological and behavioral nature. Sinigrin and other mustard oil glucosides serve a defensive function against certain insects.

Insects rarely eat aquatic plants, more often eating subaquatic ones. Aquatic plants (for example, species of Nymphaeaceae, Salviniaceae, Azollaceae, Araceae, etc.) have their specific herbivores, but not many. It seems that the variety of deterrents (more than 50 alkaloids) reduces the chances of specialization (Ostrofsky and Zettler, 1986). This is one of the reasons why it is so difficult to find specific herbivores to use to control introduced aquatic weeds.

Various species within a natural family of plants generally show a remarkable similarity in chemical composition — e.g., Apiaceae with essential oils; Solanaceae with alkaloids, such as solanine or tomatine; Brassicaceae with myrosin or mustard oil; Polygonaceae with oxalates; Rubiaceae with various glucosides; brooms with sparteine; St. John's wort with hypericin, an attractant for some moths and beetles; beans and other legumes (Fabaceae) with cyanogenetic glucosides; Rutaceae with aromatic ethereal oils, etc. These substances are mostly selected by taste, but some of them have a fragrant odor (e.g., Rutaceae, Apiaceae, Lamiaceae, Verbenaceae, Oleaceae such as *Jasminum* spp., etc.). Volatile substances such as terpenes found on trees also attract scolytids and other ambrosia beetles. A-terpineol is a specific attractive substance of bark beetles.

Insects that feed on Brassicaceae (Cruciferae) — Lepidoptera, Coleoptera, and aphids, for instance — are attracted by mustard oils and their glucosides. Mustard oils also attract small braconid wasps (Hymenoptera) that are parasitic on many phytophagous insects. With plants one can think of a triangle: the vegetal, the phytophagous insect, and the parasitoids which kill the pest and protect the plant, a topic that has been covered extensively in the literature.

These substances (mustard oils) also stimulate spore germination of the parasitic fungi *Plasmodiophora* spp. on cruciferous plants It must be noted, however, that the oviposition preferences of adult insects seem to be narrower than the choice capacities of the larvae. Since 1910, researchers have undertaken many experiments to explain this difference. The earliest classical experiments were made by Verchaffeldt in 1910, using the white cabbage butterflies *Pieris brassicae* and *Pieris rapae*, which feed exclusively on Brassicaceae, Capparidaceae, Tropaeolaceae, Salvadoraceae, and Resedaceae. All these plants have special glucosides (isothiocyanates) in common. Verschaffelt succeeded in getting the caterpillars to feed on plant leaves of different families such as Leguminosae by impregnating the leaves with diluted mustard oil. Also, it must be noted that even if the pierids feed mostly on Brassicaceae and chemically related families, in certain areas of the world they choose the species of the genus *Cassia*, a Caesalpinaceae. The species of the genera *Catopsilla* and *Callidryas*, for instance, are among them. This is an excellent example of combined oligophagy, perhaps due to a mutation. Similar experiments have been done with the host-specific leaf beetle, *Gastrophysa viridula*, the dock species, and with other phytophagous insects, always with the same results.

The role of both alkaloids and chlorophyll in food selection cannot be neglected. Experiments using synthetic media made from these materials have shown that even if a polyphagous species (an *Heliothis*, for instance) willingly accepts a relatively simple medium, oligophagous species are much more difficult to feed and require a more carefully compounded synthetic medium. "Green" odor and green leaf volatiles are often important in the final choice of a caterpillar. Those volatile components are rather complex and are the result of the metabolism of lipids. Green or other colors are often important in the selection of the feeding site in certain insects.

By inserting an electrode into each end of an insect's antennae, one can record antennograms. Insects feeding directly on leaves (e.g., grasshoppers, locusts, aphids, beetles, caterpillars) respond to green leaf volatiles such as hexanol and hexanal, which are the "green" odors. This leaf extract often has to be added to an artificial medium (see Figure 5.1E).

Sometimes the role of smell in the immediate surroundings of the plant seems predominant over taste for food selection, since the caterpillars rarely make mistakes even without previous gustative trials. The sense of taste is, nonetheless, important. For a long time, it has been thought that its importance was negligible, as the caterpillar could not distinguish the four taste qualities but only whether the food was agreeable or unpleasant. Taste distinction is, on the contrary, very well developed in some adult butterflies which not only are nectariphagous insects, but also sometimes accept pollen, putrid food, fruits, urine, tears, blood, or excreta as food. I remember, when I was in Kivu many years ago, a lepidopterist kept human excreta in his refrigerator on pawpaw leaves. He then placed them every morning in his garden to attract the splendid species of *Charaxes* butterflies. Many species of Nymphalidae and Papilionidae (for instance, in New Guinea highlands, the superb *Graphium (Papilio) weskei*) are attracted by fruit, rum, and even human urine. This butterfly is often collected in mountains drinking on the wet sand of torrents. The caterpillars of this butterfly are still unknown, and even if the adult has been seen several times ovipositing on a tree, the plant species was not identified. The adult is common and laboratory tests could solve the enigma of its life cycle.

Schoonhoven (1969) analyzed the reactions of chemoreceptor cells of oligophagous caterpillars to distinguish among the receptors sensitive to sugar (glucose, sucrose, or both), salt, alkaloids, glucosides, inositol, sinigrin, amino acids, and anthocyanins. His results indicated a greater complexity than previously thought. Their sensory receptors are localized on the sensilla styloconica of the maxillae (see Figure 5.2B,C).

Schoonhoven's studies of food selection among phytophagous insects show three selection stages: (1) discovery of the plant, (2) gustative trial, and (3) consumption. These studies made with more than 20 different caterpillars have clearly shown that each species has a characteristic receptive system, and no two are identical, which explains trophic preferences and specializations. Monophagous species would differ from a polyphagous species because it tolerates fewer secondary substances and detoxifies a only few of them.

Chemosensitive phenomena certainly govern the host plant selection of St. John's wort by species of *Chrysolina*, of mint species by *Chrysolina* and *Cassida*, of *Solanum* species by species of *Leptinotarsa*, *Platyphora*, *Lema*, and others. Rees (1969) identified chemoreceptors and mechanoreceptors (trichoid sensilla) on the fifth tarsomeres of the legs of these species. After their arrival on the host plant, apple tree aphids detect a flavonoid (phlorizin) which is found on the leaves and the twigs of the tree. This is probably the substance which attracts the insect. That locusts and grasshoppers nibble a plant before laying their eggs is a well-known fact, and the process, in relatively eclectic species, shows a certain discriminative power.

Nutritive substances used by sucking insects seem to be similar to ones eaten by other insects. Aphids can be bred easily on synthetic nutrient media.

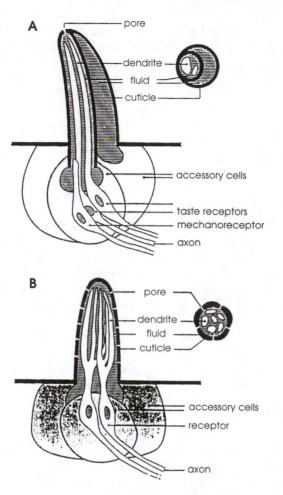

Figure 5.2 Physiology of plant selection, mines, and pollination. **(A)** Schematic drawing of a longitudinal and transverse section of an insect taste seta. The sensillum is innervated by two chemoreceptors and one mechanoreceptor. (Adapted from Schoonhoven, 1990.) **(B)** Schematic drawing of a longitudinal and a transverse section of an insect olfactory seta.The sensillum is innervated by two bipolar neurons. (Adapted from Schoonhoven, 1990.) **(C)** Cross-section of an upper surface mine. (Adapted from Hering, 1951.) **(D)** Cross-section of a lower surface mine. (Adapted from Hering, 1951.) **(E)** Cross-section of an epidermal mine of a species of *Phyllocnistis*; mine only in the upper epidermis. (Adapted from Hering, 1951.) **(F)** Cross-section of an interparenchymal mine (*Phytomyza affinis* Fall. on *Cirsium* sp.). The undamaged cells of an uppermost palisade parenchyma are of a lighter green. (Adapted from Hering, 1951.) **(G)** Pollination system of a species of orchid of the genus *Gongora*. A bee which penetrates into the flower meets a gliding surface and falls toboggan-like along the column (gynostene), where it strikes the viscidium (retinacle) which adheres to the insect's abdomen. The bee flies away and takes with her the pollinia. (Adapted from Arditti, 1966.) **(H)** Trigger mechanism of a species of the orchid genus *Catasetum* which provokes pollination once released. (Adapted from Arditti, 1966.)

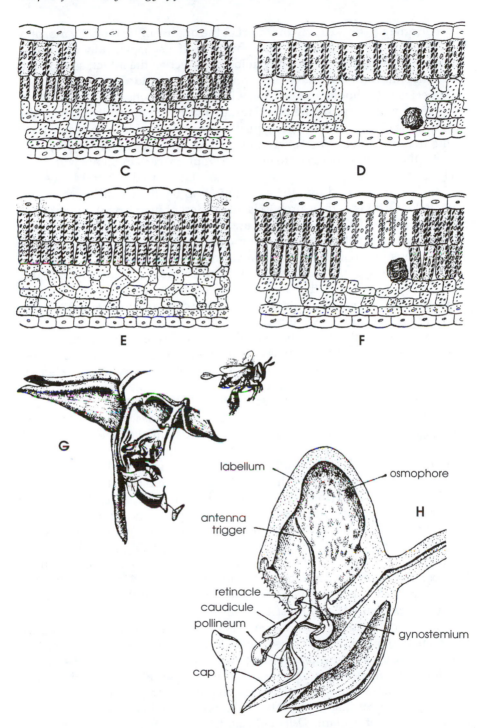

Figure 5.2 (continued)

Synthetic media show clearly the effect of the quality of food on the morpho-logical expression of aphid polymorphism, itself also genetically controlled. The age of the plant also has a physiological effect on the insect, as de Wilde and Schoonhoven (1969) have proven with the Colorado potato beetle.

Species of Orthoptera which are, with few exceptions, polyphagous bite into plants without prior chemical stimuli and try all at random. Taste also prompts the choice of the breeding place, as the adult feeds on the same plant as the nymphs. Species of Tettigoniidae chew the plant before laying eggs. Despite the polyphagous nature of the Orthoptera, certain plants with latex, such as species of *Calotropis* (Asclepiadaceae), are used as food by certain species but are rejected by others. Aphids also try their food by short and repeated insertions of their proboscis into different places on the plant or on different plants. Outside olfactory and gustatory senses, a rather vague chemical sense must also exist among insects. In fact, the location of these sense organs is variable and not always well known.

According to some authors, the protein, amino acid, carbohydrate, fat, sterol, mineral salt, and the B-group vitamin contents of green plants are relatively constant for all plants, although toxic substances do vary. Plant fats are not necessary for insect nutrition.

Aphids and mites attack plants only during certain seasons, even in the tropics. It seems that acidity (pH) of sap varies, which may provide tempo-rary resistance for the plant. No formal proof shows that seasonal changes in plants are responsible for migrations to different host plants, as in the alter-nation of generations in aphids. Specificity of host plants could be due to substances such as alkaloids, cyanogenetic glucosides, flavonoids, terpe-noids, essential oils, saponines, tannins, and acids acting as chemical stimuli, either positive or negative. Those are the secondary substances of various authors. Hering (1950) believed that proteins were responsible for that selec-tion, but this theory has been discarded after new biological findings.

The sense of smell (that is to say, the reactions of the olfactive apparatus to a specific stimulus) is generally more developed in adult insects than in their larvae (see Figure 5.1). Research on the Colorado potato beetle smell perception has been conducted in various countries using a wind tunnel. It has been shown that the distance from which the larvae of this beetle can see the host plant is about 2 mm but the adult and the larvae can perceive only vaguely its shape. It seems clear that the perception of the host by a larva, which is practically blind, must be primarily olfactive. The antennae also play a major role along with the maxillary and labial palpi. Ocelli are in-volved in phototropism, and these together constitute most of the perception of light intensity. Water vapor tension, the nature (texture) of the substratum, and the physiological conditions of the larvae also take part in chemotro-pism. As Bernays and Chapman (1994) mention, there is really very little information concerning the distance from which the insects are able to per-ceive the odor of a plant. *Delia antigua*, the onion fly, perceives the odor source 100 m away. Mixtures of odors from various plants (intercropping) can seduce or suppress the signal at least for the cruciferous flea beetles. Caterpillars perceive odor gradients very well though their olfactory receptors on their head.

The gustative sense of the Colorado potato beetle larvae seems to be localized mostly in the mouth cavity, which differs from that of flies and butterflies, for instance. On the internal surface of the labium (or on the labium itself) of the beetle larvae are found placoid and trichoid sensilla, the gustative receptors. Tangoreceptors (touch) are found more or less over the entire body. Finally, the antennae and the palpi contain basiconic and styloconic sensilla in definite numbers and positions. According to certain authors, there are olfactive receptors, but only because no liquid is involved (see Figure 5.2).

Schoonhoven (1974) also mentions sensilla on the galea, maxillary, and labial palpi of the Colorado potato beetle larvae, and those sensilla are certainly taste and phagostimulant receptors. The tarsal setae of *Leptinotarsa decemlineata* contain receptors for sugar, salts, and alkaloids (Figure 5.2). One of the St. John's wort leaf beetles, *Chrysolina brunsvicensis*, has tarsal chemoreceptors which are sensitive only to hypericin. Other receptors are stimulated by salts, but not by sugars. Those leaf beetles that do not eat *Hypericum* spp. do not react to hypericin, which surely indicates that it is the selective secondary substance.

Sensilla of the maxillae of species of *Pieris* caterpillars are sensitive to mustard oil, and to hypericin for the *Chrysolina brunsvicensis* beetle. This could also be one of the mechanisms for detecting specific attractants. Schoonhoven also shows that the majority of the chemoreceptive cells of the oligophagous insects including Lepidoptera are localized among the styloconic sensilla of the maxillae.

Wei-Chun (1969) has found some synergistic effects between feeding stimulants for the caterpillars of *Pieris brassicae*. A chemoreceptor is present in the medium sensillum styloconicum of the maxillae stimulated by many specific feeding inhibitors (alkaloids). Attractants and deterrents each play a role responsible for the final selection of the cruciferous feeders, which is probably the reason why among them (butterflies, moths, flea beetles) there is a selection among the species of Brassicaceae (Cruciferae).

The consumption of leaf matter by a phytophagous insect varies in quantity in proportion to its size. For example, studies of a small species of a polyphagous flea beetle, *Systena blanda*, show that one individual eats about 0.3 cm of beet leaves per day. Neither density nor temperature modify the process. When they are grouped, adults of the flea beetle *Macrohaltica jamaicensis* are able to skeletonize the enormous leaves of *Gunnera insignis* (Gunneraceae) in Panama, leaving only veins and a few spoiled bits of parenchyma. Colorado potato beetles eat a lot of potato leaves when abundant. Once the adequate host plant is found, and the process of feeding is started, it does not stop until the foliage is eaten or spoiled by frass and decomposition. Cycloalexic larvae, such as the ones of species of *Coelomera* (galerucine beetles), feed on the large leaves of *Cecropia* spp. in South America.

They have a cleaner way of feeding as do most of other species practicing cycloalexy, a procedure indicated when the larvae feed in line up against each other in a row, or in a circle when at rest. Sometimes some *Cecropia* spp. leaves also are skeletonized by certain species of *Coelomera* beetles.

As far as we can judge, all leaves contain the quantity of nutritive substances needed for growth and development for insects. Therefore, the big question is what are the differences among plants that make them attract some insects more than others. These differences are subtle. Odor, taste, toxins, and nutritious substances are some of the factors accountable for their choice. Essential oils and alkaloids are responsible for the taste and the smell of the plant, together with other chemicals.

When an insect attacks a plant, it is necessary for the growing period of the plant to coincide with the feeding stage of the insect and vice versa. Those conditions not being fulfilled is what Painter (1936) called "evasion or pseudo-resistance". This peculiarity is utilized to devise so-called integrated control agricultural techniques, and something as simple as a difference in time between the date of sowing seeds and the hatching of insect eggs can provide good crop protection.

Hsiao (1969), studying the Colorado potato beetle and the alfalfa weevil, found that chemical stimuli influencing host searching are attractants, repellents, stimulating signals, and rejecting stimuli. Parasitoids also use stimuli or volatile chemicals emanating from the plant as aids to host location (Godfray, 1994). This is a third-level element in insect-plant relationships. By attracting parasitoids, a plant can defend itself against its defoliators. Many parasitoids feed on flowers and on extrafloral nectaries, which are commonly also visited by mites, ants, and many other insects, including, in certain countries, Thysanoptera. The main function of flowers is to attract pollinators, but they also attract parasitoids, and they in turn could have played a small role in floral evolution. Parasitoids comprise between 10 and 20% of all insect species, and their role could be more important than previously thought. According to Mattiacci et al. (1995), cabbage plants respond to *Pieris brassicae* caterpillars by releasing a mixture of volatiles that make them highly attractive to parasitic wasps (*Cotesia glomerata* and others) which attack the caterpillars.

In summary, the conclusions reached by Painter, Dethier, and others are that sight, phototropism, geotropism, anemotropism, and hydrotropism play significant roles in insect orientation for both oviposition and feeding, and that the elements which attract insects at a short (or long) distance are essentially chemical. Insect behavioral activity therefore is as follows: (1) an orientation towards the food, (2) a response to biting or sucking, and (3) continuous feeding.

The same odor does not necessarily stimulate both orientation and feeding, as has been proven in tests using click beetles (Elateridae). Kogan (1977), summarizing numerous experiments, describes the selection process for phytophagous insects as being a chain of events between plant stimuli and the responses given by the insect. According to this author, five stages can be distinguished: (1) finding the host plant in its habitat (by sight and smell), (2) discovery of the individual host plant, (3) recognition of the host plant, (4) acceptance of the plant, and (5) chemical acceptability as food by the host.

At phase (4), several chemical substances seem to control various phases of the feeding process. For the silkworm, various compounds existing in the mulberry leaf are associated with the first bite, then with ingestion, and finally with increased feeding. These stimulants are chiefly secondary metabolic compounds exuded by the plant epidermis and provide the stimulation necessary to trigger all of those processes. The wax of the leaves contains chemicals also playing their part in this trigger action. Some plants also emit volatile oils through their leaf glands and stomata that also affect whether the insect eats the parenchyma or not.

Recent research, namely electrophysiological studies on food perception among insects, has demonstrated that odor and taste seem intermixed and are decoded inside the central nervous system as "host" and "nonhost". Some classifications have correlated the stimuli produced by the plants and the insect responses during the selective process. Those secondary chemical factors of the plants have been termed "allelochemicals" by Whittaker and Feeny (1971). They define these as "chemical products by which the organisms of one species affect the growth, health, behavior, or biology of an organism of another species," or "a substance that affects individuals or populations of a species, different from the source." The latter is more concise but not much more clear. Such a relationship can exist between plants and insects. Pheromones (ectohormones) are essentially chemical messages between the individuals of single species and are very different from allelochemicals, which affect individuals or populations of a species different from the one producing the chemical. Also involved are allomones, which are chemical agents of adaptive value for the producing organism, and kairomones, which have an adaptive value for the receptive organism. Table 5.1 is slightly modified from Kogan (1977), Dethier (1960), Beck (1965), and Whittaker and Feeney (1971) and shows the distinctions between the main groups of chemical factors and their effects.

In principle, secondary substances produced by the plant that are responsible for the trophic selection by insects are toxic towards other animals and plants or are at least repulsive to them. Insects which feed on those plants either detoxify the chemicals or sequester them as a defense mechanism against predators. These chemicals assure biochemical protection (allelopathy) of the plants against pathogens, herbivores, or competition by other plants. They are known as allelopathic substances. There are however, cases, as mentioned by Whittaker and Feeny (1971), for which the allelopathic pheromones, or the nutritional compounds produced by a species, also react as kairomones for another species.

Certain insects succeed in avoiding toxic or repulsive substances by various means, including the consumption of withered leaves devoid of

Table 5.1   Main Groups of Chemical Factors Produced by Plants
(Allelochemicals) and Behavior or Physiological Effect on Insects

Allelochemicals	Behavioral/physiological effects
Allomones (repellents)	Give an adaptive advantage to the producing organism (plant)
Repellents	Orient the insect outside the plant
Locomotor stimuli	Start or accelerate movement
Suppressants	Stop biting or piercing
Preventatives	Prevent continuation of feeding and oviposition
Antibiotics (growth inhibitors)	Prevent normal growth and larval development; reduce longevity and fecundity of the adult
Kairomones (attractants)	Give an adaptive advantage to the receiving organism (insect)
Attractants	Orient insects toward their host plant
Arrestants	Slow down and stop movement
Excitants	Produce biting, piercing, or oviposition
Nutritional stimuli	Provoke continuation of feeding

these materials (certain aphids, some microlepidoptera) or the selection of
nontoxic parts (sapsucking Homoptera). Also, it should be noted that aphids
prefer young or old leaves to the mature ones. Though often polyphagous
and with alternate cycles, aphids clearly show botanical preferences that are
rather difficult to analyze. The choice of young or old leaves is probably
sometimes linked with their cyclic development.

Interesting to note about *Leptinotarsa decemlineata*, the Colorado potato
beetle, is that without its normal host plant and the adopted potato (*Solanum
tuberosum*) itself, it accepts or rejects many species of the genus *Solanum*.
Normally, the beetle is attracted by various species (in various genera) of
Solanaceae, but according to certain authors it can also live under experimental conditions (not in nature) on Asteraceae (*Lactuca sativa*, lettuce or salad
greens) and Asclepiadaceae (*Asclepias syriaca*, a milkweed). The first plant is
a "neutral" one, but the second is highly toxic, as are all the plants of that
family.

Perhaps there is some kind of pre-adaptation in the species of the genus
Leptinotarsa similar to other related genera, such as species of the genus
Labidomera which normally develop on species of *Asclepias*, and species of
Doryphora which in turn feed on *Prestonia* spp. (Apocynaceae). Species of
Elytrosphaera, apterous mountainous species, feed on species of Asteraceae
and Solanaceae. Species of *Elytrosphaera* are closely related to species of
Leptinotarsa. According to Hsiao (1986, 1988) and Jacques (1988), several
species of *Leptinotarsa* feed selectively on Solanaceae, Asteraceae, and

Zygophyllaceae, but none, so far, on Asclepiadaceae. The food plants of the Peruvian species remain to be discovered. The Central American species live on *Solanum* spp. shrubs exclusively (Jolivet and Hawkeswood, 1995). Despite the fact that species of several other Chrysomeline genera eventually feed on species of Apocynaceae and Asclepiadaceae, most of the species of *Doryphorina* are solanaceous feeders.

Arnett has shown that *Leptinotarsa lineolata* is very host specific on *Hymenoclea monogyra* (Asteraceae) in Arizona and cannot be forced to eat other plants, even those mechanically acceptable to the larvae. In any case, this shows that the natural and normal preference for species of *Solanum* for most of the species of *Leptinotarsa* is not due to nutritive substances but to a specific attractant. The researches of Fraenkel show that a strong trophic selection is directed by the presence or absence of secondary chemical substances and not by nutritive substances.

Hsiao (1978) has shown that the Colorado potato beetle is an indigenous oligophagous species in North America and may adapt to ten indigenous Solanaceae other than the natural host, *Solanum rostratum*. Races, varieties, or biotypes, whatever we may wish to call them, may actually develop and form those new oligophagous races with extended or restricted food preferences. The geographical isolation of the insect and the original host plant is surely an important element. Hsiao insists that this is different from intraspecific variation in the case of polyphagous species where the formation of biotypes is adapted to the host and is produced sympatrically. According to Hsiao, the beetle populations from North America show a great deal more interpopulation variations as compared to European beetle populations which were introduced accidentally in 1922 in the West of France and descended from a very small group of insects, perhaps from a sole couple or an unique female, perhaps from a unique batch of eggs.

According to Mitchell (1974), host plant selection by the Colorado potato beetle appears to be largely mediated by alkaloids found in solanaceous plants and feeding deterrents caused by steroidal alkaloids. Tomatoes can be repulsive to some strains of the beetle and not to others that are able to accept tomatine or may even be attracted to it.

Taxonomic researches on the species of the genus *Leptinotarsa* by Jacques (1989) completely challenges the long-criticized classical research of Tower (1906, 1918). During our researches on Chrysomelid host plants (Jolivet and Petitpierre, 1973, 1976), we found a very clear evolution of certain oligophagous genera from a basic, primitive choice (*Timarcha* spp. from *Galium* spp. to *Plantago* spp. in Europe, *Chrysolina* spp. from Lamiaceae to *Plantago* spp., Scrophulariaceae, and Ranunculaceae, etc.). A common denominator exists between species of *Timarcha* and *Chrysolina* — *Plantago* spp., but a species of *Chrysolina* never has been found feeding on species of *Galium* or a species of *Timarcha* on Lamiaceae. Feeding experiments should be attempted, at least, within a species such as *Chrysolina staphylea*, which is potentially oligophagous. However, very often, the common denominator is missing or probably extinct, having developed along several lines during the course of evolution. Petitpierre and myself (1973, 1976) have also found a certain relationship existing between the chromosome formulae of species of *Chrysolina* and their trophic preferences, which can only confirm the taxonomic relationships between their several ecological distinct groups or subgenera. The evolution of all these related genera of Chrysomelidae can be explained by an ancestral polyphagy and a subsequent specialization, or by a deviation from a basic monophagy on species of Ranunculaceae, Plantaginaceae, Rubiaceae, Lamiaceae, Salicaceae, Myrtaceae, etc., and botanical relationships cannot easily explain the insect's choices. Some preferences hardly can be explained, as, for instance, the fondness of willow and alder insects, such as the galerucine genus *Agelastica* (*A. alni, A. coerulea*), for species of Rosaceae.

It is also well known that toxic secondary substances which protect the plant also protect the monophagous insects which feed on it, if the insects sequester the toxins. Numerous are the examples, such as the toxic flea beetles from the Kalahari desert in southwest Africa which are used by bushmen for making arrow poison (Jolivet, 1967). Species of danaid butterflies and species of *Poekilocerus* locusts feed on *Calotropis procera*, a very toxic Asclepiadaceae, and the small red moth *Tyria jacobaea* feeds on *Senecio jacobaea* along with one species of *Longitarsus*. They all acquire the toxicity of their host plant. In the case of the Kalahari flea beetles, the food chain is rather complex, because their mimetic carabid beetles (species of Lebiinae, which feed on the alticines) also have acquired the toxicity of their prey and consequently of the plant itself. In conclusion, as Fraenkel said, the defensive agent of the plant becomes equally a protection for their specific herbivorous predator and, at least theoretically, those substances can act as repellents for other insects and predators.

The last problem remaining concerns the ecdysones and the juvenile hormones or their analogues synthesized by plants (phytojuvenil hormones and phytoecdysones). Those could act as repellents and may even be transmitted venereally from the male to the female, at least in some Hemiptera, such as *Pyrrhocoris* spp. In any case, they are elements which prohibit the development of certain insects. It is true that the plants that possess those analogues are more or less resistant to pests and that a kind of coevolution has probably developed between the potential host plant and certain insects. Insect development is also inhibited by certain deficiencies in essential proteins in the host plant or the increase of tannin content in some.

Insecticidal properties of plants were known as far back as Biblical times. Extracts from many of species of plants have been tested and found to have insecticidal or juvenile hormone properties. For instance, Saxena et al. (1977) have tested the essential oils of *Acarus calamus* (Araceae) for insect chemosterilization ability. In this plant, asarone is not a juvenile hormone, but it may represent a new type of antigonadal agent. Many plants contain natural insecticides, such as species of *Melia, Pyrethrum, Lonchocarpus, Derris, Ginkgo,* and others, and they also have their own guild of parasites immune to their poisons. One reason why ferns are relatively immune to insect feeding is because of the phytohormones they generally contain.

Let us now return to the phenomenon of larval conditioning, the famous Hopkins Host Selection Principle, which was discussed previously. That principle, often criticized, seems to fit perfectly some known cases among the Lepidoptera species. Feeny (1977), referring to recent work, writes that adult females generally prefer to lay eggs on the plant species on which they have fed at the larval stage. This may help a population to concentrate at a given time on that plant species to which it is most adapted and which is the most readily available. It is, however, difficult to explain Hopkins' principle. Can one speak of "phenotypic flexibility", if that term means anything, or of selection among races genetically pre-adapted? It is very difficult to tell what is going on, but new research with modern techniques using species of Lepidoptera as subjects might shed some light on the matter. Synthetic media permit a modified insect diet, which also requires further investigation. One thing certain, however, is that, given a choice, species of *Manduca* and *Heliothis* caterpillars always choose the plants on which they originally started to feed, even if it was only 24 hours earlier.

Also, it is to be noted, as shown in an experiment by Schoonhoven (1967), that conditioning to a synthetic medium can greatly modify the trophic preferences of the oligophagous species. There is a total loss of host plant specificity after the experiment. The caterpillars of *Manduca sexta* (Sphingidae, a common hawk moth) which normally feed only on solanaceous plants show, after cultivation on a semisynthetic medium, an ability to eat other plants such as dandelion (Asteraceae), cabbage plants (Brassicaceae), and plantain (Plantaginaceae). Caterpillars are generally more adaptable than beetles, but their normal selection tendencies reappear quickly. Several days of feeding on their normal host plant are enough for *Manduca sexta* to recuperate its old selective habits. Schoonhoven, in various papers (1968; 1969a,b,c; 1970; 1972a,b), has summarized the chemosensorial basis of host plant selection by insects, particularly caterpillars. It is impossible to give details here about the wealth of research which has evolved rapidly during the last few years, particularly the studies using electrophysiological methods. These techniques permit the precise location of the chemoreceptors of the antennae and the mouthparts. An electroantennogram (EAG) or an electropalpogram (EPG) response reflects the overall response of the olfactory or gustatory receptor cell population (Visser, 1983). In caterpillars, the sense of smell is roughly localized on the antennae and the maxillary palpi. Taste, on the other hand, is situated partially on the maxillary palpi but mainly on the maxillae or on the epipharynx and hypopharynx inside the buccal cavity. Even when two species share the same host plant, the sensory reactions can vary and they can be attracted by different secondary substances, although in most cases, the chemical substance involved is the same (e.g., associations of Brassicaceae and butterflies or moths and flea beetles). (See Figure 5.1.)

Interesting results have been obtained by removing the maxillae of an insect. Complementary electrophysiological data show the location of particular sensory areas. The loss of maxillae permits these insects to accept foods not ordinarily eaten, that is, the loss of food selectivity for the insect. It even permits the consumption of various inert substances. Agar-agar is eaten by *Manduca sexta* and filter paper by the Colorado potato beetle and silkworm when their maxillae are excised. When phagostimulants and repellents are combined, they produce synergical reactions among insects. More details on the topic can be found in Schoonhoven (1968, for example).

As we have noted previously, the difference between monophagy and oligophagy can be quite subtle. Only certain chemical compounds, combined with the presence or absence of some secondary chemicals in a plant, could allow such selection. However, a thorough analysis within this chapter is difficult, as this is a phenomenon as complex as the entire topic of food selection among phytophagous insects.

According to Moody (1978), tropical floras have nearly twice as many species of alkaloid-bearing plants as temperate floras. That is normal when the tropics contain many more varieties of plants than do temperate forests. Also competition there is more severe; insect generations do not actually stop development during dry months, if there are any. The plants are always under pressure, and they have to be well protected against aggressions of every nature.

According to Edmunds and Alstad (1978), coniferous trees must maintain genetic heterogeneity through outcrossing, thus confronting their insect pests with a continuously changing array of toxins (Matthews and Kitching, 1984). If coevolution really exists, insects should produce a "stepwise reciprocal selective response", which is not always the case or can be delayed considerably, even through a very long evolutionary time. All plants defend themselves with various mechanisms and chemicals. If ferns are devoid of alkaloids, they produce hormone mimics. Species of *Cycas* are very toxic to most of the herbivores, and *Aristolochia* plant species produce aristolochid acids sufficiently different so that the selection of species of birdwing butterflies is strictly controlled. Generally, species of Apocynaceae and Asclepiadaceae do not seem as selective for their herbivore pests, but the latex and bitter cardiac glucosides in these plants eliminate a great number of nonspecialized insects. They seem to be relatively immune to herbivory in their native habitat and totally immune where they have been introduced.

Conifer feeders among moths seem less host specific than moths feeding on angiosperms (Holloway and Herbert, 1979), presumably because angiosperms have evolved a greater number of secondary plant substances. Ferns, coniferous trees, cycads, and ginkgoes all have developed many effective ways to protect themselves against the nonspecialists.

5.5   Talking trees

"Do plants tap SOS signals for their infested neighbors?" This is a question raised recently by Bruin et al. (1995), but their theory, originally published in papers from 1979 to 1983, so far has really never been proved. According to some research, the defenses of neighboring trees rapidly increase in response to airborne chemicals released by attacked trees, but some skeptics have answered: "Trees don't talk; do they even murmur?" (Fowler and Lawton, 1984). (See Figure 10.1.)

It is a fact that individual plants are interconnected, namely by the root systems (haustoria) or via mediating organisms (mycorhizal fungi); also, water and air can transfer chemicals from one plant to another. We know very well the resinous atmosphere over a spruce or fir forest. Walnuts (*Juglans* spp.) produce juglone, an allelopathic red crystalline compound released in the air, and more than 70 of the species of trees in North America emit appreciable quantities of hydrocarbons into the atmosphere, including a-pinene, limonene, and isoprene. We smell the scent of citrus trees and of many other plants at a distance; these scents are readily perceived by the human nose. Insects, obviously, also perceive these scents, which can be either an attractant or a repellent, not only for these herbivores but also for their guild of parasitoids, hyperparasitoids, and predators. There is also competition between neighboring plants (allelopathy), and volatiles could be transmitted by air as well as by soil.

The general idea that defenses of neighboring trees rapidly increase in response to airborne chemicals released by attacked trees was first published by Baldwin and Schultz (1983) and Rhoades (1983) and several others. They suggest that a damaged tree will produce ethylene, a gas somewhat similar to certain phytohormones. When an insect has voided too much excreta on the leaves or on the ground near by, this causes an increase in the quantity of tannin and phenol inside the leaves, produced as a means of defense against the herbivores. When caterpillars feed, a biochemical response is triggered in the plants that within as little as 30 minutes deters further feeding by any caterpillars. The production of phenolic alkaloid or protein-ase inhibitors is the response of the plant. Such ideas raised skepticism and according to Fowler and Lawton (1984), seemed to touch on the realms of fantasy.

We do know however, that signals which attract parasitoids are supposed to be sent by a plant when it is attacked by an insect. Volatile substances produced only in case of attack would be selective for certain predators and parasitoids of the plant and act as an SOS signal sent out by the plant.

The matter is still under discussion. It seems that caterpillars are indifferent to damage of leaves, and some even prefer the damaged leaves themselves (Fowler and Lawton, 1984). Then how can we explain the chemical changes in the plants? They might be directed against microorganisms rather than against caterpillars. It seems, however, that attacks by insects on oaks, maples, and other trees increase the phenols, tannins, resins, fiber, and other plant defenses. Is that so efficient?

Brackens, which are normally protected by an array of chemical weaponry such as cyanide, thiaminase, tannin, lignin, silicates, and sesquiterpenes, are still attacked by herbivores including leaf miners such as some tropical leaf beetles. They are also protected by pugnacious ants thanks to their extrafloral nectaries. We can argue that without this arsenal, ferns would be completely eaten by herbivores, and that there are some which bypass their best protections. Phytoalexins, those low-molecular antimicrobial compounds, seem to influence a plant's resistance to disease, but there is no proof of any response to attacks on a neighboring tree.

Wind tunnels have been devised to test the theory of talking trees. Undamaged plants are exposed to a continuous air stream from either infested or noninfested plants. Those experiments have not yet proved irrefutably if information is transferred through the atmosphere from damaged to undamaged plants.

According to Bruin et al. (1995), "Damage-induced volatiles may represent traits that are promoted by natural selection in the plant population." The authors believe more carefully designed field experiments are required, and the doubt persists. The research done by respected scientists remains as evidence, and perhaps trees really do talk to each other. But what do they talk about? That remains to be fully understood.

And are there plants that click? An Australian scientist, in 1981, found that drought-stricken plants make noises when they are thirsty! Clicking sounds are caused by the vibration of the water vessels inside the plants. Ultrasonic chirps, similar to the sound of popcorn popping, increased as stress intensifies. This could be heard by bark beetles which are ready to invade ailing trees. In this way insects could tell an unhealthy tree from a healthy one, and the insects seem to prefer chirping branches to chirpless ones. Too good to be true, or when the trees cry, do the beetles hear?

chapter six

Carnivorous and protocarnivorous plants

Carnivorous plants and myrmecophilous plants represent two metabolic ways in which plants have adapted to poor soil, and they are so named according to the ways they use nutritive substances derived from animals. Another way plants can be adapted to life in poor soils is by the use of symbiotic relationships between their roots and mycorhizal fungi, an adaptation not involving animal interventions. Carnivorous plants trap insects and other arthropods, even (but rarely) birds and mammals, in a specially modified leaf adapted to trap these organisms. Myrmecophilous plants make use of the excreta and cadavers abandoned by ants in specialized chambers and are protected by the ants. The flowers of both plant types are fertilized by flying insects, a necessity even for plants associated with other arthropods.

One unique plant has combined carnivory and myrmecophily. A plant from Borneo, *Nepenthes bicalcarata* (Nepenthaceae), is an extraordinary plant which captures insects in its pitcher. It is fertilized by bees and other insects which are not trapped in the pitcher, and it harbors ants in veins of the leaf beneath the pitcher itself.

Recently, a new distinction has been made between the classical carnivorous plants, terrestrial or aquatic, and the newly discovered protocarnivorous plants. New discoveries have been brought these to light in the last ten years. The species of one genus, *Ibicella*, were overlooked by Lloyd (1994) in his classical memoir. (See Figure 6.1.)

6.1 Carnivorous plants

The definition of a carnivorous plant is "a plant that is capable of catching and digesting insects or other small animals." Generally, these plants produce their own digestive enzymes, with the few exceptions cited in the literature (*Heliamphora* spp. and some others), and make use of the digestive power of decaying bacteria. However, these exceptions are referred to as carnivorous species because the genus *Heliamphora* belongs to the Sarraceniaceae,

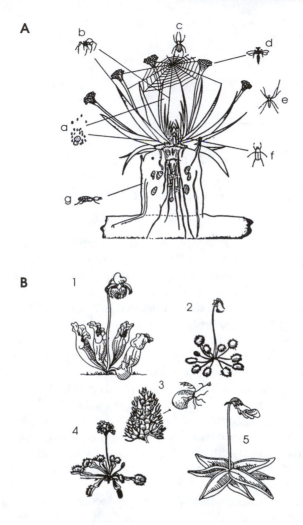

Figure 6.1 Protocarnivorous and carnivorous plants. **(A)** A protocarnivorous plant, *Paepalanthus bromeloides* (Eriocaulaceae), with its biocoenosis of insects and spiders, in Serra do Cipo, Brazil. (Adapted from Figueira and Vasconcellos-Neto, 1993.) **(B)** Several classical carnivorous plants: (1) *Sarracenia* sp. (Sarraceniaceae); (2) *Drosera* sp. (Droseraceae); (3) *Utricularia* sp. (Utriculariaceae); (4) *Dionaea* sp. (Droseraceae); (5) *Pinguicula* spp. (Lentibulariaceae). **(C)** *Darlingtonia californica* Torr. (Sarraceniaceae), in California. (Adapted from Engler and Prantl, 1909.) **(D)** *Sarracenia purpurea* Linné (Sarraceniaceae), from the east coast of the U.S. (Adapted from Engler and Prantl, 1909.) **(E)** *Heliamphora nutans* Benth. (Sarraceniaceae), from the highlands of Tepuys, Venezuela. (Adapted from Engler and Prantle, 1909.) **(F)** A protocarnivore, Venezuelan: *Brocchinia reducta* (Bromeliaceae), from Tepuys, Venezuela. (Adapted from Juniper et al., 1989.) **(G)** Schematic drawing of a leaf of *Nepenthes* sp. (Nepenthaceae), southeastern Asia. (Adapted from Fessler, 1982.) **(H)** *Cephalotus follicularis* Labill. (Cephalotaceae), from southwestern Australia. (Adapted from Hutchinson, 1959.)

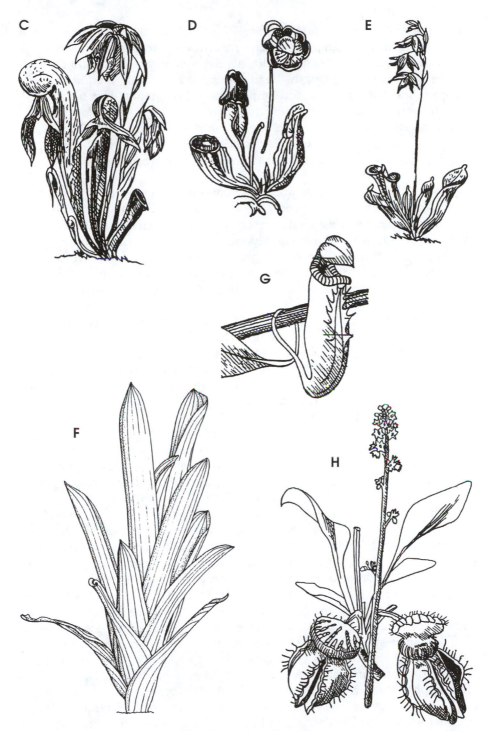

Figure 6.1   (continued)

an important American family of real carnivorous plants. Species of the *Roridula*, a South African plant which very efficiently traps insects by making use of a resinous secretion, do not directly digest their prey, but rather absorb by the roots what is left of the dead insects. It seems that the *Roridula* sp., which is one that catches insects with a resinous substance, does not possess any digestive enzyme, but instead relies on a tiny bug (*Pameridea roridulae*) living on its leaves to digest its meals for it. The plant then absorbs the nourishing substance, the insects' excretions (Ellis and Midgley, 1996). Absorbing urea through the leaves seems much more efficient than absorbing it through the roots.

We will see later on the affinities between the carnivorous plants and those known in the fossil record. We have fossil pollen and seeds, but no carnivorous plant has been fully preserved, even though acarophilous plants are known as fossils in Australia. Myrmecophilous plants are also unknown as fossils. Probably the carnivorous plants (aside from the protocarnivorous) are monophyletic, but there is still some controversy about this. Myrmecophilous plants are polyphyletic and originated independently in many families of plants.

Due to their affinities with species of *Byblis*, a relic of Gondwanaland, the two species of *Roridula* are maintained here as true carnivorous plants. This classification is subjective, but I had to decide on one of them. As said before, it seems that all the carnivorous plants are related and the cladists have recently proved that the idea, already proposed by several taxonomists (Emberger, 1960). The carnivorous plants are listed below family by family. Also, their trapping systems are discussed in detail.

There may be more than 530 carnivorous plants among the Angiospermae (the flowering plants), but many others are on the way to carnivory or suspected to be carnivorous. There appears to be at least 250,000 species of flowering plants (certain authorities say 300,000) living on the planet Earth. How do we actually classify botanically carnivorous plants?

These plants belong to two branches of dicots: the Choripetalae and the Sympetalae (Lentibulariaceae and Martyniaceae) and protocarnivorous plants (Bromeliaceae and Eriocaulaceae), which are monocots. Any phytotelmata (plant reservoir) also is more or less carnivorous, although all of their digestion is done by bacteria and no special structural adaptation is shown for trapping insects. Moreover, they rear insects inside their water reservoirs. Another way to divide the groups of the true carnivorous plants is between those that are active trappers and those that are passive trappers (see Table 6.1), but some species may be both active and passive. The true carnivorous plants have been assigned to 18 genera distributed in eight dicot families: Nepenthaceae, Sarraceniaceae, Cephalotaceae, Biblydaceae, Droseraceae, Dioncophyllaceae, Martyniaceae, and Lentibulariaceae.

The species of two families are restricted to Australia, namely, Cephalotaceae and Biblydaceae, but the species of the South African *Roridula* are actually now considered to be a Byblidaceae. The species of Sarraceniaceae

Table 6.1 Trapping Systems of the Main Carnivorous Angiosperm Genera

Trapping system	Genera
Active	
Trap with closing system (snap)	*Dionaea, Aldrovandra*
Suction trap	*Utricularia*
Passive	
Pitcher plant (pitfall)	*Brocchinia, Paepalanthus, Nepenthes, Sarracenia, Darlingtonia, Heliamphora, Cephalotus, Catopsis*
Adhesive droplets	*Pinguicula, Drosera, Drosophyllum, Ibicella, Byblis, Triphyophyllum, Roridula*
Pseudodigestive tract	*Genlisea*

Adapted from Heslop-Harrison, Y. (1978). *Sci. Am.,* 238(2): 104.

are American; those of the Nepenthaceae are mostly Asiatic but there are also some in the east African islands of Seychelles and Madagascar; the Dioncophyllaceae are African species; and among the Droseraceae and Lentibulariaceae the distribution is large and varies according to the genera.

Only the species of the genera *Drosera* and *Utricularia* are really ubiquitous. The genus *Drosera* is represented by 80 to 90 species living in cold peat bogs of the Northern hemisphere and the cold peat bogs of the tropical mountains of America, New Guinea, the Sahelian zone, in temporary swamps in South Africa, and also in Australia, where their diversity is the greatest. *Utricularia* species are generally aquatic but can be epiphytic in the Venezuelan tepuys or sometimes terrestrial, or may even live in the suspended aquaria or phytotelmata of certain epiphytic Bromeliaceae in South America. The carnivorous fungi are listed in Table 6.2; they do not capture insects, but feed mostly on nematodes.

The taxonomy of the carnivorous plants is rather simple. For the classical botanists, they belong to two related groups, the Sarraceniales (Sarraceniaceae, Nepenthaceae, Dioncophyllaceae, Droseraceae) and the Rosales (Byblidaceae and Cephalotaceae). The Martyniaceae fall separately and are placed with the Lentibulariaceae among the Tubiflorae.

Despite these relationships and the fact that these plants have evolved towards having mechanisms for the capture and digestion of animals, the trapping systems vary within a family, and sometimes within a genus. Species of the two genera, *Pinguicula* and *Utricularia,* are very different — one being terrestrial and the other generally aquatic or subaquatic. However, both genera belong to the same family. Among the species of Droseraceae, the differention is great between the species of the genera *Drosera, Dionaea,* and *Aldrovandra.*

Generally speaking, the environment of those plants (carnivorous or protocarnivorous, as well as myrmecophilous) is made up of poor soil,

Table 6.2   Main Genera of Carnivorous Plants

Family	Genus	No. of known species	Georgraphical distribution
Nepenthaceae	*Nepenthes*	65	Southeastern Asia to Sri Lanka and Madagascar
Sarraceniaceae	*Sarracenia*	9	Eastern North America
	Heliamphora	5	Guyana, Venezuela
	Darlingtonia	1	Northern California, southern Oregon
Cephalotaceae	*Cephalotus*	1	Southwestern Australia
Byblidaceae	*Byblis*	2	Northwestern and southwestern Australia
	Roridula	2	South Africa
Droseraceae	*Drosera*	90	Ubiquitous
	Drosophyllum	1	Southwestern Iberian peninsula, Morocco
	Dionaea	1	Southeastern U.S.
	Aldrovandra	5	Southwestern Europe, Africa, India, northwestern Australia, Japan
Dioncophyllaceae	*Triphyophyllum*	1	Western Africa
Martyniaceae	*Ibicella*	3[a]	Tropical South America
Lentibulariaceae	*Pinguicula*	30	Holarctic
	Utricularia	275	Ubiquitous
	Biovularia	2	Cuba, Eastern South America
	Polypomphalyx	2	Southern Australia
	Genlisea	10	South America, West Africa
Mushrooms	Numerous w/traps	20[b]	Ubiquitous
Total	15 genera of dicotyledons	523	

[a] Probably only 1 carnivorous.

[b] Or more.

Adapted from Lloyd (1942) and Heslop-Harrison (1976).

although the habitats are heterogeneous. Species of *Sarracenia, Darlingtonia, Drosera,* and *Pinguicula* are found in swampy areas, poor in organic matter, over species of *Sphagnum,* or even some in the dryer heathland. Several species of *Nepenthes* are epiphytic (8%), and the rare *Heliamphora* species (six in all) live with *Brocchinia* spp. in moss forests in bogs on the Venezuelan mountains (tepuys). Many species show astonishing adaptations; in Australia and Africa, *Drosera* spp. are found in granite bogs, in the Sahel along brooks or springs, and on Brazilian mountains directly on rocks along streams. In some places these plants dry up completely during the dry season. According to their botanical classification, we find the following carnivorous plant groups.

6.1.1 Nepenthaceae

The pitcher plants number about 71 species, with 100 more natural and artificial hybrids. The species of the unique genus *Nepenthes* are essentially tropical. They are found in Madagascar, the Seychelles islands, Sri Lanka, Assam, Southern China, India, Vietnam, Malaysia, New Guinea, Northern Queensland, and New Caledonia, but not on mainland Africa.

Twenty species of *Nepenthes* grow in Borneo. A unique species in the Seychelles has sometimes been placed in the genus *Anurosperma*. Since the year 1800, the British have cultivated in greenhouses *Nepenthes rajah* and two other species considered as very rare in nature today. The center of distribution of the genus is Borneo, but in the middle of the eighteenth century in Madagascar the first species was discovered.

Most species of *Nepenthes* are herbaceous plants, some more or less lignified, growing in swampy areas or as vines climbing along the trees. The pitchers develop intercalary invaginations below the tendril tip, prolongations of the central vein of the leaf. The pitcher has a lid above the opening which is closed only at the young stage when digestive fluid is still sterile. The pitcher has lips growing around the opening, and in that area are many extrafloral nectaries. At some distance in the interior, other glands are disposed deep in the internal surface. Insects are attracted by the nectar and the brightly colored pitcher, as well as by its fragrance. These insects slip into the water where they drown and decompose. The remaining nutrients are absorbed by the plant. The plants are able to use these products of digestion thanks to the action of enzymes secreted by specialized glands in the pitcher. *Nepenthes* spp. plants require high humidity and temperature to produce pitchers. Some pitchers reach 30 cm in length (*Nepenthes rajah*) and are large enough to hold two liters of fluid. In these pitchers, birds, frogs, or even small mammals are drowned and digested.

Several species of *Nepenthes* are epiphytic. Some have pitchers of two sizes, others (*N. ampullaria*) have two kinds of leaves, some with tendrils and without pitchers. Some terrestrial species of *Nepenthes* live in high altitudes above the moss and tree-fern forests, such as in New Guinea, where the night can be cold (5°C). Others climb along the trees up to 15 m high. *Nepenthes* spp. are dioecious, i.e., there exist both male and female plants.

6.1.2 Sarraceniaceae

6.1.2.1 Sarracenia

The *Sarracenia* are met in the sunny swamps and roadside drainage ditches of the North American Atlantic coast. These swamps are flooded part of the year and in some regions may be frozen in winter. Only one species among the eight known reach arctic Canada — *Sarracenia purpurea*. This species has been naturalized in Ireland, Scotland, and Switzerland.

The plants of *Sarracenia* spp. are herbaceous with rosette leaves. They sometimes reach a length of 68 cm. Each leaf is transformed into a narrow pitcher on the ventral side with a flat "wing" of green photosynthetic tissue. There is a lip above the opening as in species of *Nepenthes*. The pitcher opening has numerous nectariferous glands, below which is a slippery area, then the inverted hair area, and in the bottom water collects. There the trapped insects are drowned. The pitchers are often brightly colored.

These plants are passive carnivores, i.e., the insect is attracted by the pitcher by its smell, its color, and its nectar. It penetrates the pitcher, and after the slippery zone it passes into an area lined with inverted hairs which prevent its escape. At the bottom is secreted into the collected water an enzyme which digests the imprudent visitor. Very often, the smell of the pitcher becomes repulsive because of the rotten insects laying inside.

The natural and horticultural hybrids of *Sarracenia* spp. are numerous and often very beautiful. Eight species have produced 28 interspecific hybrids. The trap is very effective and even in Europe in glass greenhouses the pitcher is full of flies and bees. When there are too many insects only part are digested and the rest decay.

Diptera, Hymenoptera, and other insects attracted by smell and color are easily and efficiently captured. Transparent windows towards the summit of the pitcher are also agents of capture, since the insect, willing to escape towards those windows, is trapped a second time. Cheers (1983) says that he has often observed with *Sarracenia* spp. mosquitoes trying to suck the "blood" from the red veins of the plants.

6.1.2.2   Darlingtonia

A unique species (*Darlingtonia californica*) of this genus is found in northern California and southwestern Oregon from sea level to 2600 m. It can reach 1 meter high but is generally shorter. The plant was named the Cobra lily because of its general aspect, its hood, and the "forked tongue" in front vaguely resembling a snake.

Darlingtonia californica differs from *Sarracenia* spp. in having a pitcher hood with a distinct flap. These pitchers also have reverted hairs, as in the previous genera, to repel the insects against climbing back. Insects are attracted to the plant by odor, color, and extrafloral nectaries. Some nectar is secreted as a trap on the "forktail", inside the lid of the pitcher. Transparent windows in the hood add to the insect's confusion. Similar windows exist in several orchids and are used during pollination to confuse the pollinator. Often we can hear the buzzing of Diptera and Hymenoptera trying to escape through these false windows. At the bottom, digestion starts, partly by enzymes and partly by bacteria.

6.1.2.3   Heliamphora

The species of *Heliamphora*, six in all, with several varieties, are the rarest of all carnivorous plants. They grow on the top of the tepuys, mostly in Venezuela but also in neighboring parts of Brazil and Guyana. These plants are very primitive, as both sides of the pitchers are not completely fused, thus permitting the drainage of surplus water. It is very humid in the highlands, and rain is frequent. There is also a siphon from the bottom through which more water is expelled. The leaves are waxy and seem velvety. On these plateaus, protection is needed against the enormous difference between a freezing night and a burning hot day.

The passive trap of this plant is similar to one species of *Sarracenia* in that they both have inverted hairs inside the pitcher. The small lid is generally bright red. The flowers are white and later turn red, similar to the pitcher and the lid. The pitchers can reach a length of 30 cm. These plants are often very abundant in these mountain swamps.

According to Mabberley (1987) and Cronquist (1981), the species of *Heliamphora* have many primitive features and resemble species of *Nepenthes*. These are certainly primitive plants developed at the same time as the origin of the *Sarracenia* and *Darlingtonia* species. The species of *Heliamphora* spread over the mountain tops and separated later. They are very close and the speciation seem to be very recent in this genus.

6.1.3   Cephalotaceae

The beautiful plant *Cephalotus follicularis* grows in southwestern Australia, near the town of Albany. The "Albany pitcher plant", as it is sometimes called, was discovered by Robert Brown in 1801 and is restricted to swampy coastal tracts and slow streams filled with sphagnum moss. The pitchers on these drought- and fire-resistant plants develop from July to January. They are known to resist trampling, but even so it is still in danger of extinction.

The ligneous rhizome, a kind of subterranean root, annually produces two kinds of leaves. The inferior leaves of the rosette produce the pitchers, while the superior leaves are flat and green. The pitcher is similar to one of the species of *Nepenthes* and catches the insects in the same way. Ants cross the edge of the pitcher looking for the nectariferous area, lose their footing, and drop into the pitcher filled with water and digestive juices. Some people believe it is the reflection of light on the water in the pitcher seen through the lid which attracts flying insects. Nectar and odor must also play an attractant role. At any rate, the trap is very efficient.

6.1.4   Byblidaceae (Roridulaceae)

Two species of *Byblis* live in Australia and two *Roridula* in South Africa (Rosales). Both genera differ a little; hence, they were placed in separate families, but according to Cronquist (1981, 1988) they are closer to one another than to anything else and he reunited them into one single family. They have also affinities with Nepenthales and Pittosporaceae (see *Cheiranthera* with four Australian species).

The species of *Byblis* are Australian herbs. They are suffrutescent (a plant in which many of the branches die after flowering, leaving a persistent woody base) and can reach 60 cm high. Unlike species of *Drosera, Byblis* species are passive; their capitate glandular hairs do not move. These plants grow in sandy swamps of western and northern Australia (Queensland). They die back each summer and later revive, growing up from the roots. The rainbow plant, as these are known, has not been studied very much, and their digestive enzymes are little known. Despite its resistance to drought, its survival is jeopardized.

The two species of *Roridula* are pollinated by small Heteroptera adapted to avoid the resinous hairs of the flower. They are showered with pollen when they touch the sensitive anthers which suddenly spring upright. Many other plants do the same (species of *Berberis, Mahonia,* etc.). It seems that the prey captured by the plant is not digested; rather the prey drops to the ground. After decay of the prey, the nutritive elements are absorbed by the roots. The leaves are covered with numerous glands, secreting an exsudant close to balsam, a resinous compound compared sometimes to rubber. This material traps any insect that passes by.

Several insects and spiders move freely on the leaves and escape being trapped. There they suck the juice of the captured insects. The structure of these glands is peculiar and, in some aspects, comparable to species of *Drosera,* to which the species of *Roridula* are also related. The problem of the nature of the carnivority of this plant is subject to further study, but it seems that the plant lacks any digestive enzymes. At least it seems probable that this is true carnivory and that it is related to other carnivorous plants, as it does capture prey.

6.1.5   Droseraceae

The Droseraceae belong to the Nepenthales and comprise four genera and around 90 to 106 species distributed throughout the world. The leaves are spirally arranged or in whorls, each with gland-tipped hairs (species of *Drosera* and *Drosophyllum*), or they develop an active trap as in species of *Aldrovanda* (aquatic) and *Dionaea* (terrestrial). In species of *Drosophyllum,* the hairs are not irritable as in species of *Drosera.* The gland-tipped red or greenish hairs of *Drosera* spp. are capable of movement when irritated. They hold and digest insects. In the unique species of the genus *Dionaea* — *D. muscipula* (or Venus fly trap) — the lamina is made of two hinged lobes with cilia and three sensitive hairs. The leaf folds upon stimulation of one or two of the triggers. Some species of *Mimosa* also fold their leaves rapidly, but they lack triggers such as these three sensitive hairs. *Dionaea* spp. leaves secrete digestive enzymes from their surface glands. The plant can get "tired" after several closures, especially if the stimulation is not coming from edible prey.

The 90 to 100 species of *Drosera* comprise a cosmopolitan genus. They grow on rocks, in swamps, and along banks of streams and even in a semi-desert environment during the short rainy season. The shape and size of the leaves vary according to the species. These leaves are covered with glandulous hairs containing many vessels and terminating with red, inflated heads, and they secrete a gelatinous and brightly colored fluid. Sundews, as members of the genus *Rossolis* are called (the name coming from those bright adhesive droplets), attract insects by their color, their odor, and their brilliance. Flies, butterflies, and other insects are captured by those tentacles which, when touched, bend onto the prey, secrete a digestive enzyme, and, after killing the prey, reopen to catch more.

Half the species of *Drosera* are Australian. *Drosera gigantea* from the Australian continent reaches a height of 1 m, and there are also large species in the Cape area. Several species possess a tubercle (*D. erythroriza* and *D. wittakeri*), and *Drosera pygmaea* is one of the smallest species, measuring only 2 cm. There are several dwarf species on Brazilian mountains and the tepuys, along with some medium-sized ones as well.

Only one species of *Drosophyllum* (*D. lusitanicum*) is known. This species lives in the south of Portugal, Spain, and northern Morocco. The leaves, which are somewhat woody and marcescent, contain two kinds of glands: pedunculate glands secreting a glutinous liquid and sessile glands secreting digestive juices only when stimulated by nitrogenous compounds. Insects drop on those glands and are rapidly caught. After a time of struggling, the insect drops inside the trap and is asphyxiated. The enzyme digests the insect rapidly. The pedunculate glands do not move but are also capable of secreting an enzyme.

Dionaea muscipula is the most famous of the carnivorous plants. It has short rhizomes and carries a rosette of leaves near the ground. Darwin named this the Venus fly trap, "the most extraordinary plant of the world." The generic name *Dionaea* is another name for the goddess Venus.

Willis and Airy-Shaw (1973) described the mechanism of capture of this plant: "Each leaf has a lower and a upper blade. The former can be regarded as a winged petiole; the latter has a quadrangular shape and the margins project as long teeth placed close together. The two halves of this part of the leaves are bent upwards so as to present a flat V-form in cross section. The edge of each leaf is green; the inner part of the surface is covered with reddish dots, which are digestive glands. Unless stimulated, no secretion is carried on. On each half of the leaves are three long hairs, the trigger hairs jointed at the base so they fold downwards when the leaf closes." Slight touch of the hair or hairs causes an immediate closing. The teeth cross one another. Then, the leaves close tightly together and digest the insect; after that, they reopen.

Aldrovanda vesicularia is a rootless aquatic plant with whorls of leaves, each with a trap such as *Dionaea muscipula* uses for insect capture. Its distribution is from Central Europe to Asia, Africa, and Queensland in Australia. The plant is probably distributed by birds. In Asia, it is found in India and Japan, where its distribution is somewhat erratic. The plant has perennial rhizomes and bulbs. It is menaced by water pollution. The leaves are almost a translucent green or reddish and are terminated by setiform segments — false setae (bristles) — which surround a trap made of two blades which remind one of the *Dionaea*. The flowers are isolated on a pedicel which overpasses the leaves above the water surface. Roughly speaking, each leaf is made of a stem and a double blade which capture and digest small animals, as does *Dionaea muscipula*, but their prey is much smaller.

6.1.6   Dioncophyllaceae

This family is comprised of three genera. Only one genera, *Triphyophyllum*, has a species that is carnivorous: *T. peltatum*. The discovery of the carnivorous potentialities of this plant was not published until 1979.

Triphyophyllum peltatum lives in western Africa (Ivory Coast, Liberia, Sierra Leone) where it seems to replace species of *Nepenthes* and is localized in the primary forest, or at least what is left of it. It is a plant in regression and will become extinct pretty soon; as yet, it has not been cultivated.

The plant has three kinds of leaves, but only the filiform leaves are glanduliferous. These leaves carry pedunculate or sessile glands which capture flying insects and other small arthropods. The glands are extremely sophisticated and very much similar to those of of species of *Drosophyllum* and *Drosera*. When mature, the glands carry hydrolytic enzymes. Stimulation of the plant by the prey increases the secretion. A remarkable fact is the periodicity of the secretion, a periodicity which is not unique among the carnivorous plants but which also exists among several species of *Sarracenia*, *Cephalotus*, and even *Dionaea*, where the flattened petiole (phyllode) increases late in the season.

6.1.7   Martyniaceae

The family has sometimes been compared to Pedaliaceae (Mabberley, 1987). Willys and Airy-Shaw (1973) consider the Martyniaceae as having three genera and 13 species, all tropical and subtropical in distribution and all living in dry or in coastal regions. The family is, however, very close to Pedaliaceae and Bignoniaceae.

The genus *Ibicella* has three species in tropical South America and Mexico. *Ibicella lutea* is entirely covered with glutinous hairs capable of capturing insects, mostly gnats and other small Diptera. *Ibicella lutea* produces beautiful yellow flowers attractive to pollinators. The plants absorb albumineous substances which come in contact with its leaves.

6.1.8   Lentibulariaceae

The Lentibulariaceae comprise around 354 species in five genera. There are all insectivorous. They live in or near water or some in dry habitats. *Pinguicula* is the only genus with species with roots, and they live on humid soil. Also, several species of *Utricularia* are subterrestrial, namely those living in the tropics.

There are approximately 46 Holarctic species of *Pinguicula*, with some species invading the West Indies, Mexico, Chile, and other Neotropical regions. These are small, carnivorous, terrestrial herbs with leaves in rosettes, viscid above, with secretory glands that trap small insects. The glands are either sessile or pedunculate and secrete an adhesive liquid. Rain washes the insects against the sides of the leaves which are slightly curved up. When these leaves are stimulated by nitrogenous material, the edges curve more against the prey. The sessile glands secrete an enzyme, digest the prey, and absorb the products of the digestion. After that, the leaf unrolls again. In England, the plant is named "butterwort" because the leaves are oily. *Pinguis* means "fat" in Latin, and the French name "grassette" has the same meaning. The beautiful flowers vary from white to red, purple to yellow. Some are rather big, growing to 20 cm for certain species, for otherwise small plants. According to specialists on the genus *Utricularia*, it is composed of 180 to 300 species (the genus has been revised recently). The species are distributed world wide but generally remain in temperate waters. Several species are epiphytic within the phytotelmata of Bromeliaceae and other plants (*Orectanthe* spp., Xyridaceae, from Tepuys) of the Neotropical realm. Some are epiphytic among mosses on trees or terrestrial, with bulbuous roots, but most of them live rootless in water.

Terrestrial species have normal leaves and more or less subterranean traps on the stolons or the roots; these are probably only partially functional. However, they always live in a very moist environment, and water surrounds them. The leaves of aquatic species are divided and have bladders with trap doors, where small animals are captured and readily digested. *Utricularia* spp. traps are no bigger than a match head but they are very sophisticated and readily kill their prey. In all forms there is no clear distinction between roots, stems, or leaves. When the trap doors are triggered by small crustacean, for instance, the animal is sucked into the bladder and the trap door recloses. *Biovularia* is more or less a synonym of *Utricularia*. Two species are described, each with a trap similar to that of *Utricularia purpurea* from Brazil.

Two to four species of the genus *Polypomphalyx* live in Australia. The genus is close to *Utricularia*, but the traps are dimorphic in size and in structure. Their complicated trap has been studied by Lloyd (1942), but it functions the same as those of species of *Utricularia*.

There are between 16 and 35 recognized species of *Genlisea* living in tropical America, the West Indies, tropical Africa, and Madagascar. The species of *Genlisea* are floating, rootless plants. The fruit is globular. Their leaves are either blades with chlorophyll or tubular, branched and with twisted traps. These two kinds of leaves come directly from the rhizome. The flowers are similar to those of the species of *Utricularia*, but the way the prey is captured is totally different; it uses the technique of the eelpot, as Darwin called it, but in a much more complex way. Only the flowers are above the water; the rest of the plant is submerged. The size of the trap varies between 2.5 and 15 cm according to the species. The "digestive tract" shown by these species is really extraordinary. The "mouth", with two lips, has little acute cells (rather like "teeth") to prevent the sucked-up animal from escaping. Digestive and mucus glands are spread along the wall. The spines are directed towards the inside, preventing the captured animal from coming out, the same as for the other carnivorous plants. The central bulb is, in a way, the "stomach" of that strange plant for which only an "anus" is missing to complete the resemblance to an animal. Peristalsis waves or contractions passing along the length of a digestive tract are also missing. We will come back to the function of that true vegetable intestine. Returning to the entrance slit (mouth), this opens into the two spiral arms, which in turn, facilitate the access of prey.

Mushrooms as carnivorous plants (fungi) are not mentioned here as they capture protozoa and mostly nematodes. They do not attack small crustaceans or any other arthropod.

6.2   Protocarnivorous plants

We separate the protocarnivorous plants from the real carnivorous species because they capture insects and spiders in their pitchers, attracting them by color or odor, but they do not produce digestive enzymes. Digestion of the prey is done by bacterial action only, inside the water reservoir of the plant. As with pitcher plants, these plants have commensals around their leaves and aquatic larvae living with total impunity inside the pitcher (mosquitoes and various flies). Several species of *Heliamphora* and a few species of *Sarracenia* do not secrete enzymes or else have lost that power, but they are kept with the carnivorous plants because they belong botanically to carnivorous families and genera. It is important to remember that all carnivorous plants are dicots and all protocarnivorous plants are monocots. (See Table 6.3.)

6.2.1   Catopsis berteroniana

The insect trapping properties of this plant were first mentioned by Fish (1976) in southern Florida (see also Ward and Fish, 1982; Frank and O'Meara, 1984). *Catopsis berteroniana* grows in full sun at the top of red mangroves in Florida. This is a Bromeliaceae, with a yellow-green color, living in an upright habit. A slippery, white, ultraviolet-reflecting powder exuded by the leaves is considered to attract insects and prevent their escape. Many Bromeliads, most of them epiphytic, provide leaf bract tanks suitable as a breeding place for mosquitoes. *C. berteroniana* also harbors larvae of species of *Wyeomya* (Frank and O'Meara, 1984).

Table 6.3 Genera of Protocarnivorous Plants

Family	Genus	No. of known species	Geographical distribution
Bromeliaceae	*Brocchinia*	18[a]	Tepuys, Venezuela
	Catopsis	19[b]	Florida
Eriocaulaceae	*Paepalanthus*	485[c]	Serra do Cipo, Minas Gerais, Brazil
Total	3 genera of monocots	4	Neotropical

[a] Only 2 protocarnivorous.
[b] Only 2 protocarnivorous.
[c] Only 1 protocarnivorous.

6.2.2   Brocchinia *spp.*

Two other bromeliads, *Brocchinia reducta* and *B. hechtioides*, also are, according to many, carnivorous since they entrap insects. They live in swampy savannas on the top of the Venezuelan tepuys. Their erect, yellow leaves have a cylinderical shape with a waxy slippery powder in the inside surface. The liquid emits a nectarious smell, and the leaves contain trichomes able to absorb the digested amino acids. Very large bogs containing *Brocchinia* spp. are widespread in these mountains. Mostly ants and small insects are captured by these plants in nature. In greenhouses, the variety of prey is much reduced.

B. reducta is abundant in Roraima and many other tepuys, including the Auyan-tepuy. It is found at an altitude of between 900 and 2200 m. *B. hechtioides* is clearly related to *B. reducta*, is reported to be saxicolous and terrestrial in the savannas, and is found at altitudes between 1500 and 2125 m on many tepuys. The species of *Heliamphora*, grow in the same areas on moist, sandy, acidic soil, exposed to the full sun of the summits and to the cold of the night. There it also rains most of the year. Several species of *Utricularia* grow as epiphytes between the leaves of species of *Brocchinia* and *Orectanthe*.

6.2.3   Paepalanthus bromeloides

Paepalanthus bromeloides, an Eriocaulaceae, is the last protocarnivorus plant found in the Serra de Cipo in Minas Gerais in Brazil. There are about 485 species of this genus in tropical America, and so far this is the only one with the shape of a bromeliad with a central water resevoir capable of capturing insects (Figueira et al., 1993, 1994; Jolivet, 1993). Actually, other species such as *P. robustus* have the appearance of a bromeliad but lack a central reservoir. Probably other species have evolved towards a bromeliad shape in the Serra de Diamantina in Brazil, but they are still to be found. The aspect of a typical species of *Paepalanthus* is completely different from that of a bromeliad, with their erect flowers and small leaves.

P. bromeloides carry a biocenosis (of insects and spiders) on its leaves and flowers. The central reservoir is mucilaginous, and a waxy powder cover on the leaves reflects the ultraviolet light, which attracts insects. The insects and spiders drop the remains of their feast and their excreta into the central reservoir, where they are absorbed by the trichomes after digestion by bacteria. Two theses have been written in Brazil regarding this plant (Castro, 1986; Figueira, 1989), as well as two papers (Figueira and Vasconcellos-Neto, 1993; Figueira et al., 1994).

6.3   Other plants on the way to carnivory

No other real protocarnivorous plant is actually known in the world, but many genera and species are on the way toward carnivority. It can only be guessed that the absorption of prey occurs either inside water tanks by bacteria (species of *Musa, Dipsacus, Arum*) or by direct assimilation of the nitrogenous products through the stem or the leaves. It does not seem that *Caltha dionaeifolia*, a Ranunculaceae from Terra de Fuego, Argentina, and Chile, is carnivorous (Joel, 1985). However, this plant lives in the same swampy habitat that species of *Pinguicula* and *Drosera* inhabit and possesses leaves in the shape of those of *Dionaea muscipula*. Almost no observations have been done in the field, and in the laboratory only on dry specimens. Many plants, such as species of *Gaylussacia* and *Vellozia* in Brazil, have a glutinous petiole capable of capturing ants, probably as a protection for the flower. Several species or varieties of solanaceous plants, such as species of *Solanum, Lycopersicum, Nicotiana*, or *Petunia*, have adhesive hairs which are glutinous and toxic. There even exist quadrilobate hairs on *Solanum polyadenium* and *S. berthaultii* capable of capturing insects. Only species of *Aleurodes* escape from the traps, thanks to the powdery wax which covers them. Species of *Silene* and the species of many other genera also have glutinous hairs. These plants do not digest the insects they capture, but they do kill them.

Certain species of Orchidaceae, Asclepiadaceae (*Calotropis*), and Araceae capture Hymenoptera temporarily for pollination. Species of *Kniphofia* (Liliaceae), mountainous African flowers, capture bees in Europe, but do not release them, probably because they lack the ability to do so. They are not yet adapted for this; they have not coevolved. The African tulip tree, *Spathodea campanulata*, a Bignoniaceae, attracts insects with extrafloral nectaries. Their inflated calyx is full of secreted water. Outside of Africa (in Brazil, for instance) species of *Melipone* and *Trigona* bees drown and die in the liquid. The bees rot rapidly but are not really absorbed because the flower drops from the tree too soon. I did not see the same phenomenon in Panama, where probably another cultivar has been introduced from Africa.

In temperate climates, *Salvia glutinosa* has adhesive leaves, but the Ericaceae *Befaria racemosa* has buds with an exterior surface of petals, sepals, and fruit covered with a viscuous adhesive material on which insects, including even bees, are captured in numbers. Pollination takes place inside the flower which produces no glue. The chemical is as strong as flypaper, and Eisner and Meinwald(1983) suspect that, outside of the protection against herbivorous insects and ants, this system allows the plant to become a facultative or secondary insectivore. Nitrogenous decay products from the captured insects would leach into the ground and feed the plant. Species of *Roridula* would be carnivorous the same way, producing not glue but resin. Carnivorous plants could be plants which started by capturing insects and then slowly developed the ability to digest them on the spot. *Befaria racemosa* is restricted to the coastal plains of Florida and Georgia.

Several Poaceae (Gramineae) such as species of *Setaria*, *Molina*, and others capture insects with hooked hairs on their reproductive ears. Birds are impaled on the spines of the mountain bromeliad, *Puya raimondii*, in Peru, and very often birds are entangled in its leaves. It is said that nutrients from this prey, as well as their droppings, may be absorbed by the plant (by the roots, even by the leaves?).

The genus *Lathraea* (Scrophulariaceae) is Eurasian and is comprised of seven species. They are parasitic on roots of tree species of *Fagus*, *Populus*, *Salix*, *Corylus*, and others. They possess thick rhizomes bearing four rows of tooth-like scaley leaves, each hollow with glands in side chambers. Dead insects are often captured and digested. These glands resemble very much those of the true carnivorous plants. *Caltha dionaeifolia*, mentioned previously, also seems to have glands on its leaves, but according to Joel (1985) these are stomata and do not have trigger hairs; however, this remains to be verified.

Finally, several seeds with a mucilaginous cover also may be carnivorous, such as *Capsella bursapastoris* (Brassicaceae). Their natural glue contains protease which would help seeds with little food reserves to survive on poor soil. It is said that Pasteur, the French microbiologist and botanist, upon seeing this plant for the first time, said, "It is similar to my purse (bursa); it is completely flat." The flatness of the seed is compensated for by its ability to gather extra nitrogenous substances.

6.4   Phytotelmata or water reservoirs

The phytotelmata are natural water reservoirs specially adapted in plants such as bromeliads in the Neotropics, Musaceae in Asia, pitcher plants, bamboos, tree holes, and many other water containers. Generally, the reservoir is a breeding place for aquatic larvae of dragonflies, flies, mosquitoes, beetles, tadpoles, etc. Even certain damselflies are exclusively adapted to tree holes in the tropics, along with many mosquito species, and each one of these abundant species is permanently in competition with the others inside of these small water bodies.

Carnivorous and protocarnivorous plants harbor abundant aquatic fauna not digested by the liquid which even eventually prey on other captured insects. Bromeliad malaria exists in the Neotropical forests and is due to a species of *Anopheles* mosquito breeding in the epiphytic small tanks. In Ethiopia, and elsewhere, banana leaf axils are responsible for species of *Culex* and *Anopheles* breeding therein and thus the transmission of several diseases including malaria and filariasis. In the Pacific islands, filariasis is due to species of *Aedes* breeding, in the larval stage, in the empty shells of coconuts.

A book edited by Frank and Lounibos (1983) is devoted to phytotelmata. In it, the definition of phytotelmata as given by Fish is "a structure formed by terrestrial plants that impound water, such as modified leaves, leaf axils, flowers, stem holes, or depressions, open fruits, and fallen leaves." Twenty-nine plant families respond to that definition and almost all aquatic insect orders occur in phytotelmata. The relative habitat specificity of phytotelmata-inhabiting insects may be attributed to chemical immunity against enzymes of carnivorous plants and other special adaptations.

When the great Neotropical damselflies, *Mecistogaster* spp., drop their eggs with their very long abdomen into tree holes, they do it by throwing the eggs into the hole and water, but they are careful not to enter the hole themselves. The ballistics are almost perfect. Several biocenosis exist, namely for species of *Nepenthes*, with maximum variety in southeastern Asia, particularly in Borneo, which is probably the center of distribution of the genus. This fauna is resistant to the digestive juices of the plant and the dilution of the contents of the pitcher with water improves that tolerance. The pitcher fauna is much smaller in the Seychelles and in Madagascar, the western limits of the plants' distribution.

Some absorption of nitrogenous substances derived from the reservoir fauna of the Bromeliaceae is probable but always due to bacterial decomposition, and it is very small, indeed. The plant is not dependent on that to survive. The two mountain species of *Brocchinia* and the Floridian species of *Catopsis* are exceptions among bromeliads and are specially adapted for the capture of terrestrial insects, namely ants by species of *Brocchinia*.

New discoveries about carnivorous plants seem improbable, and the only surprises could come from tropical mountainous plants, well-known botanically but poorly studied biologically.

6.5   Conclusions

According to Joel (1988), carnivorous plants do not develop Bakerian mimicry, a deceptive mimetic system attracting insects into a pseudoflower by their color, odor, and extrafloral nectaries accompanying the traps. On the contrary, they would maintain mutual, rather than deceptive, relations with insect communities. Insects benefit from nectar and at the same time pay for the benefits by sacrificing a small portion of their communities to the plants. That hypothesis would mean that many insects, after drinking nectar, do escape the traps. This idea, as interesting as it can be, though, is purely semantic. It does not maximize the goal of the system: to obtain nutrients for plants growing on a poor soil, although these plants have spent a lot of energy in devising their very sophisticated traps.

The 500 or so carnivorous plants have very different kinds of traps and some are very complex. It is not our goal to study them all here. Many books and papers have been published on the topic. These traps, however, are either active or passive. In the first category fall the active snap traps of species of *Dionaea* and *Aldrovanda* which catch insects by folding the two parts of the leaf and setting the trap by using trigger hairs. The sucking traps of species of *Utricularia*, *Biovularia*, and *Polypomphalyx* are water-filled bladders. These too, are very sophisticated. Their prey is absorbed inside very quickly. Also, we must add to this list the pseudodigestive suction trap of species of *Genlisea*.

Among the passive traps are the pitcher plants (the species of Nepenthaceae, Sarraceniaceae, Cephalotaceae, and the protocarnivorous bromeliads). The flypaper type of traps, using adhesive droplets, can be completely passive (species of *Byblis*, *Drosophyllum*, *Roridula*, *Triphyophyllum*, *Ibicella*, and several rare species of *Pinguicula* and *Drosera*). On the contrary, most of the adhesive traps of the species of *Drosera* and *Pinguicula* are active; the edges of the leaves of *Pinguicula* spp. and the glutinous hairs of *Drosera* spp. curl above the trapped insects.

The digestion of insects by carnivorous plants is done through enzymes secreted by special glands on the leaves inside the pitcher (Juniper et al. 1989). The system is very sophisticated, as after digestion, proteins are absorbed by the plant leaf, often by the same glands which were secreting enzymes. Numerous enzymes are secreted, but species of *Heliamphora* and the protocarnivorous plants are exceptions, as the digestion is done by bacteria.

Acids secreted by the digestive glands of *Dionaea* facilitate the absorption of the amino acids after digestion. It is similar to an animal stomach where acids also aid in the process. There are two different types of digestive glands in species of *Cephalotus*. A brief review of the role of the enzymes and the digestive process is given in Jolivet (1987). Generally, because of the lack of chitinase, the insect exoskeleton and wings remain in the "botanical stomach", an expression from Darwin in his classical book (Darwin, 1875).

Animal proteins increase plant vigor, and the plants absorb not only nitrogen but also mineral salts including phosphorus rare in their environment. In the case of the species of *Genlisea*, the prey moves along the digestive tract, the walls of which are covered with special glands that help in moving liquid along the "intestine". Protozoa, crustacea, and insect larvae comprise their main food, the remains of which are accumulated in the utricule, which lacks a "botanical anus".

With several carnivorous plants, namely the pitcher plants, when digestion is complete, the fluid is reabsorbed. Secretion can last several hours, and
the remaining volume can exceed the plant's capacity to re-absorb it. Reabsorption fails, and the leaf starts to rot, a victim of botanical indigestion.
We could say that the pitcher plant has bitten off more than it can chew. Such
botanical indigestion can also happen with other carnivorous plants (for
example, species of *Pinguicula, Utricularia*, and *Dionaea*) when a big insect
produces oversecretion of digestive fluid over a period of several hours.
Normally, reabsorption takes longer than secretion for the average plant.

The pitchers of species of *Nepenthes* are contaminated by bacteria when
the lid opens. Also, their digestive fluid contains hydrochloric acid exactly as
in a human stomach. Certain carnivorous plants, such as species of *Pinguicula*, also capture atmospheric pollen and digest it. Young spiders do the
same, which sustains their early life. The number of captured insects varies
according to the species of plant, but it can reach as high as seven per plant
(in *Drosera* spp.) and in the thousands in species of *Sarracenia*.

Several botanists (Emberger, 1960) have tried to link all carnivorous
plants to the same origin. It is certain that all are more or less related, unlike
the myrmecophilous plants, which surely are polyphyletic.

About 500 species of angiosperms belonging to eight families are known
to be carnivorous, and four belonging to two families are protocarnivorous.
The last are on the way to carnivory but have never developed specialized
enzymes to digest their prey. Ultraviolet patterns in the traps of carnivorous
plants attract insects. Certain traps have ultraviolet patterns attracting insects, as do flowers (Joel et al., 1985). Both carnivorous and protocarnivorous
plants have ultraviolet trapping systems, and insects, particularly Hymenoptera, are extremely sensitive to "black" light.

No fossil carnivorous plant has been found, but the trait seems to have
developed at the end of the Jurassic or early Cretaceous with the advent of
ants and butterflies. Only pollen and seeds of carnivorous plants have been
found, in the Miocene of North Borneo, for instance.

Probably the carnivorous syndrome and the different types of traps have
evolved several times since the lower Cretaceous, as the group is not widely
polyphyletic as far as we can understand. Several authors (Croizat, 1960:
Heads et al., 1984; Emberger, 1968) have been in favor of a certain monophyletic classification among these plants. A review of the topic is found in
Juniper et al. (1989), but recent work based on nucleotide sequence (Albert et
al., 1992) indicates that carnivory and trap diversity have arisen independently in different lineages of angiosperms.

Carnivory is not the most economical way for plants to secure nutrients
(Benzing, 1987), but a parallel can be drawn with myrmecophilous plants.
This shows how disparate life forms have changed, under evolutionary
processes, existing characters into novel combinations to achieve new functions.

According to Juniper et al. (1989) and Albert et al. (1992), carnivory involves morphological features associated with attraction, retention, trapping, killing, and digestion of animals and adsorption of nutrients. That definition involves the protocarnivorous and the *Heliamphora* species lacking proper digestive enzymes. *Heliamphora* spp. are primitive and probably the ancestors of other species of Sarraceniaceae. *Brocchinia* species and others are on their way to carnivory. Carnivorous plants do not have a unique origin, and as Darwin (1875) hypothesized, carnivory results from natural selection operating on a pre-existing variation. Perhaps, in the future, *Caltha dionaeifolia* will "learn" to fold its leaves and will secrete enzymes, but this is one of the many mysteries of the evolution.

chapter seven

Myrmecophily and ant-plants

The relationships between plants and insects are very diverse and complex. Ant-plant relations are a result of evolution that has allowed plants to survive on a poor soil with the help of the insects. Contrary to that of carnivorous plants, this is a true mutualism; ants, in exchange for lodging, provide the plants with pruning of weeds and climbers, defense, and nitrogen enrichment. Mite-plants, or acarophytes, seem to exist and to be more common than previously believed. The reality of this association, rejected by Schnell (1966), is accepted by biologists working on the topic today, mostly Americans and Australians. Ant relationships with plants are very diverse and include such things as seed dispersal, ant gardens, etc. All these aspects will be discussed below.

7.1 Myrmecophytes, or ant-plants

There are about 540 insectivorous plants as opposed to around 465 myrmecophilous plants or myrmecophytes (see Figure 7.1 and Table 7.1). New myrmecophilous plants are discovered every year, while the list of carnivorous plants seems almost complete. Myrmecophilous plants are surely polyphyletic. The interrelationships the ants have found are entirely different from one plant to another. Even within species of the same genus (*Macaranga*, for instance), the system is different in Africa, Asia, and even between the southeastern Asian species (see Figures 7.1 and 7.2).

I cannot list here the 465 species, or even the 52 families with myrmecophilous genera. I refer to Jolivet (1986; 1987; 1991; 1992; 1993a,b,c; 1994; 1995; 1996a,b) for details about the species involved. I list here only some very characteristic genera of the three tropical areas: America, Asia, and Africa. America is the richest for the variety and the quality of the relationships found, Asia follows, and then Africa is the poorest. The detailed biology of certain genera has not been fully studied in Africa, and a lot remains to be done with several American and Asian genera.

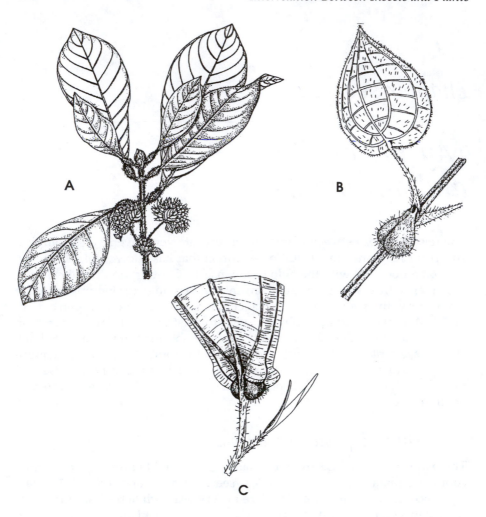

Figure 7.1 Myrmecophilous and acarophilous plants. (A) *Hoffmannia vesciculifera* Standley (Rubiaceae), occupied by a species of *Solenopsis* ants, from the Highlands of Panama. (Adapted from Dwyer, 1980.) **(B)** *Tococa coronata* (Melastomataceae) inhabited by azteca ants, with paired ant pouches and entrance pores. (Adapted from Benson, 1985.) **(C)** *Clidemia tococoidea* (Melastomataceae) with paired stem domatia inhabited by azteca ants. (Adapted from Benson, 1985.) **(D)** *Clidemia hammeli* Almeda (Melastomataceae): (1) representative leaf (abaxial surface); (2) enlargement of lower leaf surface showing pocket domatia; (3) dorsal and ventral view of the mite, *Ololaelaps* sp. (Adapted from Almeda, 1989.) **(E)** *Cecropia glaziovii*; In = internode; LS = leaf scars; MC = Müllerian food bodies, which lie on the trichilium or cushions of setae; P = petiole; Pr = prostoma; TB: terminal bud. (Adapted from Dumpert, 1981.) **(F)** *Acacia sphaerocephala*, the hollow thorns (T) of which are mostly inhabited by ants, which gnaw entrance holes (or find natural openings) in the young thorns; BB = Belt bodies or trophosomes; N = extrafloral nectaries. (Adapted from Schimper, 1888.)

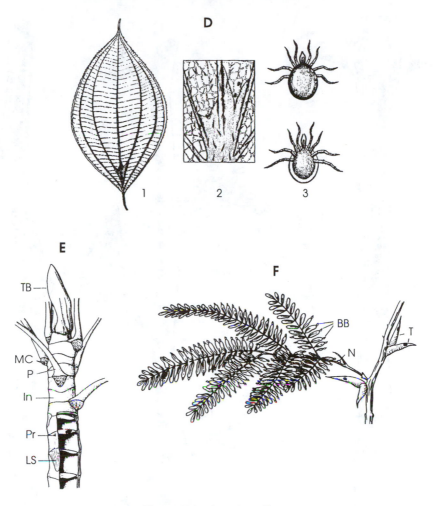

Figure 7.1 (continued)

Table 7.1 Myrmecophilous Plants, Trophosomes, Food Bodies

Plant	Ant	Food body	Composition
Acacia	*Pseudomyrmex*	Belt bodies	Proteins, lipids
Cecropia	*Azteca*	Müllerian bodies	Glycogen, lipids, proteins
Macaranga	*Crematogaster*	Beccarian bodies	Lipids
Ochroma, etc.	Various spp.	Pearl bodies	Lipids
Piper	*Pheidole*	Delpinian bodies	Proteins, lipids
Phymatodes, Lecanopteris	*Iridomyrmex*	Sporangium bodies	Lipids
Barteria	*Pachysima*	Glanduliferous teeth	Proteins, lipids

Adapted from Buckley (1982).

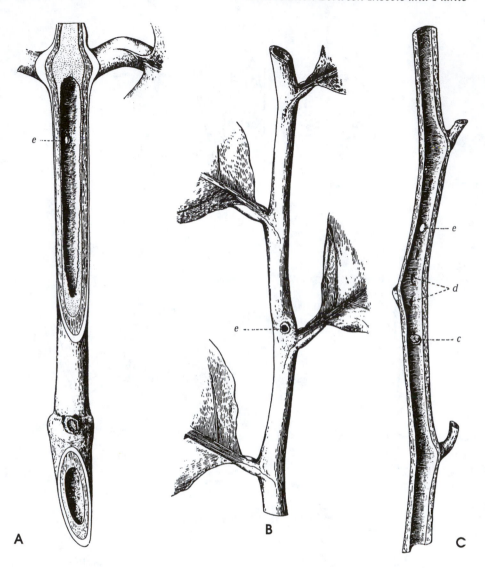

Figure 7.2   Myrmecophilous plants. **(A)** *Nauclea* sp. (Naucleaceae). Portion of stem is inhabited by ants; sectioned longitudinally in the domatium area; found in Zaire. **(B)** *Barteria deweyrei* de Wildemann and Durand (Passifloraceae): external view of a portion of a lateral branch inhabited by ants, found in Zaire. **(C)** *Barteria deweyrei*: longitudinal section of this branch showing a coccid (c) fixed on the inner walls and a small depressions (d); (e) is an orifice leading into the domatium; found in Zaire. **(D)** *Vitex staudtii* Guerke (Verbenaceae): longitudinal section of a stalk inhabited by *Viticicola tessmanni* (Stiz); e = entrance to cavities; le = accessory exit; l = lateral galleries excavated by ants. Found in Zaire. **(E)** *Myrmecodia schlechteri* Valeton (Rubiaceae) growing on Casuarina in central highlands of Papua New Guinea; abundant in Goroka. (Parts A, B, C, and D adapted from Bequaert, 1922; Part E adapted from Huxley and Jebb, 1993.)

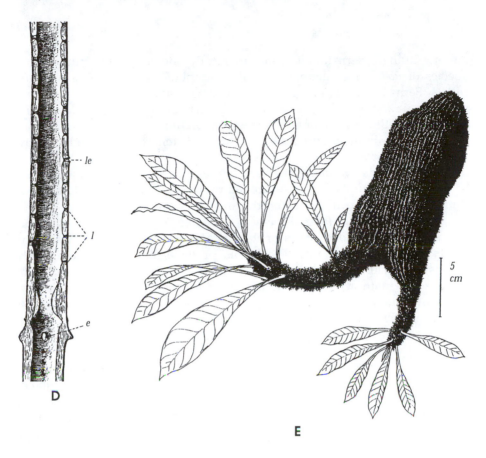

Figure 7.2 (continued)

Myrmecophilous plants are plants which live in symbiosis with ants. They provide shelter to these ants, and they often feed them (trophosomes or food bodies, extrafloral nectaries, pearl bodies). In exchange, the ants protect the plants against herbivores (namely, species of *Atta* or leafcutter ants in South America), destroy the climbers or vines (lianas and surrounding plants), and feed the plants with dead ants and excreta, at least that is the theory. In reality, the theory is not so perfect. The ideal is probably reached by the Central American *Acacia* tree species, the Neotropical species of *Cecropia*, the southeastern Asian species of *Macaranga*, and several other plants, such as certain *Piper* species which also provide trophosomes, or Delpinian bodies. Also, we can add species of *Myrmecodia* and *Hydnophytum*, which are epiphytic on Asian trees but do not provide any trophosomes, having instead viscid and edible fruits. Some plants devoid of extrafloral nectaries provide ants with sugars thanks to the coccids they raise inside their natural dwellings.

7.1.1   American plants

Among 20 angiosperm families, one fern family, and one conifer family (with species of *Podocarpus*) there are peculiar adaptations to ants in American plants. Ferns have special rhizomes harboring the ants, which feed on this sweet tissue and probably also on sporangium bodies (species of *Solanopteris, Polypodium*). Several American species of Orchidaceae and Bromeliaceae harbor ants — in the pseudobulbs (the inflated stem) of the orchids and by the leaves being constricted and closed at the top (e.g., species of *Tillandsia*) in the bromeliads. Normally, the Bromeliaceae plants are opened at the top to receive rain, but the *Tillandsia* are fed pretty well by the ant excreta and thus do not need open leaves and the ants can avoid rain.

Several species of *Piper* (Piperaceae) shelter *Pheidole* ant species in the petiolar cavities (folded edges of the petiole) and in the stems, probably partly dug out by the ants. The plants produce food bodies (Delpinian bodies) rich in lipids inside the petiolar cavities which feed the ants and their larvae. The main activity of the ants is to remove the eggs of invading herbivores and parasitic vines. The growth of the food bodies is linked to the presence of the ants or symbiotic clerid beetles, which eventually feed on the ant brood when they coexist (Letourneau, 1990). The beetles learned to decipher the code for increased trophosome production.

Cecropia (Cecropiaceae) is a very peculiar genus of trees comprising nearly 100 tropical American species, of which only a few are myrmecophobic. These trees harbor ants in the leaf stem and feed them with Müllerian bodies growing on the stem below the petioles, over a hairy cushion (trichilium). The species without ants are found on mountains, islands, and sometimes on plateaus, such as, for example, *Cecropia hololeuca* in the Brazilian Atlantic area. One or two species live in the Amazons. *Azteca* ant species live inside stems. The queens penetrate through a small depression called the prostoma. She pierces a hole (stoma), taking about 2 hours, and then lays eggs inside the cavity. After this, she closes the hole, which will be reopened later on by workers issuing from the first eggs. *Azteca* ant species will patrol the tree and rear coccids inside the trunk. There is no pleometrosis, i.e., the colony is founded by one queen only. The workers and the queen kill all the other queens, normally at the beginning, with only one per node. *Cecropia* spp. trophosomes contain glycogen, or animal starch.

Triplaris tree species (Polygonaceae) along the Amazon and its tributaries have hollow internodes each with prostoma. They are inhabited by ferocious species of *Pseudomyrmex* ants. Coccids are reared inside the stem, but no trophosomes are apparently produced by the plant.

Among the American myrmecophytes, several *Acacia* species (Mimosaceae) from Mexico and Central America surely represent the most perfect myrmeco-philous plants. The trees produce yellow food bodies (Belt bodies) on the tip of the petioles and sweet enlarged nectaries on the petiole. The big thorns are inhabited by ant colonies which either dig into the tender tissue or penetrate through a natural opening, this mostly when the thorn is empty. The ants defend their tree with efficiency and clean the tree and its surroundings from vines and weeds. Acacia trees are inhabited by species of *Pseudomyrmex* ant colonies, monogynous or polygynous according to the species. Several Central American *Acacia* are without ants, being protected by the toxicity of their leaves. Janzen (1967, 1969) has studied the narrow relationships between ants and *Acacia* species.

Thirteen tropical America genera of Melastomataceae have species that are partly myrmecophilous. Normally, they have paired or unpaired foliar or petiolar pouches such as species of *Tococa* and *Maieta*. *Pheidole* ant species seem to be efficient in protecting the plants, improving the plant's fitness, and safeguarding the fruit. The pouches grow without ants in greenhouses. Similar billets exist among the species of several African plant genera and in Central America in *Besleria formicaria* (Gesneriaceae) and *Hoffmannia vesciculifera* (Rubiaceae). *Besleria* species harbor species of *Pheidole* and *Solenopsis* ants in Panama, and species of *Hoffmannia* harbor only *Solenopsis* ant species. In Panama, we found that *Besleria formicaria*, a mountain plant, harbors the queen, workers, and brood in one pouch and the larvae of a species of Diptera (Mycetophilidae Keroplatinae), *Proceroplatus belluus*, a predator, in the other. The larvae capture ants on a silk thread (Windsor and Jolivet, 1996; Aiello and Jolivet, 1996).

In the Neotropical region, the myrmecophilous syndrome appeared not only in the tropical forest but also in semiarid areas of Mexico, Central America, and Paraguay (in *Acacia* species), but only in Mexico do *Acacia* species produce food bodies (or Belt bodies) on their leaves. There are more ant-plants in South America, but I have listed here only samples of the most characteristic genera.

7.1.2   Asian plants

This is the richest kingdom for myrmecophytes after America. Outside of several ferns, there are two monocot families (Palmae and Orchidaceae) and 24 dicot families exhibiting more or less important signs of myrmecophily.

Among the epiphytic ferns, several Polypodiaceae (species of *Phymatodes* and *Lecanopteris*) have, in their rhizomes, cavities inhabited by ants with internal roots absorbing nitrogen debris. The sporangia carries fat bodies removed by ants as food.

Among the Palmae, the rattan palms (species of *Korthalsia*) have foliar bases which form a closed shield, narrowly enveloping the stem; among several species, that organ is dilated into a swollen base, rounded or oblong and hollow, named ochrea or ocrea (a carton structure), often pierced and occupied by ferocious *Camponotus* ant species. Few herbivores feed on species of *Korthalsia*, but some leaf beetles seem to be immune to these ant attacks.

Several Asian Orchidaceae show galleries occupied by ants in the base of the swollen pseudobulbs. Probably the orchid *Grammatophyllum* species gets their nitrogen from excreta and dead ants accumulated inside. One carnivorous plant, *Nepenthes bicalcarata*, from Borneo, sometimes harbors ants in its hollow leaf. This case seems unique and has been confirmed recently.

Among the Euphorbiaceae, several species of *Endospermum* and *Macaranga* trees harbor ants in their hollow branches. *Macaranga* species replace *Cecropia* species as parasol trees in southeastern Asia and produce food bodies (Beccarian bodies) in various places. The trophosomes are spread above the young leaves between the veins (*M. formicarum*). Food bodies and sweet secretions of coccids are practically the only food of the ants. In some *Macaranga* spp., the ants also live under the lanceolate, erect and persistent bracts of the bud. Sometimes the inside of the bracts carries abundant white food bodies. Occasionally, the ants bring their larvae inside the bracts. The ants, primarily the *Crematogaster* species but also species of *Dolichoderus*, protect the plant very efficiently against herbivores and vines.

The pitcher plant, *Dischidia rafflesiana* (Asclepiadaceae), also is a very interesting ant-plant. Some specialized leaves becomes a true pitcher, with a curved-in edge and being about 10 cm deep. Inside grow adventitious roots coming from the stem and the adjoining petiole. All nitrogenous fermentation is absorbed by the internal roots. The internal surface is waxy, and the absorption of ant debris and excreta is entirely due to the roots. *Dolichoderus bituberculatus*, *Crematogaster brevis*, and *Pheidole cordatus* can live separately or together inside the pitcher. Sometimes the pitcher can be double (*D. complex*) and the system seems destined to reduce water loss for a xerophytic species. The inside ascidium is a possible refuge for the ants against flooding.

The evolution of *Dischidia* spp. toward myrmecophily is progressive from species with elongated fleshy leaves to species with concave leaves with roots appressed against their support, then to species with concave leaves full of roots appressed against each other, such as a lamelibranch shell, then further to species with simple, then double pitchers. There is a tendency towards myrmecophily among the species of *Dischidia* and related genera. The ants start to colonize the plants with concave leaves appressed against their support, such as a trunk or a rock.

Among the Rubiaceae, the Hydnophytinae are among the most extraordinary epiphytic ant-plants. Huxley and Jebb (1991) designated this subtribe, which was previously studied in detail in 1884 and 1885 by the Italian botanist Odoardo Beccari.

Myrmecodia spp. are epiphytic on many trees, from the mangrove forests to primary and secondary upland forests, in the highlands as well as on the highest mountains of New Guinea. At high altitudes in New Guinea, they grow directly on the ground. The genera *Hydnophytum*, *Myrmecodia*. *Anthorrhiza*, *Squamellaria*, and *Myrmephytum* comprise all together around 95 species; almost all are epiphytic. Few of them are uninhabited by ants. The galleries of leafy stems (hypocotyle axis) are not dug by the ants but grow naturally by internal differentiation. Some are used by ants as a breeding place, others as latrines and depots for cadavers, and the nitrogenous products and dead ants are absorbed by the chamber walls, the warts, and, in some rare species (such as *Myrmecodia schlechteri* from New Guinean highlands), by internal roots. The air-conditioning of the pseudotubercule is highly sophisticated with small pores and alveolate cells. Normally *Philidris* (=*Iridomyrmex*) *cordatus* inhabits the stems, but many other ants can live inside. They are very efficient defenders against any mammalian intrusion, but *Philidris* spp. tolerate a certain number of Lepidoptera (Lycaenidae) caterpillars, weevils (*Oribius* spp.), and other herbivorous insects. Inside, in the wart cells with refuse, there are mites, nematodes, and fungi. The subtribe is distributed in all of tropical southeast Asia on the plains and in the mountains of New Guinea and Queensland, Australia.

Fruits and seeds of species of *Hydnophytum* and *Myrmecodia* are distributed by the ants, namely *Philidris* spp. and perhaps by birds. The mesocarp is viscous when ripe and adheres to tree trunks, where it germinates and is immediately colonized when the first hole appears in the hypocotyle. The fruit is a drupe, white or red in color. The last may indicate a possible ornithochorous distribution. The water reserve of the pseudotubercule allows the plant to survive in an epiphytic situation through a long dry spell.

Asian (and Oceanian) myrmecophytes are diverse and somewhat sophisticated. The species of certain genera such as *Macaranga*, *Myrmecodia*, and *Hydnophytum* are very well adapted to harboring and feeding ants.

7.1.3 African plants

The myrmecophytes in Africa are less diverse and generally less adapted to ants than in America and Asia. According to certain botanists, the African plants have suffered long periods of dryness and that is the reason why the differentiation is not so advanced. However, efficient protection is given to trees by ants in the case of eastern *Acacia* species, *Barteria* species, and others. Food bodies probably will be found one day in some of those plants. Several myrmecophytes feed their ants with pith and sweet secretions, but so far no trophosomes have been mentioned in the literature. East African *Acacia* tree species harbor, with the ants in their stipular thorns, a very diverse insect fauna that includes beetles and is highly diversified without any equivalent in Central America. Thirteen families of ant-plants have been recorded in tropical Africa, compared to 22 in America and 27 in Asia.

In Africa there can be distinguished stipular formations (in species of *Macaranga* and *Acacia*), foliar formations, pouch-like structures at the base of the leaf blade (*Cola* spp. and *Scaphopetalum* spp.), caulinar formations or hollow stems (species of *Vitex* and *Clerodendrum*), and caulinar formations swollen in the internodes or between the nodes or at the base of certain branches. Finally, also in Africa, the main structures permiting myrmecophily are present, with the exception of the swollen hypocotyles of the epiphytic

Asian Rubiaceae. In Africa, no fern, no Bromeliaceae, and no Orchidaceae shows any sign of myrmecophily, and it is very doubtful that this condition will be found in the future.

Acacia tree species in eastern Africa (Mimosaceae) have hollow stipular thorns as do their relatives in Mesoamerica, but they do not have Belt bodies. They generally have extrafloral nectaries. It seems that two species grow in southern Africa and one in Madagascar, but the real ant trees are only distributed in Kenya, Uganda, and the surrounding countries of eastern Ethiopia, Tanzania, and Namibia. Several ants inhabit those thorns, and *Crematogaster* ant species are very aggressive and very efficient defenders. Among the species living with ants are several leaf beetles (species of Clytrinae, Cryptocephaliinae) which seem to be symphilic inside the thorns which have yellow trichomes appeasing the ants. Other Chrysomelid beetles (*Syagrus* species, Eumolpines) seem to be synoekete only, but all the other insect species have their biology yet to be studied. Probably this beetle penetrates into the thorn at the egg or larval stage with the help of ants, where it grows and dies. Perhaps they live long enough to survive the end of the ant colony but that remains very doubtful, and the hole dug by the ants is too small for the beetle to escape. In any event,that aspect of the biology of ant guests is totally unknown.

Similar species of *Acacia* exist in West Africa, but they have small stipular horns and are without ants. East African *Acacia* species transplanted to Dakar, Senegal, have big horns, but they remain empty. The local ants do not try to visit them, even if natural holes appear in the side of the stipules.

Leonardoxa africana is a leguminous tree (Caesalpinaceae) from Guinea and Congo. The internodes of the young branches are often swollen with alternate pores, dug or not by the ants but existing at a predisposed place (prostoma). According to McKey and Davidson (1993), who studied the tree, two ants live in the swollen internodes: *Petalomyrmex phylax* (Formicidae) and *Cataulacus mackei* (Myrmicinae) in Cameroon. The second ant could be a parasite of the association. The plant provides lodging and extrafloral nectaries, and *Petalomyrmex phylax* chases the leaf beetles and the caterpillars but seems inactive against vines.

Species of *Macaranga* trees (Euphorbiaceae) are modest compared to their Asian counterparts. They do not produce Beccarian bodies as do the Malaysian species. They have extrafloral nectaries, possess large-sized stipules twisted into urns, and harbor ants (*Crematogaster africana*). Their myrmecophily exists but is very weak.

Species of *Cola* and *Scaphopetalum* (Sterculiaceae) are more interesting. *Cola* species in Africa have their branches and leaves covered with numerous elongated, stiff, brown or red-brown hairs, a myrmecophilic character common with many other ant-plants.

The large leaves of *Cola marsupium* carry two pouches at the base of the lamina. These pouches are made of a swelling of the blade on the superior (adaxial) face and open at the inferior (abaxial) face by slits. *Engramma kohli* and *Plagiolepis mediorufa*, two very shy species of ants, frequent the pouches. If not efficient guards, they can clean these leaves of the eggs of insect invaders. The pouches of *Cola marsupium* strongly remind one of those of several species of South American Melastomataceae, Gesneriaceae, and Rubiaceae. Such similar conditions have evolved separately on different continents, and different plants often share these same peculiarities: dense hairs separating the ants; violet colored pouch inside the domatia; use of one pouch for a latrine, the other for the brood; and other modifications that these systems can produce. Coevolution of ants and plants often seems the only plausible explanation for these relationships.

As is true for species of *Maieta* in South America, *Scaphopetalum thonneri* (a Sterculiaceae similar to a *Cola* species) has uneven foliar pouches alternaterating from one leaf to the other, on the right of the main vein for right leaves and on the left for left leaves. The pouches are 25 to 50 cm long and 6 to 8 cm broad. They open through a narrow slit above. The pouches have stiff, bristly, red-brown hairs and are inhabited by two ant species of the genus *Engramma*.

Barteria fistulosa (Passifloraceae) is a shrub with hollow or swollen stems at the base. The Zairian natives do not like to cut the tree because of aggressive ants coming out immediately upon the slightest touch of the branches. *Pachysima aethiops* and other ants live inside the branches, and their way of "raining" over an intruder reminds one of *Triplaris* tree species of the Amazon and their ferocious occupants, *Pseudomyrmex* ant species.

It seems that the *Pachysima* spp. dig the pores themselves, but the pith inside the branches disappears by lysis action and is not taken away by the ants. The sting of *P. aethiops* is very painful and remains so for hours. An old *Barteria fistulosa* plant is occupied by a single colony of ants, probably resulting from the fusion of various nests as occurs in nests in cecropia trees in South America.

Trees inhabited by *Pachysima* ant species are generally vigorous even if on rare occasions some insect invaders are tolerated. The ants drink at the extrafloral nectaries but also raise coccids inside the stem. They feed on coccids, pith, pollen, mycelia, etc. The ants also clean the ground around the tree in a radius of 5 to 6 m by nipping the tender shoots of young plants as they invade the area. The same phenomenon is known with ants on the American species of *Cecropia, Acacia, Triplaris,* and many other myrmecophytes.

The openings (stomata) of *B. fistulosa* are opposite of the leaves, which would mean that a prostoma or zone of lesser resistance exists. The insect-plant association seems very positive for both partners.

Among the Verbenaceae, the African genera *Clerodendrum* and *Vitex* have many species that are truly myrmecophilous. In Africa, and sometimes in Asia, the internodes of certain species are hollow and inhabited by ants.

Extrafloral nectaries are generally abundant and attract many insects belonging to many orders. The old stems lose pith, but species of *Clerodendrum* with caulinar swellings are really adapted to myrmecophily (*Clerodendrum formicarum, C. fistulosum*).

As stated above, many species of *Vitex* are myrmecophilous — for example, *V. grandiflora, V. thyrsiflora,* and many others. Their obligatory guest is the aggressive ant *Viticicola tessmanni* living in stems that are hollow and without swellings. The openings of the cavities exist in pairs at the nodes between the insertion points of the leaves. The symmetry of the entrance pores of *Vitex aglaeifolia* is so perfect that through one pore one can see the opposite one.

Uncaria and *Nauclea* (Naucleaceae) are genera which contain several myrmecophilous species. *Crematogaster* spp. live inside the internodes. *Nauclea* spp. produce slight swellings on these internodes.

Several genera of African Rubiaceae are myrmecophilous in different ways. Species of *Heinsia, Canthium, Cuviera,* and *Vangueriopsis* present caulinar swellings inhabited mostly by *Crematogaster* ant species. Other shrubs (*Grumilea* spp.) have inhabited stipules and also contain *Crematogaster* species.

7.2   Theories of myrmecophily

The ant-plant association is sometimes loose and probably results in the tendency of any ant to occupy any available cavity. However, sometimes the association is so perfect that we must find an explanation compatible with our concepts of evolution. In regard to the American species of *Acacia* and *Cecropia* and the Asian species of *Macaranga* (all trees with trophosomes), epiphytic Asian Rubiaceae, Hydnophytinae, some American ferns, and so many others, one must ask the question, "How, since the end of the Jurassic, could such perfect adaptations have evolved?" (See Figure 7.2.)

Francisco Hernandez (1651), who discovered myrmecophilous Mexican species of *Acacia*, and Rumphius (1750), who discovered myrmecophilous species of *Myrmecodia* and *Hydnophytum*, had a very simple explanation: the plants produce the ants by spontaneous generation. Later on, botanists asked more questions, and their ideas varied slightly one from the other according to the period of their study. These scientists incorporated into their studies most of the predominating ideas of the time. Unfortunately, most of them are totally rejected today. I refer again to Jolivet (1996) for more details and for further references.

7.2.1   Theory of Spruce and of Buscaloni
            regarding immersed lands

Richard Spruce explored the Amazon (1849–1864), and when he returned to England presented to the Linnean Society a memoir, which was published only in 1900 after the death of the author because it was judged too Lamarck-ian. However, the basic ideas of Spruce were that myrmecophous plants evolved to harbor the ants because the ants had no other choice in areas flooded permanently or temporarily and no other place to establish a nest. This theory was picked up again by Buscaloni and Huber (1900), 30 years later.

The general idea, an evolution brought about due to floods preventing ants from building an earthen nest, is not without merit, because ants can establish a nest in trees or between leaves of trees such as Old World species of *Oecophylla*. Also, ants have colonized plants in subdesert areas (Central America and East Africa in species of *Acacia*). Epiphytic *Myrmecodia* spp. grow in dry regions, and the epiphytic position is generally associated with dryness. There are many other cases where myrmecophily is not linked with swampy areas, permanent or temporary, or the banks of rivers. Myrmecophytes grow on poor soil, humid or not, and survive as do the insectivorous plants providing a supplementary nitrogen supply is brought by the ants.

7.2.2   Finalist theory of Belt, Müller, and Delpino

The association of ants with plants is, in most cases, a pure symbiosis with mutual benefits. Belt, Müller, and Delpino studied this symbiosis, and Delpino (1886), for instance, noted that the function of extrafloral nectaries is to place ants and wasps in a position to guard the plants to prevent the tender parts of the plants from being destroyed. For Delpino, the hypothesis was self-evident.

As expected, many botanists and zoologists rejected the idea, including Wheeler (1913). Some even tried to prove that the extrafloral nectaries were useless and that ants are often shy and do not protect the plant. For them, it was not symbiosis, but just a phenomenon, an accident. Delpino's theory is not a theory, really — more a statement of fact — and probably does not involve any evolutionary theory.

7.2.3   Lamarckian theory of Beccari

Beccari (1884) attributed the origin of myrmecophytes to the heredity of acquired characters, i.e., to Lamarckism. Francis Darwin himself (1877) wrote that the food bodies had been produced by the chewing of the ants coming about after a long evolution hereditary. These characters had been prefixed and would grow later on with or without the presence of the ants. Needless to say, Beccari had very few followers, but Hocking (1970, 1975) invoked the Baldwin effect in the origin of the stipular thorns of the east African species of *Acacia*. I have never been satisfied by this strange theory, a sort of bizarre adaptation of Lamarckism to Darwinism and somewhat a hybrid of both.

7.2.4   Cecidian theory of Chodat and Carisso

This theory attributes the inhabited thorns of the species of *Acacia*, fistulous stems of the myrmecophytes, and ant pouches of many herbaceous plants to the formation of galls. The theory, revived by Jeannel (Alluaud and Jeannel, 1922), is unfounded, because these myrmecophilous formations are very different from galls and in the absence of ants (in nature or in greenhouses) still develop their peculiar structures.

7.2.5   Pre-adaptation of the structures: theory of Schnell

This idea, originally proposed by Cuenot, was finally adapted by Schnell (1970) in France in his papers and books on myrmecophytes. According to this theory, the structures inhabited by ants pre-exist before the arrival of the ants and are not at all induced by them The ants, then, would be pure parasites of the plant as are lice or fleas, and if they protect them sometimes, it is purely accidental. This theory is acceptable in the case of some species of *Clerodendrum* such as *C. fallax*, unmodified, and where the stems become naturally hollow and are occupied by the ants. There are several other similar cases among the myrmecophytes, but these have to be fully rejected for plants such as species of *Cecropia*, *Hydnophytum*, Mexican *Acacia*, *Macaranga*, *Piper*, and many other Rubiaceae, Melastomataceae, Boraginaceae, etc. When the adaptations are perfect, with extrafloral nectaries, trophosomes, prostoma, inflated pouches, or nodes, this explanation has no value.

7.2.6   Natural selection and coevolution

Finally, we reach the notion of coevolution, when "two or more populations interact so closely that each serves as a strong selective force on the evolution of the other, resulting in reciprocal stepwise adjustments." When this mutual effect between ants and plants has benefited the plants much more than the ants, only a theory of natural selection and coevolution can satisfy the biologist, even if such a mechanism is difficult to conceive. According to Janzen, a fixation of characters useful to the survival of the plant in poor soils (mangrove, epiphytes, dry savannas, flooded areas) has developed a coevolution between ants and plants. The finality of the association, where experiments have been done, is positive for plants and for ants, both of which find advantages in the association. This topic has been discussed in detail in Jolivet (1996).

7.3   Acarophytes or mite-plants

This subject has been debated much more than the ant-plants theory. Acarodomatia were discovered in 1887 by Lindstrom, i.e., 13 years after Belt had published his symbiotic theory of myrmecophily. These formations could be axillary formations of veins, folds of the limb margin, hair tufts, etc.

The problem of the existence of acarophytes has been mostly studied by Jacobs (1966) and by Schnell (1963, 1966, 1970), who plainly rejected the symbiotic theory and spoke of a certain "pre-adaptation" of the plant but considered these as totally useless in function. Schnell also based his ideas on some morphological structures of the leaf sometimes repeating themselves along the veins. Acarodomatia cannot be confused with acarocecidies, which are also inhabited by various mites, which are induced by them and produce hyperplasias.

All recent researches, even experimental ones, seem to prove the usefulness of the association. These mites seem to feed on fungal spores and eggs of various insects; some are fungivorous, others are carnivorous and feed on insects, and still others are herbivorous. In Australia, such formations have been found on fossil plants dating back 42 million years, from mid-Eocene (O'Dowd et al., 1991) or even much older, the angiosperms dating from the end of the Jurassic or perhaps the end of the Trias. These structures pre-exist for mites, are hereditary, and are even used in identification keys. Mite-plants are probably symbiotic, and the system seems to work the same as it does for the ant-plants. The cleaning power of the mites cannot be overlooked. Rozario (1995) also sees a form of mutualism between plants and mites in Bangladesh. Also, acarophilous plants can be found in temperate and tropical climates.

A mite, *Oribatula tibialis*, lives on the leaves of *Pinguicula longifolia* (Lentibulariaceae), a carnivorous plant, in Spain (Antor and Garcia, 1995). The mite moves without being trapped by the mucilaginous droplets of the leaf surface. *P. longifolia* provides shelter and food for the mite, while the plant may benefit because of the fungivorous and scavenging activities of the mite. This is a very peculiar mite-plants association because the cavities used by the mites on the leaf are similar to acarodomatia.

Small insects such as thrips (Thysanoptera) sometimes are associated with trees such as species of *Eucalyptus* in Australia and can live, as do mites, in close association with the plant and feed on extrafloral nectaries. Of course, in this case this is pure parasitism (see Figure 7.1D).

7.4 Seed eating and seed dispersal (myrmecochory) by ants

Harvester ants (Myrmicinae) collect seeds in arid and semiarid areas and belong to species of several genera (*Pogonomyrmex*, *Veromessor*, and others). They collect seeds from their host plants or directly from the ground under the plant. These they store for the dry season in special chambers (their grainaries) in their underground nests. These so-called "biting" ants (because they have a stinger on their abdomen) feed on these seeds. They usually eat the plumule and the radicule at time of germination, which of course kills the seeds. However, if the seeds become wet, the ants take them outside the nests, spread them out to dry, and then take them back again inside the nest. The Bible and the Talmud both mention these ants and their wisdom. It is evident that plants which germinate around those nests were not planted intentionally. They are only the seeds lost or rejected during harvest by the ants. It has been shown that seeds of violets cut into by the mandible of ants germinate quicker than the ones not carried by them. Probably both the cutting and saliva together help in the germination process.

Myrmecochory, specifically, is the transport of seeds by ants, with the ants eating the only edible part of the seed (elaiosome). These ants do not kill the seeds, which still can germinate far from the place where they originated. Seed dispersion, when the seed is not eaten, can benefit the plant by having its seeds deposited in situations favorable to survival. Dispersion of seeds by ants happens in temperate and tropical areas but seems prevalent in dry areas such as Australia, the Sahel zone, and South Africa; it is also often carried on by harvester ants, even though normally they kill the seeds.

In temperate countries myrmechorous plants are mostly herbaceous plants found in lower strata. Trees are anemochorous (seeds disseminated by wind), but berry bushes in the arbustive strata dominate. These seeds are mostly disseminated by birds — for example, the epiphytic (hemiparasitic) mistletoe (Loranthaceae) with red or white berries. In reality, the distribution of the Loranthaceae and relatives is more complex in the tropics, and there are cases of autochory or self-explosion of the seeds. They are generally glutinous and viscid and are distributed either with the saliva or the excreta of the birds.

In the tropics, many epiphytes, mostly the myrmecophytic ones, produce sweet and sticky berries, white or red (Hydnophytinae), which are disseminated by ant species of the genus *Philidris* (=*Iridomyrmex*) and practically none by birds. These ants are so aggressive against any vertebrate attempting to touch the plant that we could hardly imagine that birds are able to reach the berries. Many myrmecophytes are myrmecochorous epiphytes. Let us cite as an example the Asian species of *Dischidia* (Asclepiadaceae), which also is distributed by species of *Philidris* ants, but not in the same way.

Some ants in temperate areas, such as *Formica rufa* and related species, carry seeds to use in building their nest. The most favored seeds are ones with an elaiosome or oily body or ones which exude oil to cover the test of the seed.

Seed distribution of aerial plants by ants is very efficient for more than just the myrmecophytes. Eleven different types of these seeds disseminated by ants have been described. The number of seeds carried by one ant alone can be considerable (almost 40,000 for only one worker of *Formica rufa* in 80 days). The distance traveled can be as far as 70 m. Generally, these are seeds rich in oil, such as those of *Viola, Veronica*, or species of *Corydalis, Cyclamen, Chelidonium*, and *Melampyrum* (Ridley, 1930). The seed caruncle (the elaiosome) contains oils, proteins, and various fats. We know that phasmids, feeding on various plants, drop their eggs on the soil at random. Some phasmids in the tropics use an edible caruncle, sort of elaiosome, placed on the top of the egg to have them disseminated by ants. Sometimes, an egg is saved from fire when it is stored in the ant nest.

A classic example of elaiosome is the large celandine, *Chelidonium maius*, a weed in Europe common in hedges and garbage dumps and on old walls. Species of the ant genus *Myrmica* carry away these mature seeds which they then drop on the ground or take to old walls. Only these ants are responsible for the establishment of the plant on these walls.

Myrmecochory is extremely common in the tropics, where all strata of the forest are involved. The inevitably present *Atta* ant species are kept in check in Neotropical forests by ants with aerial nests or living in myrmecophytes. These ants defend their habitat ferociously as well as being responsible for seed dissemination of their plant hosts. We will see that the seeds of the ant gardens in South America are probably really "planted" by their ants, as already foreseen by Ule in 1902.

7.5　Ant gardens

We now reach an association of ants and plants characteristic of the Neotropical region, though it seems that they also exist in another form in Madagascar and southeastern Asia. An ant garden is a mutualistic interaction between specialized epiphytic plants, some being real myrmecophytes themselves (*Marckea* spp.), most of them not in tropical America. The ants provide nutrients enabling the plants to live as epiphytes; the plants provide a stable substrate, a frame for the ant nest. Ant gardens are really nests of cartonlike material constructed by certain species of ants around specific epiphytic roots. The ant gardens were discovered by Ule in 1902 and then studied by many specialists, zoologists and botanists alike. Some plants are obligatory hosts of ant gardens, such as species of *Codonanthe* (Gesneriaceae).

Some dominant ant species which normally occupy ant gardens are, for instance, *Crematogaster parabiotica*, *Camponotus femoratus*, and *Solenopsis parabioticus*. All of these ants live in parabiosis; the first two are aggressive, the third one harmless.

In those "balls of earth", as Ule named them, grow a multitude of specialized plants belonging to Araceae, Bromeliaceae, Peperomiaceae, Solanaceae, Gesneriaceae, Moraceae, and Cactaceae. The ants obtain part of their food from floral and extrafloral nectaries, fruit pulp, seed arils, and the fleshy coverings of the seeds, and the roots of the epiphyte, which form an integral part of the ant nest, obtain their nitrogenous substances from ant detritus, cadavers, and excreta. The ant garden varies from being the size of a nut to the size of an orange or a soccer ball.

The seeds germinate in the aerial garden and obtain nitrogenous compounds from soil brought by the wind and from detritus collected by ants. It seems that gardens result from a fortuitous association of pre-existing

Table 7.2 Comparison Between Carnivorous and Myrmecophilous Plants

	Insectivorous	Myrmecophilous
Nutritional value of the substratum	Very low	Very low
Use of animal food	Probably nitrogen	Probably nitrogen
Duration of life	Perennial (few annual)	Perennial
Distribution	Tropical-temperate (some boreal)	Tropical and subtropical
Assimilation system	Specialized cells	Generally internal roots or trichomes or warts
Situation	Aquatic surroundings, bogs, creepers, few epiphytic, mangroves	Generally flood areas, dry or humid forests, mangroves, epiphytic

Adapted from Thompson, J. H. (1981).

structures useful to both partners. Most of the garden plants are autogamous, probably because of the aggressiveness of the ants and the lack of pollinators in that strata of the forest.

According to Davidson and Epstein (1990) and Seidel et al. (1990), some of the volatile compounds on seeds of ant garden epiphytes probably play a role in ant attraction to epiphytic seeds. So the ants really, as Ule said before, plant their suspended garden.

7.6 Seed dispersal by insects other than ants

At the risk of being accused of anthropomorphism, it can be said that altruism does not exist in the plant kingdom. We have previously seen that when a flower gives up its nectar this usually ensures its pollination; when myrmecophilous or carnivorous plants provide food for their guests they receive, in exchange, protection and the nitrogen they need (see Table 7.2). When fruits are attractive to animals, seed dispersal is assured. We will review here insect dispersal of plant seeds, or entomochory.

Entomochory is mostly myrmecochory, i.e., transportation of seeds by ants, but other insects also can be dispersal agents. It is evident that during their travels many animals disperse seeds from their body fur or their excreta. This includes dispersal by vertebrates, insects, crustacea, and even earthworms. A great part of the animal world contributes to this dissemination. During the Mesozoic era, some reptiles were herbivorous, the major part really, and they must have been responsible for the dissemination of gymnosperm seeds including those of the ginkgo and later those of angiosperms. Some of these seeds have been found in the imprint of fossil stomachs. Today, though, the role of reptiles has been greatly reduced or is nil. Throughout history, the role of insects in seed dispersion must have been important, developing after the appearance of angiosperms during the Jurassic at the middle of the Mesozoic.

Insect dissemination of seeds and spores of plants takes place in three different ways: (1) by swallowing them and voiding them undigested in their excreta (endozoochory); (2) by taking them into their nests, where they eat only oil bodies (or caruncles) adjacent to the hilum in certain seeds (elaiosome); and (3) by adhesion of spores or seeds to their body (epizoochory).

The dispersion of seeds by the first method is rare but is a common way for spore dispersal. However, Darwin, in the *Origin of Species*, mentions that in Nepal he reared seven different plants from the excreta of several Nepalese migratory locusts. The spores of specific fungi depends on insects and even slugs for dispersal. This is the way *Phallus impudicus*, a fungus with a strong odor that is found in temperate regions, and its tropical relatives are dispersed. Flies attracted by the fungus' odor eat the mucus spores, but these are undigested. Instead, they are excreted elsewhere. In the excreta of one fly, specialists calculated that there were in it as many as 20 million spores.

Beetles, namely certain Chrysomelids such as species of *Diabrotica* and *Chaetocnema* play a modest role in seed dissemination, but they are more responsible for the dissemination of spores and bacteria. Ridley (1930) mentions Trichoptera as seed disseminators of marsh plants, as well as other examples involving Orthoptera and Diptera. The later two disseminate mosses (*Splachnum* and *Tetraplodon* species) and fungus spores. Collembola also carry lichen soredia. We could add more examples, but the carriers of the greatest number of fungus spores are Diptera, Homoptera, and Hemiptera. They also disseminate bacteria. It is to be noted that rye ergot (*Claviceps purpurea*), a pest of many cereals and the source of ergotism or "mal des ardents" in man, is carried by flies attracted to its sugary secretion, which is produced during a certain stage of the life cycle of the fungus as nectarinian spores. Thus nectar production is not a complete monopoly of the higher plants. In Europe, during the Middle ages, the ergotism was widespread among the peasants and as recently as 1951 an outbreak of ergotism occurred in a small village in the south of France.

Termites play a modest role in the dissemination of seeds of flowering plants. They also carry spores and sclerotes of fungi which they grow in their nests. In Transvaal, the grass *Cynodon incurvatus* is carried in the seed and fragment stage into the termite nests where it is stocked in the upper chambers of the nest. Large nocturnal termitophagous mammals, *Orycteropes* spp. for instance, break open the termite nest, allowing the seeds to germinate. The seeds of the giant flower of *Rafflesia* spp. in Indonesia are disseminated actively and passively by termites which invade their fruit and ovary. However, myrmecochory is the mainstay of seed and fruit distribution by insects.

In a recent paper, Shykoff and Buchell (1995) mention pollinator visitation patterns of the white campion in the U.S. (*Silene alba*, a Caryophyllaceae), a native of Eurasia established in the U.S. in the early nineteenth century. The spores of the basidiomycete smut fungus *Microbotryum violaceum* or *Ustilago violacea* (Ustilaginales) seem to be transmitted to male and female plants "venereally" by pollinators looking for nectar. Pollinators prefer plants with large floral displays, male to female plants, and healthy to diseased plants. Transmitted sexual or venereal diseases (TSD) are known in small numbers in insects but is doubtful in plant spores. The white campion is generally pollinated nocturnally by moths and mosquitoes. The flowers are night scented and rich in nectar.

chapter eight

A rare plant-insect relationship: epizoic symbiosis

When searching for weevils in the moss forests of Borneo and Panama, entomologists can see several individuals of certain species with greenish elytra. Probably algae, blue or green, are growing on the elytra of these long-living insects. Dr. Stockwell, at the Smithsonian Tropical Research Institute, Panama, has frequently collected greenish *Geobyrsa nodifera*. Although offering camouflage, this coloration seems to be due to the high humidity reigning at this altitude rather than a color brought about by selection.

In the Ivory Coast, I have seen reduviid nymphs (Hemiptera) cover their bodies with cast skins, feces, and various vegetable and animal debris. The larvae of Cassidinae (Chrysomelidae) often carry cast skins and feces over their abdomen, and even some Neotropical species carry a complete "nest" of feces in filaments over their bodies. This is not only camouflage, but also protection against rain or the rays of the sun, as well as a shield-like protection against possible predators.

Nothing is alive in those camouflage designs, but the system is very efficient and rather widespread among these insects. They do much the same as hermit crabs in the sea by covering their adopted shell with marine algae and sea anemones. Some reduviid nymphs use the integument of recently captured termites to cover their odor and as a bait to catch more termites near their nest.

External plant-insect relationships are rare, but they do exist — mostly among fungi and rarely among algae. Among the fungal relationships are species of Laboulbeniales and similar groups which are not really pathogens or ectoparasites on the insect cuticle, even if they often use haustoria to fix them to the insect surface. These haustoria do not absorb nutrients or water; they are for fixation only.

Green algae sometimes have been found on the wings of a desert locust in India, on a Chinese geometrid moth in moss forests, and on a Queensland

143

spider. Algal growths were also observed on some tropical locusts and Neoguinean grasshoppers. These isolated cases must be more common, but they are rarely recorded; entomologists generally throw such captures directly into the killing jar without further observation. Algae are very common on aquatic species, particularly mosquito larvae, larvae of chironomids, dragonfly naiads, and many other aquatic immatures or adults of common insects. Species of *Cricetopus* larvae (Chironomidae) are found exclusively on pads of species of *Nostoc* where they burrow, feed, pupate, and live in some kind of mutual relationship (Gerson, 1974). Endozoic symbiotic algae live in the caudal lamellae of some damselflies, outside or inside (*Chlorosarcina* spp.). Many aquatic and terrestrial insects disseminate these algae. Various marine and freshwater algae grow on arthropods without causing any harm, whether they are insects or crustacea.

Lichens also are often associated with arthropods, and lichen mimetism is very common in the tropics, not only among species of Tettigoniidae, Phasmidae, or Mantidae, for instance, but also among many other insect orders (beetles, moths, lacewings, etc.).

The predacious South American larvae of the *Hemerobius* sp. (Neuroptera) cover themselves with lichens and mosses, thus becoming an exact image of the round lichenous patches on forest trees. In Ohio, Slocom and Lawrey (1976) discovered that the larvae of the Neuropteran *Nodina pavida* (Chrysopidae) carry an accumulation of cryptogamic vegetation on their backs, a sort of camouflage against predators. This is composed of lichen soredia and thalli, pieces of bark, parts of moss gametophytes, pollen grains, fungus spores, and other debris from plants and insects. A Colombian lacewing, *Chrysopa* sp., has been found covered by soredia of several lichens. The larva excrete sticky silk to which soredia adhere. Weber (1974) has found two lepidopterous caterpillars in Australia and New Guinea covered with fragments of *Parmelia* spp. Also, marine lichens such as *Arthopyrenia halodytes* and *Verrucaria* spp. have been observed on living mollusks and barnacles.

As we have seen above, the associations between cryptogams and insects, mites, spiders, etc. are rather rare, but nevertheless exist. The phenomenon is more common in the sea with crabs and it has its equivalent on land with the epizoic symbiosis described by Gressitt (1965; 1966a,b,c; 1969; 1970; 1977) in the highlands of New Guinea.

These cryptogamous gardens have been discovered only on apterous weevils with fused elytra, such as Leptopiinae, Brachyderinae, Cryptorhynchinae, Otiorrhynchinae, and Buridinae. We also see that a species of *Dryptops* (Colydiidae) carries some lichens and algae, which means that this association may have developed on other beetles yet to be discovered. Why not on other orders of insects? Until now these association have been found only in the New Guinean rain forest.

These associations are complex, comprising acellular and vascular cryptogams and a fauna living in that mini-forest (nematodes, mites, protozoa, and very probably tardigrada). This vegetation covers mainly the cuticle of species of large weevil of the genus *Gymnopholus*, especially of the subgenus *Symbiopholus*, and many other weevils and beetles.

This vegetation is reminiscent of the algal flora which adorn pagurid crab shells in the sea. Very probably the weevil flora functions as a camouflage against possible predators at that altitude. At first, Gressitt thought that birds of paradise were the probable predators, but as those birds are becoming rarer and rarer, they do not often feed on these beetles. The most likely predators are marsupials such as species of *Antechinus*, *Petaurus*, and *Eudromicia*. The weevils do not seem to be protected by toxic blood or secretions as are many chrysomelids. It is a true symbiosis, the insects providing a favorable environment for plants, the plants giving a form of camouflage to long-living insects. Gressitt coined the term "epizoic symbiosis" for this new form of external association between a plant and an insect. Oddly enough, symbiosis has never been discovered in this form in the high tropical mountains of either North or South America, Borneo, or even in the Malaysian area where local conditions are sometimes very similar to the cloud or moss forests of Papua New Guinea. Epizoic symbiosis is also completely lacking in tropical Africa, where the high mountains are either volcanoes or old crystalline massifs such as the Mt. Ruwenzori. It is rare there to see large insects, such as galerucine beetles, above 3000 m in the moss forest. Tiny beetles are then common but not big enough to develop a specialized flora, and their life span is very short, too short to support such a flora. Alticine beetle species range in very high altitudes, but they are small and also short lived.

One must also understand that even if those associations were visible in New Guinea on the living insect, they are difficult to recognize on dead museum specimens. On those specimens, in collections, the vegetation dies, dries up, and is practically invisible.

In summary, this phenomenon concerning mostly weevils seems to be extremely rare. Indeed, it is possibly a unique occurrence in the Papua New Guinea subcontinent, without parallel elsewhere, a quite surprising fact.

This epizoic symbiosis provides certain species of Coleoptera with a living garden, mostly on their back, or even more intriguing, as a suspended garden inhabited by many tiny animals. This ecological association includes various cryptogams such as green and blue (cyanobacteria) algae, fungi, liverworts, lichens, mosses, and even in certain cases the gametophytes (protothalli) of a fern, an oribatid mite, rotifers, nematodes, and various micro-organisms. This phenomenon is limited to the moss forest at an altitude ranging between 1200 and 3600 m at the equator. From 1965 to 1977, Gressitt and his colleagues made many observations on this association, and progressively their interpretation has evolved. I personally visited Gressitt in 1969 at the Wau Ecology Research Station, in eastern Papua New Guinea in the Kaindi mountains, where I conducted many experiments. In the gut of a species of the weevil genus, *Gymnopholus*, a colleague and I found a gregarine related to one from an African mountain weevil. However, during my research in Papua New Guinea, I have rarely encountered these weevils, but I did find some in the central highland areas. Either they are rare, or they are well camouflaged. Gressitt gave many reasons for the scarcity of the species, but predation seems rare and the duration of their life can be as long as five or more years. Not only is this considerable length of time probably necessary for the establishment of a dense flora, but old specimens may eventually partially lose their plant cover. We know that the Galapagos tortoises, at least the females, partially lose their lichen cover after repeated copulation. Male specimens keep the epizoic flora almost intact.

Can we really use the word symbiosis in the *Gymnopholus* complex association with cryptogams? Yes, there is a very real camouflage for an apterous and defenseless insect, one that is very slow moving and already living in the shade of a forest with a microforest on its back. However, the debate continues, even though there is no question that the association exists and is very remarkable in itself.

Lichens live on the back of species of *Dryptops* beetles (Colydiidae) but it is on species of *Gymnopholus* and *Pantorhytes* weevils that are found the most complete series of plants. The cryptogams are numerous and varied, sometimes Ascomycetes with red, orange, and black bodies, some Fungi Imperfecti, various Cyanobacteria (Cyanophyceae), green algae, lichens of the families Physciaceae and Parmeliaceae, liverworts, mosses, and even, as already mentioned, a fern protothallus. In the middle of that flora, and probably on it, Gressitt found many mite species of the genus *Symbioribates*, an oribatid, and others. These acarina are very small (0.2 mm) and are completely hidden beneath the fungus cover. The movements and mating activity of the weevils cause fungal spores to be disseminated and the adhesive secretions of the elytra of other weevils to collect them by chance. Along with the fungi also live rotifers, nematodes, diatoms, and other microorganisms, all very small. No doubt the fauna has not been completely investigated. Occasionally, some Psocoptera are seen feeding on the algae and fungi. These host beetles are also parasitized by some mites which are often phoretic, which has nothing to do with the original complex association. These mites stay on the ventral surface, antennae, and legs of the beetles.

The species of *Gymnopholus* are found between 500 and 3600 m, where the high humidity of these mountains is evidently conducive to the development of the original flora. It is, however, above 2000 m that most of the *Gymnopholus* species can be found, or at least the ones with abundant flora on their backs.

It is true that species of the subgenera *Symbiopholus* of *Gymnopholus* harbor, almost exclusively, these associations. Species in other subgenera of the genus usually have smooth elytra and live at low altitudes. These cryptogamic gardens are miniaturized; the plants are the "bonsai" of those suspended gardens.

The morphological adaptations of species of *Symbiopholus* seem to encourage the development of these plants. Their elytra have big hollow holes, crevices, tubercles, edges, special scales, setae, and setal tufts which seem to provide places for retaining the flora. Sticky secretions in the hollows seem to provide a place to catch the spores and to encourage the growth of algae. About 28% of all the collected beetle specimens bore conspicuous lichen growths, and another 56% primary growths. Most of the species of lichens were collected on *Gymnopholus lichenifer*.

These insects are sedentary and apterous, with fused elytra. They live a long time and are relatively polyphagous. As for many forest Curculionidae in the tropics, they are found on a great number of bushes and small trees belonging to the same ecological habitats (16 families of plants in all). The larvae are endogenous and feed on roots, very probably polyphagous as adults.

The plants fix themselves on structural modifications of the elytra, thorax, and sometimes on the femora. Often the flora coincides with the hairiness of the beetle. A waxy secretion at the posterior edge of the pronotum can also help to affix the spores. Young beetles have their scales intact and are without flora, while mature specimens are covered with lichens. Nevertheless, as with the Galapagos tortoises, there is a lot of rubbing in the very old specimens and a loss of part of the flora.

These beetles appear still to be evolving. They are distributed over all the New Guinea highlands, including the Indonesian part. Each chain of mountains claims its own species. I can only compare the diversification of a genus in full evolution to the related Mascarene genus *Cratopus* in the Indian ocean, itself showing a great diversity of species. Unfortunately, the *Cratopus* species are going to be extinct before having achieved its full evolutionary development. Urbanization and insecticides are destroying their habitat in Mauritius and La Reunion.

We can expect a longer life for the species of *Gymnopholus* which are not yet subject to great pressure from humans in its mountains.

It is important to note that the cryptogams found on the weevils are found also on the leaves and trunks of trees in humid moss forests. The hypothesis that their protection is the same for the weevil as for the decorated shell of the pagurid crab is attractive. Indeed, it may be true. If not, let us say that this association is remarkable, even if it is only the colonization of a preadapted chitinous structure in a favorable humid environment.

In conclusion, at the present time, epizoic symbiosis of insects has been identified nowhere else, at least in the terrestrial insect world. There is only a rough analogy between it and the blue algal associations in the hairs of the Neotropical sloths and the pyralid caterpillars and moths which are also established there. However, species of Lepidoptera seem to feed only on the desquamated skin and cutaneous secretions of the sloth, as do the mallophagan lice. If they were feeding on algae, the association would be more complex. Anyway, it has been proven that these caterpillars once lived only on the excreta of the sloths, on which the moth imagoes still lay eggs during the sporadic defecation of the animal. More than a camouflage, the algal cover of these mammals appears, as in Papua New Guinea weevils, to be the result of the permanent humidity of the rain forest and the quasi-immobility of the beast. However, there is no doubt that the greenish cover of the sloth contributes to its protection against eventual predators.

Parasitism by algae and fungi is known to occur in human and animal tissues. Algal infections are quite rare in man, although certain fungi are highly pathogenic to man, being saprophytic on skin and mucous membranes, often in competitive equilibrium with the bacterial flora. Such is the case of athlete's foot (tinea), caused by *Candida albicans*, and many others. Other cryptogams have a neutral or somewhat symbiotic role, as for example, the sloth-algal association. It is also known that scabs on the shell of aquatic turtles are invaded by blue and green algae, filamentous or encrusted, and the giant Galapagos tortoises carry lichens on their shell. Many fish also have algal-infested gills, the algae living where there is an optimum amount of oxygen. Several lizards and crocodiles have algae growing on their skin. It is probable that this was also the privilege of aquatic plesiosaurs and mosasaurs, as well as subaquatic dinosaurs. The Sirenians, which carry cirripeds over their skin, as do whales, probably also have algae over their bodies, but they seem relatively immune.

The association of algae with aquatic invertebrates, including insects, is not so rare. Sometimes it is a true symbiosis. Examples of this are the marine turbellarian worms, such as *Convoluta roscoffensis* associated with the alga *Phytomonas convolutae*; the giant marine mollusks of the genus *Tridacna* with algae on their gills; the gastropod *Elysia viridis*, which crawls on plants and uses free chloroplasts from marine algae; several freshwater hydras with chloroplasts in their bodies; marine anemones (*Anthopleira* spp.); and some sponges, each with algal associations. Unicellular organisms using chlorophyll are represented by chlorophyll-containing rhizopod and flagellate protozoa. It is because of this that their position in the animal or plant kingdom was long debated. Actually, the situation varies according to the classification system proposed (five kingdoms or less). Carl Woes of the University of Illinois keeps them at the base of the Eucaryotes.

If the protection of sloths by their algae living on their hairs is problematic, the *Pagurus* crab species (hermit crabs) seem to be well protected by the garden of algae and living anemones planted over their adopted shells. Terrestrial cenobites such as coconut crabs (*Birgus latro*) have lost that kind of protection. Cenobites live under stones in a naked shell, and coconut crabs have produced a new carapace on their still-asymmetric, tender abdomen. Cenobites have adopted a nocturnal life similar to that of wood lice, and *Birgus latro* lives during the day in a rat hole near the beach. It goes back to the sea only for reproduction.

Eccrinales, fungi that live in the digestive tract of insects, are nonpathogenic. No algae are known inside the gut of arthropods. *Oscillospira* spp., depigmented Schizophyta, close to cyanobacteria or blue algae, have been found only in the intestines of mammals including man.

Finally, the only comparable relationship similar to epizoic symbiosis of beetles is the camouflage of the hermit crabs in the sea. The difference is that these crabs often seize the algae and plant them over their shells. In the case of the New Guinean weevils, the dissemination of the flora is done by spores planted in the mucous holes of the elytra by the wind and perhaps also venereally transmitted during copulation. The chances of dislodging part of the flora are remote. Among the old specimens, though, mating seems to do so from the back of the females.

chapter nine

Galls and mines

Much has been written about galls. Even though the morphology of the gall has been well studied (along with the guild of parasites, the development of gall-making insects, and plant tissue under the influence of the larva), we are still in total ignorance of the mechanism for the development of the gall as a benign tumor. Hypotheses are many, but the chemicals responsible have not been properly identified. We do not even know how and if the genome of the metastasizing cell has been modified. Galls and mines have in common only the fact that they contain an insect inside the plant tissue, where the insect feeds and is protected (see Figure 9.1). Probably this represents two different solutions for the animal for protection, more or less efficient, against predators and parasitoids.

9.1 Galls

The definition of galls given by the Torre-Bueno's glossary of entomology (1989) is as follows: "an abnormal growth of plant tissues caused by various organisms which irritate the plant and possibly lead to the production of some type of growth hormone." Some hypotheses, such as a proposed change in genetic information supplied to the cells by the gall inducers, are recent but not entirely satisfactory. Such an explanation accounts for the gall formation, but nothing can be proven until one has succeeded in creating a completely artificial gall on a plant.

Another term for gall is *cecidia*. Galls caused by mites are *acarocecidia*; by nematodes, *nematocecidia*; by fungi, *mycocecidia*; and by bacteria, *bacteriocecidia*. Insect galls, which are the most common of all, are called *entomocecidia*. Viruses, or mycoplasmas (mollicutes), can induce a sort of gall eventually. The shapes of the galls are extremely variable in the tropics and sometimes are very colorful, strange in form, and vary in size.

Mani (1964) gives the following definition of a gall: "Galls are pathologically developed cells, tissues, or organs of plants that have risen mostly by hypertrophy and hyperplasy under the influence of parasitic organisms such as bacteria, fungi, nematodes, mites, or insects. They represent the growth

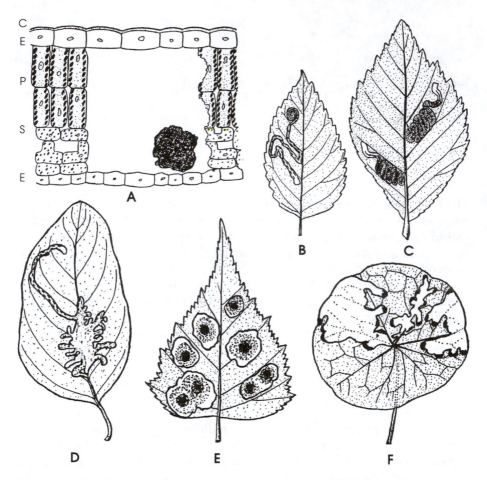

Figure 9.1 Mines and galls. **(A)** Cross-section of a full-depth mine; C = cuticle; E = epidermis; P = palisade parenchyma; S = spongy parenchyma. In the center is a heap of frass from the leaf-miner. **(B)** Mine of *Liriomyza eupatorii* (Diptera) on *Galeopsis* sp. **(C)** Mine of *Nepticula viscerella* (Lepidoptera) on *Ulmus* sp. **(D)** Mine of *Phytomyza xylostei* (Diptera) on *Lonicera* sp. **(E)** Mine of *Nepticula argentipedella* on *Betula* sp. **(F)** A bridge species for many miners and other phytophagous insects: *Tropaeolum majus.* Mine of *Scaptomyzella incana* and *S. flava* (Diptera) for transference from Centrospermae and Brassicaceae to Leguminosae. (Parts A, B, C, D, E, and F adapted from Hering, 1951.) **(G), (H),** and **(I)** Development of a (lysenchyme) gall: (G) egg of *Diplolepis rosae* imbedded into the epidermis of petiole; (H) the formation of primary larval cavity by lysigene (fusion) action; (I) larva sunk within the tissue (Hymenoptera: Cynipidae). (Parts G, H, and I adapted from Mani, 1964.) **(J)** A gall made by the wasp *Andricus quercuscalicis* on an acorn of the oak *Quercus robur.* **(K)** Longitudinal section through the gall and the centrally located larval chamber. (Parts J and K adapted from Schremmer, 1976.) **(L)** Structure of typical pouch gall of *Eriophyes similis* on *Prunus spinosa.* **(M)** The spherical pouch gall of *Eriophyes macrorrhynchus* on *Acer campestris* with the elongate unicellular setae in the cavity. **(N)** *Eriophyes* sp.: typical gall mite, ventral view. (Parts L, M, and N adapted from Mani, 1964.)

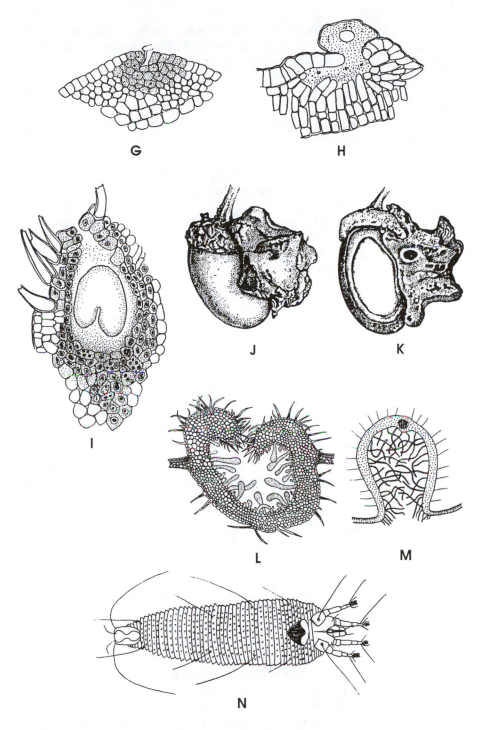

Figure 9.1   (continued)

reaction of plants to the attack of the parasite and are in some way related to the feeding activity and nutritional physiology of the parasite." This definition also does not explain the process by which a gall is born. Acarodomatia are not galls. Mani also proposes that they are a part of the morphology of the leaf, inherited by plants and occupied by fungivorous or carnivorous mites.

A gall can be only the reaction to an insect ovipositor, which deposits one or several eggs in the plant tissue. Galls are developmental and growth abnormalities produced by the plant in reaction to an irritation or a secretion. Galls are also "a deviation on the normal pattern of plant growth produced by a specific reaction to the presence and activity of a foreign organism (animal or plant)" (Bloch, 1965).

Oak galls produced by cynipids (Hymenoptera) are the best known in temperate climates and, because they are rich in tannin, the most exploited. Other cecidia also produce tannin on tamarisk in Morocco (acarocecidia) and in sumac (entomocecidia).

Galls develop on any part of the plant: leaf, stem, root, fruit, or flower. They are extremely variable in size, shape, and color. Some are a very beautiful red or yellow in contrast with the green of the leaf. Some galls exist on cryptogams, including lichens, and on various phanerogams, mostly on dicots. Insects and mites produce 80% of the known arthropod galls.

Galls were once considered to be harmful to the plants, but now the tendency is to consider the gall as advantageous since it protects the plant against possibly even greater damage — complete necrosis of the parasitized tissues. Indeed, it seems better for the plant to encapsulate the parasite and to provide it with food than to have the parasite moving around destroying the tissue. It is quite similar to the encapsulation of certain nematodes by animal tissues, but the nematode dies and the gall-maker escapes. The formation of the gall, which is a hypertrophy or a dysplasy of a tissue, restricts the damage not only to a certain region of the plant, but also to a particular tissue during the plant life cycle. It is a defensive and protective measure which seems to be the best way to combat the invasion. A gall has been likened to a false fruit, but since it has never been totally reproduced in the laboratory, the analogy is not certain.

Some galls are so well designed that they open automatically at the proper time to let their inhabitant escape. Galls induced by wasps, weevils, and other insects with biting mouth parts do not open that way; instead, the animals bite holes in the gall walls.

Before discussing the various types of galls, it should be mentioned that certain species of small ants invade the oak galls after they have been evacuated by their prior inhabitants in the oak tree. As the colony grows, the ants also leave and migrate to a larger habitat. Also, when considering ant and gall relationships, the case of *Myrmecocystus mexicanus*, an ant that lives in dry areas of North America, must be mentioned. These ants store honeydew in living honey pots, the specialized workers or "repletes". These ants suspend themselves from the roof of the galleries of the ant nest and store honey in their stomachs, up to eight times their weight. This honeydew comes partly from aphids, but mostly from secretions from galls on oaks caused by certain species of gall wasps. This is a very specialized relationship between galls and ants. We also see that galls, while providing shelter and food for one species, can also supply surplus food for others.

In those galls produced by mites, mostly species of Eriophyidae, it is very difficult to tell where the gall starts and the acarodomatia ends. The responsibility of mites, as we have seen in a previous chapter, is not always evident in the production of the various lodgings. Anyhow, acarodomatia exist and are also very common in plants.

Around 20 families of insects classified in seven orders induce galls or occur in close association with galls. About 13,000 cecidogenous insects have been recorded, which represents a small part of the living insects. The principal gall inducers are found in the following orders (Dreger-Jauffret and Shorthouse, 1992): Thysanoptera, Hemiptera (Tingidae), Homoptera (Cercopidae, Cicadellidae, Psyllidae, Aphididae, Eriosomatidae, Adelgidae, Phylloxeridae, Coccidae, Kermidae, Asterolecaniidae, Diaspididae, Eriococcidae), Lepidoptera, Coleoptera, Diptera (Nematocera and Cyclorrhapha), and Hymenoptera (Symphyta and Apocrita). Most gall-inducing insects are host and organ specific. We will treat here essentially those galls caused by insects, particularly those induced by Coleoptera (weevils), Hymenoptera (Tenthredinidae, Cynipidae, Lepidoptera (Tortricidae), Thysanoptera, Homoptera (Aphididae and Coccidae), and Diptera (Cecidomyidae). The subject has been dealt with so often in the older literature, as well as in recent papers, that now we are able to summarize the subject. Sometimes we forget that the large species of Sagrinae, a leaf beetle with bright colors and enormous femora, are gallicolous in the larval and pupal stages, mostly living inside the stems of leguminous bushes. They are similar to the Bruchidae, to which they are related, but the bruchids pierce and live in seeds and never cause galls. The archaic species of the Sagrinae are pantropical insects, dull brown in color. They exhibit a gondwanian distribution (Australia, Madagascar, Argentina, and Brazil). The species of old genera must have the same biology as the recently evolved *Sagra* species. They form a simple type of gall which is a transitional structure between boring and gall making. The sagrine gall is extremely lignified, generally infecting stems but occasionally penetrating roots. Groups related to the sagrines, such as the Megalopodinae, are only borers in herbaceous plants. No other gall makers are known among the leaf beetles.

The gall is certainly a tumor, but a mild (benign) form. Generally the gall does not grow before the larva inside hatches and starts eating the surrounding tissues. The growth of the gall stops when the larva becomes a pupa.

Galls may be swellings, cupules, and spheres. The insect egg and the larva which hatches later can be either inside the plant tissue or at its surface.

Each combination of insect and plant produces a typically different type of gall.

Certain galls are well known: the pineapple galls of spruce (*Picea* spp.), with swollen leaves overlapping each other; those produced by adelgid aphids (*Adelges laricis* and *Sacchiphantes abietis*); the oak galls made by a cecidomyid fly; the round galls (apple galls) of oak made by a cynipid wasp; and the hairy galls of roses produced by another cynipid. Galls are called tumors because of the way they are produced by an increasing growth of tissue, cell hypertrophy, nucleus alteration, and more rapid division of cells.

Experiments have shown that gall-making is due to the introduction of an auxin-like substance produced by an animal, generally the newly hatched larva. This larva secretes the substance through its skin (as in the *Mikiola* sp.), or via its salivary glands during the time it feeds on plant tissue. Exceptions are the species of *Pontania* (Hymenoptera: Tenthredinidae) and gallicolous species on the species of *Salix*. The galligenous substances are produced at the adult stage, during egg laying. Other more elaborate hypotheses about gall production have been proposed.

Insects are one of various biological factors able to induce plant growths by using a wide variety of externally applied chemical, physical, and biological stimuli (Rohfritsch, 1992). Tissues growing around the larva provide shelter and food. Plant wounding and ovipositional fluids are important growth-promoting factors for sawflies, whereas cynipid eggs and young larvae possess a strong lytic action. Salivary enzymes and light wounding seem to be responsible for the gall-making of other insects. Synthesis of tannins and lignins is part of the gall-making and are also induced by insects. In the case of the cynipids, the insects seem to control the gall and not the plant. This gall, which is not, despite appearances, a fruit, develops its own polarity and symmetry.

Quoting Rohfritsch (1992), "Gall insects modify minute areas of their host plants by soliciting unusual gene expression from adjoining cells in such a way that new developmental events result." Insects seem to reveal new morphogenetic potentialities of the host plant, but between us, we do not know much more than we did 50 years ago, as the chemicals responsible have not yet been properly isolated.

Research on the relationships between a cecidomyid fly, *Wachtliella persicariae*, and *Polygonum amphibium*, a plant of humid surroundings, have shown clearly that the tissue growth making up the gall is entirely due to the feeding activities of the insect (Dieleman, 1969). This comes about as a change in the normal development of the leaf, caused by the increase of auxins, growth substances in the gall tissue. This gall is necessary for the insect's development.

An old and classical experiment by Molliard, who inoculated healthy poppy pistils with extracts of squashed *Aulax papaveris* larvae (Hymenoptera: Cynipidae), resulted in the induction of gall formation. Unfortunately, the biochemistry of the galligenous process remains obscure.

Saliva of species of the aphid genus *Phylloxera*, a sad memory in Europe because of its damage to grapes, contains numerous substances which are supposed to accelerate or inhibit growth, including various amino acids. Normally the growth is produced only if the galligenous substances contact young cells or tissues. But in the case of galls produced on the leaves of roses, in organs already developed, the cells return to the embryonic state, a bit similar to carcinogenic cells. If we remove a larva from its gall, gall development stops immediately, and the plant does not react any longer to the inductive substance.

The number of galls in the world is enormous. Many in the tropics have not even been recorded. In Europe alone, 200 kinds of galls have been recorded on three species of oak, cataloged according to morphology, location, and inducing insects. Oak apples are well known and have a pleasant appearance, first yellow and later on turning red. These galls are always situated on the margin of the lower surface of the leaf.

The histological structure of galls is variable, sometimes rather complex. Structurally, there is an epidermis, with or without stomata and often also with hairs, which do not exist on the original epidermis of the unaffected parts of the plant. This epidermis covers an abundant parenchyma which is sometimes a bit sclerified. The plant organ (a leaf, for instance) is entirely modified.

The gall is also provided with abnormal phloem and xylem vessels which may be partly ligneous. Sometimes a cavity is found inside the gall close to or in communication with the outside, and inside this cavity lives the parasite. Near this may be found a layer of hypertrophied nutritive cells with fragile membranes and incised nuclei. This layer feeds the larva. The cells start to regenerate as the larva feeds. Then this layer is surrounded by a solid layer, somewhat ligneous and, hence, protective. Tannin inside the gall produced by *Cynips* spp. on oak is situated in the external parenchyma tissue. It has a protective as well as a repellent role.

Roughly speaking, the oak gall appears to be most profitable for the cynipids which live inside. The ligneous layer and the tannin protect the larvae against predators and somewhat against parasitoids, particularly the ichneumonids, although rather poorly, since the parasitoids are numerous and persistent.

The growth of some galls is so well organized that they open automatically at the right time for their inhabitants to escape (as in the case of the moth *Cecidoses eremitus*). Only insects with chewing mouth parts, as mentioned earlier, can dig galleries in order to escape, but many other ingenious systems have evolved, such as, for example, the piercing pupa of a species of fly genus *Giraudiella* living in the common reed *Phragmites communis*. Gall aphids escape the gall only after the leaf wall dries.

The chemical composition of the galls has been well studied; several of them are very rich in tannins and anthocyanins, which give them their colors. Organic acids and glucosides are important within galls, and a valuable comparison has been made between galls and fruits. Both formations seem to be vegetative organs similarly modified, one by the larval parasite, the other by the fertilized seed, a parasite not unlike the animal embryo in its mother's womb. In both cases, the plant reactions are analogous to an animal; it supplies the food and chemicals necessary for growth. A similar induced change can be caused by a mycoplasm parasite, which can completely modify the aspect of a flower, transforming all the flower's components (petals, sepals, stamens, pistils, and ovary) into green leaves. The phyllodial cotton has a green flower, as do sunflowers and many other plants. Witch's broom induced by fungi or bacteria is also a case very similar to phyllody. It is this phenomenon which proves that many living organisms can modify the directions coded in the plant host genes.

Many galligenous Hymenoptera show a complex cycle similar to that of aphids, i.e., the cycle can alternate between generations. For instance, *Andricus quercuscalicis* produces galls on oak acorns (*Quercus robur*), and the following spring parthenogenetlc females escape from the gall. Those females produce, in turn, very small galls on the stamens of *Quercus cerris*. Both males and females are produced in the galls. The female later lays eggs in the acorns of *Quercus robur*.

A more complicated case exists in that of the pineapple galls formed by the spruce aphid. The cycle takes 2 years to complete and comprises two generations. Also, the synchronization of the opening of the gall with the insect's development is remarkable.

In summary, galls have a certain morphological and chemical analogy to fruit. They seem to be the best solution for the plant in order to limit damage and displacement by parasites. The oldest gall known was described from the roots of an arborescent species of *Lepidodendron* (Upper Paleozoic to Lower Triassic from England) dating back to about 300 million years ago. According to Larew (1992), many gall inducers now attack the same plant species belonging to the same genera as they did in the past. That is true at least for species of *Acacia, Acer, Eucalyptus, Eugenia,* etc. Permian galls have always looked much as they do today, even though the primary types evolved during the Paleozoic on primitive green land plants and fungi (Roskam, 1992). Sawflies during the Cretaceous seem to have developed sophisticated galls, and gall-makers have evolved during the Tertiary, coinciding with the establishment and diversification of angiosperms (Raman, 1994). Coevolution between a gall insect species and a plant taxon seems difficult to justify, as one is the parasite of the other. If the plant defends itself, it does not gain any advantage to being the victim of insect attacks.

9.2   Mines

According to Hering (1951), mines are feeding channels caused by insect larvae inside the parenchyma or epidermis tissues of plants in which the epidermis or at least its outer wall remains undamaged, thus shutting off the mine cavity from the outside. Mines provide both living and feeding quarters for larvae. One cannot find a better definition of mines, because the one given by the Torre-Bueno's glossary (1989) is correct but rather succinct: "gallery or burrow visible beneath the epidermis of plant tissue, made by a larva." From the same book, a leaf miner is an insect that lives in or feeds on the mesophyll between the upper or lower surfaces of leaf. Hering's definition is more accurate and suits any case (see Figures 5.2C–F and 9.1A).

Normally, a mine cavity extends into the green parenchyma of the leaf, but mines also can be established inside the parenchyma of fruits, stems, or roots. Often, the leaf veins remain intact and the skeleton of the leaf is preserved. From his study of leaf miners, Hering conceived his classification of food selection among insects, various phytophagous arthropods, and even parasitic fungi. Mines exist on terrestrial and aquatic plants and can even preserve the green coloration of the leaves, as in the case of certain caterpillars. Larvae of Coleoptera, Diptera, Hymenoptera, and Lepidoptera are the main organisms responsible for leaf mines and other mines among ascomycetes, lichens, mosses, ferns, gymnosperms, and mono- and dicotyledons.

According to Hering, 118 European plant families are attacked by many types of mines, 95 by Lepidoptera, 85 by Diptera, 37 by Coleoptera, and 11 by Hymenoptera. In Europe and the Mediterranean basin, mines so far have not been reported among 30 families of vascular plants from Lycopodiaceae to Acanthaceae. They may be discovered later on, mostly in the tropics, but such statistics give an idea of the distribution of mines among plants.

It is true that the mining habit affords the larva greater protection against enemies, but parasites have adapted to leaf miners and share living conditions in the mine with the producer. No Tachinidae have been reported among the miners, because the female fly does not possess an ovipositor able to penetrate the cavity. There are, however, parasites of eggs, larvae, or pupae of various origins. There are even hyperparasites, parasites of parasites of miners among Chalcididae and Proctotrupidae (Hymenoptera). Braconidae, Chalcididae, and Ichneumonidae are the most common parasites. Larvae of Cecidomyidae are also found in mines, but their parasitism is low. Parasitic fungi also attack leaf-mining insects.

Ants are also predators, even if the mine is generally a good protection against them. Larvae of Hispinae in South America often mine the leaves of *Cecropia* tree species (Cecropiaceae) and are practically free of *Azteca* ant species attacks. Species of Chrysopidae can eventually suck out the larvae since they bite open the leaf epidermis. Birds are the worst enemies of leaf miners and often peck out the larvae and pupae from the mines.

Actually, it is very difficult, if not impossible, to identify some miners from the characteristics of their mines. It is often necessary to rely on examination of the larvae and pupae. According to Hering (1950), the mining habit arose polyphyletically. A mine is a hot house with all the humidity and heat necessary to develop a complex of associations: insect, parasites, inquilines, fungi. This guild deserves to be better understood. Detailed investigations of the influence exerted on the plant tissue by frass is also interesting. The frass provides a medium for fungal growth but also can be absorbed by the plant as a source of nitrogen. This aspect has also never been investigated. Ant-plants get part of their nitrogen from ants by way of their excreta. The plant absorbs it either with or without internal roots. Carnivorous plants also derive their food from digested debris and dead insects absorbed by the leaf.

Kahn and Cornell (1989) investigated the interplay in the U.S. of the native holly leaf miner *Phytomyza ilicicola*, leaf abscission of the American holly tree *Ilex opaca*, and parasitoids — a tritrophic interaction. Leaf abscission done after pupation of the miners did not affect miner survivorship.

chapter ten

Insect mimicry and homochromy in relation to plants

When I was in Thailand in the 1970s looking at the cabbages cultivated around Bangkok, I was fascinated by the extraordinary variation of color of the caterpillars of the moth *Spodoptera litoralis* (Noctuidae). The general color was light or dark green, but also yellow, reddish, brown, and very dark. According to my Thai colleagues, that was probably genetic. Others believed in mimetism, but really no one knew the cause of that variation. Recently, a paper has been written about the color morphs of the caterpillars of *Eumorpha fasciata* (Lepidoptera: Sphingidae) (Fink, 1995), which can be pink, yellow, green, or multicolored. The caterpillar feeds on various plants, and the food-plant quality seems to be a factor affecting larval color in that case. Food-plant effects on larval coloration may be widespread in the Sphingidae family, but most reports are such as for *Spodoptera* — anecdotal. *Spodoptera litoralis* feeds only on cabbage and the food-plant quality does not seem to be involved. This color variation is still unexplained but homochromy does not seem to be involved.

Mimicry is the resemblance of one species (model) to another (mimic) living together in the same area. Homochromy is a change of color to resemble the environment. So, mimicry applies mostly to a species mimicking another living species, while homochromy, as the Europeans understand it, applies to a species mimicking its environment, both inert (rocks, soil, etc.) and living (plants). I am dealing here with insects or arthropods mimicking the plants and true mimicry such as Batesian and Müllerian will not be discussed here.

Some insects, such as some of the Neotropical cassid species, abruptly change their color, probably as a deterrent to predators or as a sexual signal, but do not match the plant environment this way (Jolivet, 1994). Many of them are green or partly transparent, mimicking a drop of water or excreta.

These can be considered as homochromic. Many insects (grasshoppers, leaf beetles, bugs) are green and reach a high degree of camouflage, either physical or chemical; others are brown. Some moths or butterflies mimic the bark of a tree or the green of the leaves. In a classical example, a *Phyllium* species imitates a leaf and is nearly perfect mimicry of fungi, notches (insect damage), lichens, and even veins of the presumed leaf. Many tropical America tettigonids perfectly mimic leaves. Some of them when lying on their sides have adopted the attitude, color, and general aspect of a dead leaf in a forest path — for example, *Phylloptera* spp. in Panama which remain immobile until disturbed.

Mimetism and homochromy are the result of a very long evolution, but the phenomena were already known during the Carboniferous and the Mesozoic. Imitations of fern pinnules are frequent (paleomimetism) and have been confused sometimes (Theobald, 1937) with real insects.

The concept of homochromy with its double meaning (change [U.S.] vs. permanent copy [Europe]) brings confusion. However, it always means mimicry or camouflage of an insect (or arthropod) to resemble its resting place, a leaf, a twig, the bark of a tree, its spines, the ground, an insect, or bird excreta. It is evident that the chameleon changes its color to match its surroundings, but species of cassids do not exactly match even if several species change from gold to brown when dropping to the ground. Some rare beetles change color with changes of humidity and time of day, namely some large tropical scarabs which darken with the night.

Species of the Asian butterfly genus *Kallima* (Nymphalidae) and many others, when their wings are folded, look very much like a dead leaf with false veins, mold spots, and nicks in the edges. Many *Phyllium* spp., several other phasmids, some mantids (with moss and lichen covers, as in the Panamean *Pogonogaster tristani*), and many tettigonids exactly mimic leaves, lichens, and mosses with all their minute details. Several adult membracids and some leaf beetles (Lamprosomatinae) at the larval stage with their scatoshell mimic the spines so common along the trunks of tropical trees.

Many mantids, homopterans, caterpillars, and spiders mimic pink, yellow, blue, red, and violet flowers according to the color of the flower on which they live. The change in coloration is slow, poorly known, and probably under the influence of their food (as in caterpillar species of *Rekoa*, Lycaenidae) or is through some pigment migration or other possible causes. In Brazil, Thomisiinae spiders have a color perfectly matching the flowers in which they are awaiting their prey: yellow, purple, or white. Perhaps here one deals with different morphs choosing their color according to the flower's or, more probably, through a very slow process starting very early in the life of the spider. Weevils often mimic their support, such as the species in Curitiba bearing brown spots of rusts "painted" over its back. Parasitoid wasps (Koptur, 1989) copy the white flowers of Rubiaceae in Costa Rica.

Membracid species of the genus *Stegaspis* mimic species of *Inga* folioles (Mimosaceae). These are very common plants along Amazonian streams. The imitation is striking. A small caterpillar of a noctuid moth, *Zanclognatha theralis*, in Oak Ridge, TN, mimics the lichen *Usnea strigosa*. The larva (2.0 cm long) has evolved to resemble its lichen host. It is gray-green with light speckles that mimic the emerging papillae on the lichen thallus (Sigal, 1984).

As we have seen previously, sloths in tropical South America have algae living on their hairs, which probably increases the camouflage of the fur. Crab species of *Pagurus* (hermit crabs) collect algae from their surroundings and plant them over their shell. Reduviids nymphs camouflage themselves with debris and cast skins. Several weevils in Borneo and Panama have algae encrusting their elytra. This is, of course, epizoic symbiosis already discussed in Chapter 8. The large New Guinean weevils described there have a very long life and an exoskeleton pre-adapted to the growth of this sort of camouflaging cryptogamic flora.

In the mimicry of plants, insects and spiders are generally passive. The New Guinean weevils get their flora on their backs from aerial spores passively disseminated. Mimicry is a very long process and is the result of selection since the origin of the insects and their food plants. Insects living in lichens look very similar to lichens, and even some Central American tettigonids develop the grayish and the blackish spots of their favorite food and hiding places. I am sure that further research will help us find other cases of epizoic symbiosis in the high tropical mountains. The phenomenon cannot have evolved only in New Guinea and only with its weevils. Whether insects and plants coevolved or did not coevolve, their development since the Carboniferous shows an adaptation to plants as food but also as a means of protection. Ant-plants, galls, mines, and carnivorous plants are examples of mutual adaptations and a means of survival. Under selective pressure, more and more insects mimic plants — camouflage is one way to protect life in a very dangerous world.

Let us note that the black mutants of cryptic moths select the soot-covered trunks and the light ones the lichen-covered trees. Because of pollution, cities such as Paris and many English towns have completely lost their lichens, free or encrusted. Under natural and artificial selection, light-colored moths usually rest on fresh vegetation, while dark-winged species select tree bark or rest on the ground; a mechanism of background selection definitely exists. Bark-resting species rely primarily on crypsis to avoid detection, while ground-resting species conceal themselves from predators and take evasive action if located. *Biston betularia* melanic and typical morphs tend to behave differently (Boardman et al. 1974). Industrial melanism in England selected not only the color but also adapted to the soot-covered trees, and their behavior is the most favorable to such an environment.

Let us propose that it is by mimetism of normal eggs that several *Passiflora* vines inhibit egg laying by species of *Heliconius* butterflies. The vines possess specially modified extrafloral nectaries copying the real eggs of the butterflies (see Figure 10.1D).

Certain plants are pollinated by Bakerian mimicry. For instance, the Amazonian monoecious palm *Geonoma macrostachys* has staminate and pistillate

A

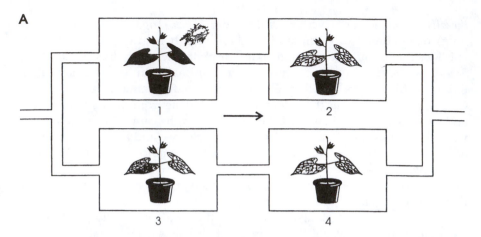

Figure 10.1 Physiology of plant selection. **(A)** Setup designed to study the influence of damage-related plant volatiles on resistance in recipient undamaged plants. Two parallel wind tunnels send a unidirectional continuous air stream with plant volatiles. 1 is the infested plant; 2, 3, and 4 are noninfested plants. (Adapted from Bruin et al., 1995.) **(B)** Chemical formula of several secondary substances of plants. (Adapted from Schoonhoven, 1969c.) **(C)** Hypothetical relationships between the degree of preference of the phytophagous insects and the basic plant substances and their degree of polyphagy. (Adapted from Edwards and Wratten, 1980.) **(D)** Above, leaf of *Passiflora coerulea* L; middle, petiole of the same leaf enlarged with trichomes transformed into false eggs of species of *Heliconius* butterflies; below, one real egg of *Agraulis vanillae* (L.) (Heliconiinae). (Adapted from Sacchi, 1988.) **(E)** Pollination of the orchid *Cymbidium insigne*, without reward, by mimicry with *Rhododendron lyi* (Ericaceae) in Thailand. (Adapted from Kjellsson et al., 1985.)

phases. The staminate flower offers pollen and scent compounds to visitors, while the pistillate does not present food rewards (Olesen and Balsley, 1990). The pistillate flower resembles the staminate in size, color, and scent. It has a staminodial tube and is pollinated by drosophilid flies, beelike syrphids, and beetles. The same occurs in the Caricaceae (pawpaws), which are also dioecious. Moths visit the male flowers (nectariferous) and the female flowers (without nectar) indifferently. Bakerian mimicry is very common among plants and is done to facilitate pollination.

As Barrett (1987) wrote, there are flowers that look very much like insects, leaves that look like flowers (some carnivorous plants), and weeds that masquerade as crop plants. Rothschild (1984) mentioned a wild carrot umbel with an inflorescence mimicking a fly: it attracts pollinators and deters predators.

In Thailand, the orchids *Dendrobium infundibulum* and *Cymbidium insigne* are pollinated by mimicry (they offer no food reward) with *Rhododendron lyi* (Kjellsson et al., 1985). Many similar cases are known all over the world. (See Figure 10.1.)

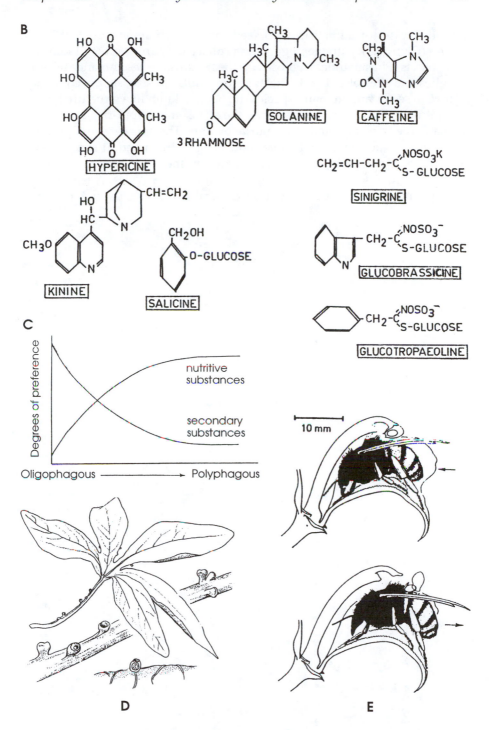

Figure 10.1   (continued)

Other cases of mimetism with seeds are known among the leaf beetle species (Clytrinae), with their eggs being collected by ants (Jolivet, 1952), and stick insects (Phasmatodea), whose eggs bear a striking resemblance to seeds (Hughes and Westoby, 1992). The eggs are similar in size, shape, color, and texture to seeds, and in many species they bear a knob-like structure known as capitulum. This structure resembles the elaiosome, a lipid-rich appendage on many seeds, an adaptation for burial by ants. The ants eat the capitula as they eat the elaiosomes. As Hughes and West (1992) noted, it is a "striking example of evolutionary convergence between the plant and animal kingdoms." These observations were made in New South Wales, Australia. The same phenomenon has been found in tropical Africa and Latin America.

chapter eleven

Natural weed control in the Holarctic region and in the tropics using phytophagous insects

Phytophagous insects and phytoparasitic fungi have been used all over the world to fight against imported weeds which have become pests in their adopted country. Classical entomological handbooks have been written on biological weed control and are referred to in the References section in this book. Actually, American (USDA) and Australian (CSIRO) teams are based in Brazil, France, Italy, etc., not to mention the British (CAB) and French (IRAT) units all over the world. These teams collect specific enemies of imported weeds. Generally, these plant parasites are leaf feeders, but the flowers, fruits, or roots of such weeds also need to be eradicated or at least controlled. Recently, a paper by White (1996) summarized the topic for North America, at least for leaf beetle species.

For many years, various teams, including CSIRO, have worked in Montpellier on the biological control of weed species of the genera *Carduus* and *Chondrilla* (Asteraceae) imported into the U.S. and Australia, respectively, and many other plants such as species of *Heliotropium*. The work consists of identifying in the Mediterranean region of France, Greece, and Turkey fungi and animal parasites of several plants, including the Old World dandelion, *Chondrilla jacea*. Then these herbivores and fungi are tested in the laboratory on various ornamental and useful plants cultivated in Australia and even on wild species of *Eucalyptus* trees that belong to a completely unrelated family, the Myrtaceae. When and if the tests are positive, i.e., if one is certain that the pathogens and parasites are selective, only then, with the greatest precautions, are these species raised, multiplied, and cleared of their parasitoids. After a reasonable quarantine, they are released in their adoptive country. Sometimes, in places such as Australia, where the quarantine laws are very strict, some insects such as coprophagous beetles are imported only as eggs which have been externally sterilized.

The Australian region has proved an ideal experimental area for this kind of research, and numerous other countries have been encouraged to begin research and consequently to introduce auxiliary phytophagous animals for plant control. It is clean and avoids the poisonous effects of herbicides which very often are not selective and destroy useful parasitoids. The Hawaiian islands, India, Mauritius, Madagascar, South Africa, New Zealand, continental U.S., Canada, and Chile have specialized teams, most often composed of an entomologist and a phytopathologist, for these studies. The phytopathologist is in charge of studying the viral, mycoplasmal, bacterial, and fungal diseases of the plants to be controlled, while the entomologist looks for phytophagous insects.

Even if the fight against some invading plants sometimes has been a success, there is still a problem with certain other weeds. Some of those for which the control is being studied and still remains uncertain are *Convolvulus* spp., *Ipomoea* spp., *Rubus* spp., *Cyperus* spp., *Xanthium* spp., *Kentrophvllum* (*Carthamus*) *lanatum*, *Centaurea calcitropa*, and *Echium vulgare*. *Matricaria discoidea* is a pest in Europe but is mostly localized in rural areas. *Solidago* spp., *Galinsoga* spp., the famous and beautiful water hyacinth *Eichornia crassipes*, and aquatic ferns such as *Azolla* spp, *Salvinia* spp., and others also are without adequate control.

Eichornia crassipes is controlled to some extent by the weevil *Neochetina eichorniae* in southern Florida. There also is a flea beetle, *Agasicles hygrophila*, which feeds on alligator weed and seems very efficient in controlling this aquatic plant.

It is much easier to fight invading plants than insect pests, provided one takes into account a series of essential rules. The most important rule is to ensure the introduced plant or animal species is not pathogenic against useful indigenous species.

One of the best successes realized so far, is the control of gorse or whin (*Ulex europaeus*) by *Apion ulicis* in New Zealand, Southern Australia, and Tasmania. It seems odd that these biological control measures have never been attempted in the "Hauts" of the island of La Reunion, near Madagascar. The plant, introduced once by a nostalgic Briton two centuries ago, is becoming a pest at the expense of grazing land of that island, as well as *Rubus alceaefolius* and *Fuchsia* spp. *Rubus alceaefolius* is a real pest having inedible berries, but *Fuchsia* spp. and *Ulex* spp. perhaps represent an esthetic aspect of the mountain summits and probably it is this which is protecting them.

The partial destruction of *Lantana* spp. bushes in Hawaii and Fiji islands by a moth (Tortricidae), a fly (Agromyzidae), another fly (Tephritidae), and a bug (Tingidae) is another example of at least limited success in the use of insects as control agents. Rounding out the list of real successes are the destruction or control of St. John's wort, *Hypericum perforatum*, a toxic weed, mostly thanks to several species of *Chrysolina* (Chrysomelidae) from western

Europe, in Australia, California, and elsewhere; the control of *Clidemia hirta* with the thysanopteran *Liothrips wrichi* in Fiji; and the control of the prickly pear cactus (*Opuntia* spp.) with the help of the moth *Cactoblastis cactorum* (Crambriidae) and several species of cochineal scab insects in Australia, India, and South Africa.

Among the successful applications of biological control is the control in Hawaii of the weeds *Eupatorium adenophorum*, *Tribulus* spp., *Emex australis*, and *Hypericum perforatum* and in Russia the orobanche or broomrape, a parasitic plant on sunflower roots, controlled by the fly *Phytomyza orobanchiae*. The orobanches number ten species, and some of them parasitize the roots of tomatoes in northern Sudan. They lack any green tissue and cannot be considered as hemiparasitic such as the mistletoe. The seeds at germination produce spiral filaments searching for hosts. In the tropics, some have been seen piercing the tarmac. In Africa, people are at a loss as to how to fight that beautiful flower (Orobanchaceae is closely related to Scrophulariaceae) because it is a terrible pest on wet season crops. It is impossible to control these plants with chemicals alone, because they are not selective enough. Similarly *Cuscuta* spp., the dodders, cause damage to trees and irrigated crops in the semi-desert steppes of Afghanistan. So far, no biological control has been tried. Strangely enough, the dodderlike *Cassitha filiformis* (Lauraceae) of the Old World tropics does not seem to parasitize cultivated plants or, if it does, has never been reported to do so. It is extremely common on native bushes and trees in Western Africa.

The problem sometimes is more complicated. The control of prickly pear cactus (*Opuntia* spp.) is aimed at the destruction of the species with spines and not the spineless ones, but specificity of an insect to similar species or varieties is extremely rare. In South Africa, both kinds of plants have been destroyed together.

I have mentioned above the genus *Tribulus* (Zygophyllaceae) in Hawaii. One species, *Tribulus terrestris*, is a noxious weed in California, where, after 50 years of herbicide application, the California Department of Agriculture finally introduced two weevils from India in 1961 which were successful in controlling the weed. The plant is very toxic to stock.

It should be noted that invading plants in the tropics are often ubiquitous because they have been carried by humans as plants or seeds almost everywhere. Many of those weeds originated in tropical America, Africa, southern Europe, or Asia. I have found, in such far-away places as Sahel (an area of west Africa, south of the southern limit of the Sahara), New Guinea, La Reunion, Vietnam, and Thailand, the weeds *Euphorbia hirta*, *Sida rhombifolia*, *Tridax procumbens*, and many others. In the highlands of New Guinea, I have been able to follow the progression of the European *Plantago major* along the roads and around the villages. It is a ruderal plant and it was probably introduced from Australia (a non-native plant there also) with seeds. Plantain seems to fill a vacant ecological niche in the suburban areas. It replaces at middle elevations the endemic plantains of higher altitudes. Some introduced plants are relatively easy to fight using food-selection specificity and introducing parasites and pathogens from the native habitat of the weeds.

Control experiments began in 1950, in Mauritius, against the Central American *Cordia macrostachia* (Boraginaceae), a small, rather common, arborescent shrub, often eaten in tropical America by cassids.

Two chrysomelid beetles were introduced in the island — one, the cassidine *Physonota lutacea* and the other, a galerucid, *Schematiza cordiae*. Both are active tree defoliators in Trinidad. Twenty five years later, while visiting Mauritius, I observed that the weed had persisted but was integrated into the local flora and was practically under control. However, only the galerucid beetle had survived; the cassidine did not become established. I have seen *Physonota lutacea* on *Cordia* shrub species in South America. The adult beetle and their larvae live in groups. The adults make noises with their heads rubbing the pronotum, probably as a warning. The beetles are very vulnerable to predators, and probably in Mauritius they found their enemy, possibly a native bird not repelled by the cassid.

White (1996) reports the successful biological control in North America of the following weeds using leaf beetles: *Alternanthera philoxeroides,* or alligator weed, controlled with *Agasicles hygrophila* (Alticinae); St. John's wort, or *Hypericum perforatum,* controlled with two species of *Chrysolina* (a third species did not survive); and *Senecio jacobaea,* an Asteraceae, controlled with *Longitarsus jacobaea* (Alticinae). Some other projects using *Aphthona* spp. on *Euphorbia esula* or *Galerucella* spp. on *Lythrum salicaria* seem headed for success. Many leaf beetles and weevils are actually used or are under study by the U.S. Department of Agriculture (USDA) in America.

During these preliminary studies of the introduction of phytophagous parasites, the species must be imported free of all its enemies. Its selectivity has to be studied studied to determine that it is harmless to cultivated plants, and it must be determined if it can adapt to the new habitat. Similarity of climates between the country of origin of the insect and its new adopted area must be carefully studied. The differences in seasons between the northern and southern hemisphere are not important. Plants and insects adapt very rapidly to the inversion of seasons, as proved by the introduction (accidental) of various species of *Microtheca* in North America (Chrysomelinae) and of *Agasicles hygrophila* on *Alternanthera* sp. in Florida, Georgia, and the Carolinas. Even if a tree needs vernalization (*Cerasus* spp., or apple trees) or an insect needs diapause, the passage from the northern hemisphere to the southern proceeds very well. For instance, *Chrysolina* spp. on *Hypericum* spp. inverted their cycle when introduced in Australia.

The fecundity of an introduced species must also be determined. One must be very careful about possible xenotrophic (allotrophic) changes and mutations that can modify its behavior and sometimes produce polyphagous and oligophagous varieties. The oligophagous hidden tendencies of a genus or a species must be carefully weighed. For instance, specialists know that the European galerucine *Agelastica alni* does not feed only on *Alnus* spp.

(Betulaceae) but show extremely rare outbreaks on Rosaceae, a normal tendency for the Far Eastern species.

When it was decided to destroy the imported blackberry *Rubus fructicosus* in New Zealand, someone suggested the use of the Buprestid beetle *Coraebus rubi*, native to southern France. The idea was abandoned because of its oligophagous habits. This insect is perfectly able to attack roses and other cultivated Rosaceae. The cure would have been worse than the illness.

The case of the "vigne marronne", *Rubus alceaefolius*, on the island of La Reunion is interesting. Introduced 200 years ago by a missionary from Indochina, who, according to the legend, confused it with the grape vine because of its leaves, the plant rapidly became a pest, particularly on the mountainside where its berries were disseminated by birds. In this case, biological control theoretically appears simple. To start with, after following strict screening procedures, one could try the native phytophagous enemies from Vietnam: the leaf beetles, *Phaedon fulvescens*, *Chlamisus latiusculus* and other species of *Chlamisus*, several eumolpines such as different species of *Basilepta*, a certain number of moths, and a fungus or a rust. They all seem to be host specific and not pests of roses. The introduction of a rust, however, is rather risky without extensive testing, and species of *Chlamisus* are rather delicate to acclimatize. *Phaedon fulvescens* is robust, specific, and able to adapt under tropical climatic conditions not too much different from the low tropical mountains of Vietnam. A virus seems to be transmitted by aphids, but more studies are needed in the native land of the plant and its future country of adoption before releasing insects or a new fungus.

Unfortunately, insect introductions are numerous; although some of them, such as species of *Phylloxera* and the Colorado potato beetle, have been accidental, many others have been deliberate. The introduction of *Papilio demodocus* on citrus in La Reunion and Mauritius was certainly a mistake, as endemic species that existed in the mountains of those islands are now near extinction. In La Reunion, *Agave* sp., *Rubus moluccanus* and *R. alceaefolius*, *Eugenia jambos*, *Lantana camara*, *Ulex* spp., *Solanum auriculatum*, *Erigeron* spp., *Fuchsia* spp., and many other plants were deliberately introduced, but they have not been effective in controlling the weeds.

The method used for selecting beneficial insects consists of selecting monophagous species and rejecting oligophagous and polyphagous ones. On the other hand, it is rare that only one phytophagous species is sufficient to destroy an invading weed. Some cases exist, however, such as, for example, the control of *Opuntia aurantiaca* and *Hypericum perforatum*, but even with these two, the main insect enemy has been helped to some extent by some other insects. Generally, one selects a whole series of pathogens and phytophagous arthropods which will attack leaves, flowers, seeds, stems, roots, and fruit. Their combined efforts may succeed in controlling the weed.

Rare failures are due to the presence of either parasites or predators of the phytophagous agents. It is well known that in Australia, a parasitoid of the *Paropsis* sp., chrysomeline on *Eucalyptus*, attacked after a relatively long period the species of *Chrysolina* imported to fight St. John's wort. Most often nonselective predators are present which rapidly adapt themselves to new prey. Another reason for failure is the change of climate between the two countries. Seasonal differences sometimes increase or diminish the number of generations of the introduced insect. *Chrysolina varians*, for instance, has doubled the number of its generations per year since its introduction in Australia. To fight *Alternanthera philoxeroides* in the eastern U.S., the entomologists had to find a cold-adapted biotype of the flea beetle *Agasicles hygrophila* in southern Argentina. The "hot" biotypes did not survive the winters of North Carolina. A moth, *Vogtia malloi* (Pyralidae), also was introduced to help the flea beetle in its destructive function.

In conclusion, let us emphasize the importance of having exact knowledge about the life cycle of phytophagous insects. No detail should be considered unimportant. Elton (1958) lists three stages at which plant and insect invaders may be repelled. First, we can prevent them at the gate by quarantine. Second, we can destroy the first bridgehead by eradication. This is a rare phenomenon, although it was a success in Brazil for *Anopheles gambiae* and partially for *Aedes aegypti*. In Bougainville island in New Guinea, in 1969, the giant African snail, *Achatina fulica*, was destroyed near the airport with flame throwers. Finally, and this is the most reasonable method, we can keep the invader population below the dangerous threshold via biological control. Fighting against noxious weeds using phytophagous control agents is one means of control which renders an introduced plant rare and harmless. Providing that rigorous screening procedures are followed, success is much easier to achieve when controlling weeds than fighting insect pests. Molecular biology seems ready to offer ways for plants to resist selective herbicides. It could also sensitize others to those herbicides. To me, that is not an acceptable solution. Poisons are poisons, and there are enough of them now over our planet.

Introducing parasitoids also can lead to disaster, as in Guam where 60% of the native lepidopteran fauna has been destroyed by imported Hymenoptera during the last 100 years. This is a limited case, but precautions must be taken when indroducing exotics for pest control.

chapter twelve

Pollination

In 1793, the German botanist C. K. Sprengel discovered the role of insects in pollination. In several exotic orchids, the phenomenon of pollination is so complex that the term "coevolution" has been given to the process that evolved. The process by which some temperate and exotic orchids are pollinated (pseudocopulation, as described later) contributed to the idea of coevolution of a very specialized type. Orchid pollination does not always take place in this manner, though, and pollination is not generally the result of a parallel. Many insects pollinate, and many flowers provide them with edible pollen and nectar.

The development of pollination was a very long process, taking place since the Cretaceous and all during the Tertiary. It seems that the earliest Cretaceous flowers were apetalous magnolias (Crepet et al., 1991). Coleoptera were probably among the earliest pollinators, though the idea is actually debated. Rosid flowers appeared during middle Cretaceous and became diversified with complete corollas. Nectaries are also present in the late Cretaceous, and these must have evolved with the Hymenoptera, namely the Meliponidae (from the Apidae). Some researchers (Grinfeld, 1975, for example) reject the cantharophily hypothesis of the first Angiospermae appearing in the Mesozoic.

The flowers are thus the product, indeed the symbol, of the last group of plants to have evolved. The flower is a structure that combines a protective envelope around the reproductive organs, sexual cells, pollen, and ovules. Flower parts are only modified leaves, which can easily be seen in the cases of phyllody, i.e., when a plant is parasitized by mycoplasmas, a primitive bacteria-like organism, and is classified among the Mollicutes. The flower — whether compound, such as those of the sunflower, or simple, such as those of cotton — transforms partially or entirely into a leafy organ when infested by mycoplasmas. The resulting green flower is evidently completely sterile and devoid of pollen or nectar.

The hermaphroditic flower of many angiosperms is composed of the perianth, or floral envelope, made up of the calyx (sepals), corolla (petals), and androecium (pistil and ovary). At the base of the pistil is the ovary or ovaries from which emerge the styles capped by the stigma. Pollination in angiosperm flowers (the prelude to fecundation, or fruitful union of male and female gametes) is accomplished by transportation of pollen grains from the anther of the stamen to the stigma. In gymnosperms, pollen from the anther is taken directly to the ovule in the cone. The fusion of the gametes in the ovary will produce an egg, which will develop into an embryo and, along with other parts, will form the seed. Insects and wind are the main pollinators of plants, helped by birds, bats, and other mechanical means such as water in certain primitive or aquatic flowers. (See Figure 12.1.)

Not only do insects pollinate the plants, but some special adaptations allow birds such as hummingbirds (Trochiliidae) in America; sunbirds (Nectariniidae) in Africa, southern Asia, and eastern Australia; sugarbirds (Meliphagidae and Trichoglossidae) in the Australian region; Hawaiian Honeycreepers (Drepanidae); and bats in tropical America, Africa, and Asia to effect this process. At least one flower is pollinated by rats in Hawaii (*Freycinetia* sp., a Pandanaceae). Species of *Banksia* in Australia are pollinated by small marsupials. Their long tongues and fine fur are covered with pollen, which is then transported from plant to plant. Some snails pollinate certain flowers (such as species of *Aspidistra* and *Rhodea*). Orchids are pollinated, according to genus and geographical areas, by Lepidoptera, Hymenoptera, birds, bats, frogs, slugs, and (for two endogenous Australian species) by earthworms. Among some plant families, such as the Palmae, there are plants that are pollinated mostly by bees (coconuts), and others that are either beetle pollinated (oil palms) or wind pollinated (date palms).

Pollination methods are named as follows: anemogamy, by wind; hydrogamy, by water; therogamy, by mammals; cheirogamy, by bats; entomogamy, by insects; and ornithogamy, by birds. Therogamy is usually restricted to rodents and marsupials. The Proteaceae (which includes the macadamia nut) in South Africa are remarkably morphologically and biologically adapted to "their" rodents by having an abundance of nectar, a special smell and color compatible with the nocturnal life of the animal, strength to hold their weight, and a large flower and petiole near the ground. The co-adaptation is almost as perfect as those between insects and certain flowers described on the following pages. There is also a temporal coincidence between the time of flowering and the nutritive needs of the animal. The same adaptation also occurs among the Australian Proteaceae and their marsupials and must exist elsewhere, namely with primates in the tropics — for instance, *Galago* sp. with species of *Adansonia, Lorus,* and *Tarsius.* If these facts are not well known, it is due to the difficulties in observing these nocturnal, shy, and rare mammals. It does not seem that pollination by the vertebrates developed much later than the insect-plant syndrome. On the one hand, there exist fossil traces of presumed relationships between insects and plants; on the other hand, the adaptation of the South African Proteaceae to their rodents is too specialized to be fortuitous. However, the tamarin monkeys in the Amazon gorge themselves with nectar from *Symphoria* tree species (Caprifoliaceae). They destroy flowers, but they do not cross-pollinate the flowers (Garber, 1988). The pendulous flowers of this tree are chiefly fertilized by bees and wasps.

Europe and temperate Asia are without any pollinating birds. The Nectariniidae reach only Israel outside of the tropics. Perhaps one reason could be the rarity of red flowers in these countries. Most insects, except butterflies, are blind to red or see it in a different way. Wild red poppies actually reflect ultraviolet, at least as far as bees are concerned, as they do not see the same colors as we do.

Hummingbirds are often migratory. In summer, they fly as far north as Alaska, where they find flowers adapted for pollination by them. The flowering time is adjusted to bird migrations and to the summertime. One African nectariniid bird flies as far north as Palestine where red flowers of the epiphytic mistletoe, *Loranthus acaciae*, are perfectly adapted for pollination by these birds. Some species of *Fuchsia*, growing at the extreme southern tip of South America, are pollinated by birds. Cultivated species of *Fuchsia* have no birds adapted for pollination; instead, they seem to reproduce vegetatively. At high altitudes, as in the Andes, a similar adaptation occurs with the bromeliad *Puya raimondii*, in the East African mountains with blue flowering giant *Lobelia* spp. blossoms, and in the Papua New Guinean mountains with species of *Rhododendron*, where they are often trees. All of these bird-plant associations are specialized, have their bird pollinators, and indicate parallel evolution.

During a stay in eastern Papua New Guinea, I was impressed during the climb of the highest peak, Mt. Wilhelm, by the diversity of size, shape, and color of the rhododendrons. A special adaptation seems to exist there between the pollinators and the flowers. At higher altitudes, red and odorless flowers are fertilized during the day by birds; at lower altitudes, white and fragrant flowers are fertilized at night by Lepidoptera, primarily hawk moths.

It must be noted that sometimes, as in tropical America, wherever Lepidoptera and hummingbirds coexist, both become competitors chasing each other. Hawk moths are chased by the birds. At high altitudes, the moth *Castnia eudesmia* chases birds away from the flowers of *Puya alpestris*. The bird is often smaller than the moth in this tropical forest. Not only do the nectarinians fertilize the flowers, but also there is a well-known case of a species of *Boerlagiodendron* (Araliaceae) attracting pigeons. The plant has fruit-like bodies, which, according to Van der Pijl (1982), are actually sterile flowers. It is difficult to know whether the nectariphagous birds are primary or secondary pollinators because they search for insects, water, food, and nectar. However, the adaptations seem very old and probably began at the end of the Jurassic with the appearance of birds and the first flowering plants.

The distribution of bats, which are color blind, and the plants they pollinate, all of which smell like ripe fruit, coincides perfectly. The adaptations seem to be very specialized and exclusive.

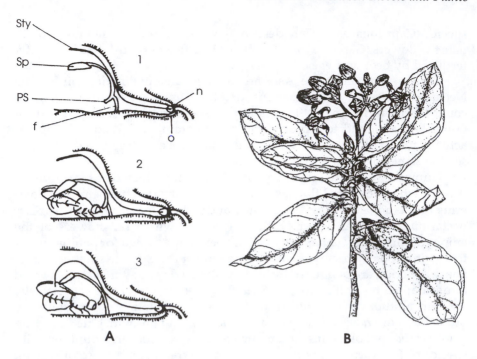

Figure 12.1   Pollination by insects. **(A)** Pollination of the flowers of *Salvia glutinosa* by bees: (1) f = filament, PS = palette, Sp = anther, Sty = style, n = nectaries, o = ovary; (2) flower at the male (pollen) stage — note pollen on bee; (3) flower at the female (ripe ovary) stage. **(B)** *Calotropis* sp. plants from Thailand. These plants are pollinated by various species of Hymenoptera, including species of *Xylocopa* which are kept temporarily inside the flower as prisoners. The plant is eaten by a danaid butterfly, some moths, and various beetles, species of *Platycorinus*. The insects that live on *C. gigantea*, including yellow aphids and locusts, show aposematic colors because the toxic properties of the plant are transmitted to the animal hosts. **(C)** Inflorescence of a giant Sumatran flower, *Amorphophallus titanus* (Araceae). The red color and fetid odors of these inflorescences attract blow flies and induce them to lay eggs on the spadix. Several species of beetles are also attracted. (Adapted from Meeuse, 1961.) **(D)** Mating reactions between: (1) *Andrena maculipes* (the bee) and *Ophrys lutea* (an orchid); (2) *Andrena maculipes* (the bee) and *Ophrys fusca* (an orchid). (Adapted from Kullenberg, 1961.)

Hummingbirds and hawk moths do not land on flowers when they are gathering honey, but hover the same as helicopters with their elongated beaks or proboscises extended. Nectariphagous bats show adaptations: elongated snouts and elongated tongues with corneous expansions at the end. Insects also have similar adaptations, resulting in a great diversity of sucking organs for gathering nectar. Many plants are dependent exclusively on certain insects, as well as birds and bats, for fertilization.

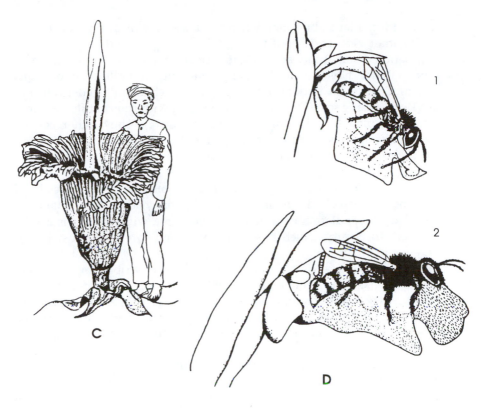

Figure 12.1 (continued)

Many pollinators fly high in the canopy of tropical forests, such as insects, bats, and birds. Among the insects are some Diptera and Orthoptera (all polyphagous), but little is known about this specialized fauna. Scattered throughout this tropical forest canopy are trees with large flowers, easily visible from a plane, which break up the green monotony of the canopy. Some forests exhibit more color than others, according to the season, the geographical area, the local flora, and the altitude. At the top of the canopy, where the temperature reaches 40°C, flowers are numerous and pollinators are abundant. At the same time, it is evident that pollinators at the lowest level of the tropical forest are rare because the shade cuts down the number of available species. At the soil level, many flowers are cleistogamous or endogenous (e.g., *Commelina* spp.).

As previously mentioned, honey gathered from rhododendrons (with about 1200 species) can be toxic to humans, primarily the honey from those species occurring in Asia Minor, but also elsewhere, as indicated by recent reports from Nepal. The honey is not toxic to bees, but numerous plants do produce a pollen that kills bees or at least makes them sick. The pollen of buttercups (Ranunculaceae) and of some shade trees is poisonous. Normally, bees avoid these plants, but often when food is scarce or during a drought, bees will gather even toxic pollen.

Pollination is sometimes acrobatic for the bee. Some orchids fertilized by pseudocopulation have the bee hanging upside down. Pollination of *Vigna vexillata* in Costa Rica is done by a hugging mechanism (Hedstrom and Thulin, 1986). I have seen *Vigna caracalla* in Brazil with fragrant flowers that have a keel coiled like a snail (the "snail flower"). They either can be autogamous or species of Apidae must pierce the corolla and turn the body in a strange fashion. Pollination, if any, must be studied further for that plant.

Much has been written on pollination and pollinators, the specialized associations between plants and insects, and the many adaptations that have evolved. Pseudocopulation of certain Hymenoptera with orchids is a relatively recent discovery which has been confirmed to exist almost everywhere, even in Australia.

The biology of bees is now well known and described in almost all biology textbooks. Their relationships with the flowers they pollinate are now known to be functions stimulated by smell, color, shape, and movement of the host plant. A motionless flower certainly attracts fewer insects than a flower shaken by the wind. Too much has been written elsewhere regarding bees to go into great detail here.

There are also species of Hymenoptera which vibrate their bodies at high frequency (between 320 and 410 hertz) and provoke the dehiscence of the anthers of several plants and the dusting of their body with pollen. Several cases are well known and analyzed. It is difficult to regard this as coevolution between the insect and the plant, because *Bombus* spp., for instance, pollinate many different flowers. Some *Bombus* spp. vibrate the flowers of various species of *Actinidia*, *Symphytum*, *Borago*, and *Polygonatum* (Corbet et al., 1988). The anthers are vibrated at known frequencies, and pollen surges out at a critical frequency and at its harmonics. It is the vibration from the bee's abdomen and thorax which are transmitted to the flower's anthers. The buzzing is merely a byproduct of the vibrations.

Vibration pollination seems to be older and more widespread than we think (Proença, 1992). Poricidal anthers do not appear in the fossil record before the upper Paleocene, but myrtalean pollen is known from the Cretaceous. Modern-day Myrtales with buzz pollination include Melastomataceae and Lythraceae. A Cretaceous origin for myrtalean buzz pollination, therefore, is proposed for the Colletidae bee family.

We have discussed previously myrmecophylic plants. We should add here the fact that the Italian form of *Apis mellifica*, introduced into Mexico long ago, quickly became a rival of the *Pseudomyrmex* ant species that protect many acacia trees there. The extrafloral nectar produced by the trees attracts bees, but the bee is always the loser in case of conflicts with the ants. The latter remain the most ferocious, since they were the first to become established on the trees and to defend their own territory (Janzen, 1975). Anyway, the recent introduction in America of the ferocious Africanized bees has probably changed the relations between the attackers and the defenders.

Cross-pollination is usually a necessity for most plants. For instance, in the case of *Salvia glutinosa* (garden sage), when a bee pollinates the flowers, the anther bends to drop pollen on the visiting bee. Later, another bee, laden with pollen, pollinates the stipe. Almost all leguminous (Fabaceae) plants (fodder plants) are autosterile, which is also true of many fruit trees.

One third of our food supply depends directly or indirectly on pollinating insects. Alfalfa (*Medicago sativa*) originated in the Middle East and is not of much interest to honey bees in our latitudes. In Europe, solitary bees such as the American species *Magachile pacifica* seem more efficient as pollinators than do many European species. Without pollinating insects, only 1% of the alfalfa flowers set seed. Also note that pollinators in western Europe, which nest in hedges, fall victim to the destruction of hedges which is also brought about by the blind use of certain insecticides and herbicides. Some races of honey bees are more efficient than others as pollinators. Their efficiency can vary from 20 to 80% among the Caucasian variety of bees, which have longer tongues than do some of the others. This is probably why, in the middle of summer in irrigated areas of Iran and Afghanistan where alfalfa is native, the plant is pollinated by local wild bees and bumble bees.

A well-known example of the need for entomogamy is the vanilla plant, a climbing Mexican orchid, which has been introduced into Madagascar and La Reunion. Only species of American bees of the genus *Melipone* are able to perform the pollination, although thanks to a discovery by a Reunion slave named Albius, it is now possible to artificially pollinate vanilla flowers, which is absolutely necessary for the formation of the pod. He did this through the process of autogamy or self-pollination. Perhaps it would have been better to introduce the melipone bees, but as for the date palm man's interference saved the day.

Later in this chapter we will discuss many similar examples, including those of yucca and figs, which, though quite common, represent a specialized and remarkable instance of adaptation. Fertilization of various species of *Ficus* is extremely diversified. Very few date trees are parthenogenetic, and cross-fertilization is necessary. Fertilization is normally effected by wind-blown pollen from male trees, which reaches a certain proportion of female trees. In order to increase the chances of successful fertilization, someone must climb up the female trunk and suspend a male inflorescence near the female one. This practice is very old, at least dating from Babylonian times. We have observed this ancient custom (a very dangerous operation, because of the spines of the trunk) in February in the Sudan along the Nile. Modern growers do this with a power sprayer using a mixture of dust and pollen. Thus, as with the vanilla plant previously mentioned, human innovation replaces insects and wind.

The discovery of sex in plants dates from the end of the eighteenth century. Despite the fact that numerous plants are sexually dimorphic, such as date trees, having both male and females, the sexual distinction is not very obvious. What visibly distinguishes a male papaya tree from a female, especially when male plants bear abortive fruits? *Cecropia* tree species can be also male or female, but the morphology of the inflorescences can rapidly inform a botanist as to their sex.

Camerarius is honored as being the first to distinguish between monoecious plants, with either hermaphrodite flowers or separate unisexual flowers of both sexes, and dioecious plants, with unisexual flowers on separate plants. He also discovered that stamens carry pollen, the male gametophyte, and that the pistil is the female organ. Through their combined action they produce fertile seeds.

Many books and journal articles have been written on cross-pollination of flowers, especially by entomogamy. This chapter is an attempt to condense all of this into a few pages. Recent research in the tropics has enriched the only data acquired in temperate regions.

A fundamental law of plant fertilization is the inferiority of self-pollination and the advantages, or often absolute necessity, for cross-pollination, i.e., the fusion of gametes produced by two different individuals. Sprengel, in 1793, and Darwin, 70 years later, both insisted that cross-fertilization is a fundamental biological law. If that is so, why then are all plants not cross-fertilized? To answer this, let's see how these processes work.

The structure of angiosperm flowers reveals a neat tendency in favor of cross-pollination which makes self-pollination improbable, if not impossible. If it does happen, however, the resulting union of gametes is usually sterile. The means used by plants for successful reproduction are various; some of them are reviewed here.

Let us remember that flowering plants (1) are most frequently hermaphroditic (flowers with stamens and fertile carpels); (2) can be diclinous monoecious (stamens and carpels on different flowers, i.e., male and female flowers on the same plant); or (3) can be diclinous dioecious, with male and female flowers on different plants. Some well-known examples of the third case are pawpaw, date palm, willow, poplar, helm, hop, cecropia, etc.

The pollen, which contains the male gametes produced by the flower's stamens, equivalent to the microspore of the cryptogams, can germinate only if dropped on the receptive surface of the stigma (or eventually through the action of enzymes in the stomach of certain insects). This receptive surface is characterized by a papillose and sticky epidermis. The ovule located beneath the stigma in the ovary is the equivalent of the macrospore of lower plants.

As an example of dioecious flowers we can consider the Rafflesiaceae. The flower is the unique external expression of the plant. The enormous unisexual flowers of *Rafflesia* species in Indonesia emit a putrid odor that attracts flies (*Lucilia* spp. and *Chrysomya* spp.) and eventually beetles. The plants are generally parasitic on the roots of species of *Tetrastigma* (Vitaceae), similar to the sacculine crab in the sea. In Sabah, the majority of plants are male, and many female flowers fail to set fruit (Emmons et al., 1991). The pollination syndrome is sapromyophily and only large flies can pollinate. The flower provides no reward for pollination; all it offers is a deceptive meat color and smell (Beaman et al., 1988).

Pollen can be classified according to the way it is distributed: anemophilous, by wind; hydrogamous, by water; or zoidogamous, by animals. Species of Poaceae and Cyperaceae, for instance, have their pollen distributed by wind, but there are exceptions, such as species of *Rhynchospora* and allies in the tropics (Cyperaceae). They have bright color bracts and attract insects (Thomas, 1984), namely bees. The Bambusae among the Poaceae are mostly wind pollinated, but not always. Some species flower annually; others are hapaxanthic and can go 20 to 30 years without flowering. A palm tree, *Corypha elata*, is also hapaxanthic, as are several other plants. Insect pollination of such plants is very rare and a real event.

Insects that visit flowers are very well suited as pollinators because pollen is a high-protein food eaten by both adults and larvae of many Hymenoptera, some Coleoptera, and some Lepidoptera. Many insects also sip nectar (especially Lepidoptera and Diptera), a watery solution generally produced by insect pollinated flowers and high in saccharose, sucrose, and fructose. Nectar exudes from nectaries, from the receptacle surrounding the ovary, from the base of petals, from stamens, or sometimes from other special organs. It is always secreted by the flower. Of course, there also exists a quantity of extrafloral nectaries that is exploited by insects, as we have seen in the examples of myrmecophilous and carnivorous plants discussed previously. These nectaries only have an attraction value for the plant and do not play any role in the pollination process itself. They sometimes just keep the ants out of the flowers. Both the act of feeding on nectar and the actual gathering of pollen result in the transportation of pollen on the insect body. This is either as a disseminated powder or by special plant structure, the pollinium (a pollen sack which sticks to the insect). By visiting one blossom after another, insects carry pollen from flower to flower and effect the cross-fertilization of separate flowers.

All pollinophagous or nectariphagous arthropods are not, however, effective plant pollinators. There are, for instance, some Hymenoptera and Coleoptera that pierce a hole in the corolla and "steal" their reward. Certain insects with a long proboscis suck up the nectar without touching the plant. Many other cases, including species of phoretic mites of the genus *Rhinoseius* which are carried by hummingbirds, are nectariphagous, but their pollination potential is problematic. Other ineffective pollinators include many Thysanoptera (thrips) and some beetles; however, these are generally exceptions.

The classical example of insects that fertilize flowers — not because of nectar or pollen or any other attractant, such as perfume or nutritive tissue, but rather to lay their eggs in the ovaries of the plant — is the process of fertilization for both *Ficus carica* and flowers of species of *Yucca*. Fig pollination is achieved by a particular galligenous hymenopteran species for each fig species. The best known is a species of the genus *Blastophaga*, a chalcidid wasp. The mature female bores into the fig inflorescence and lays her eggs in the ovaries of sterile flowers. Males produced from these eggs escape to fertilize female wasps. As they leave the male fig flowers, they are dusted by pollen which is transported to other fig trees. Theoretically, the insect stimulates only the development of the endosperm in the ovules, and it is this endosperm on which the larvae feed. In reality, the cycle is much more complex, since there are three types of flowers: male, female, and sterile.

A species of *Yucca*, a Mexican Agavaceae, is also remarkable in that it is one of the rare cases of an absolute mutual dependence between a flower and an insect. A small moth, *Pronuba yuccasella* (Tineidae), is involved. Yucca flowers give off perfume, mostly at night, which attracts the moths. The female moth has a long ovipositor and is provided with prehensile, spiny, maxillary palpi specially modified in this genus. At night, with the aid of the palpi, the female collects a certain quantity of pollen and shapes it into a ball about three times as big as her head. Then she flies to another flower and deposits some of her eggs in the ovary with her ovipositor. Next, she climbs to the top of the flower and places the pollen ball on the stigma. The ovules are then fertilized, thanks to the remarkable instinct of the moth. In addition, pollen is so abundant that there is enough for larval food as well as for the reproduction of the plant. Without this method of pollination, the ovary would wither and the plant could not reproduce itself, which would mean the moth larvae would not receive their food. However, outside of America, namely in India, the autogamous (?) production of seeds in *Yucca alsifolia* has been described, but pollination has never been observed.

In temperate regions, species of *Mahonia* and *Berberis* (Berberidaceae) have sensitive stamens that spring upwards upon contact with an insect and shower the side of the insect's head with pollen. These movements are called seismonastic, and such flower or leaf movement is found in the sensitive *Mimosa* spp., certain orchids, and many other tropical plants.

Self-fertilization is sometimes successfully realized in some plants and is necessary in certain plants with flowers which never open (cleistogamous). As yet unknown are the pollination methods of the underground, or geocarpic, flowers of certain trees, the saprophytous Australian orchid species of the genera *Cryptanthemis* and *Rhizanthella*. If cross-pollination takes place, it must be carried out by earthworms or insect larvae. In these orchids, flowers and seeds are found 2 cm beneath the surface.

Cross-fertilization is aided by various anatomical structures and special physiological mechanisms. For dioecious flowers, cross-fertilization is obviously obligatory. In most hermaphroditic and monoecious flowers, cross-fertilization is also required when the pollen grains and the pistil of the same flower cannot unite because they are separated by time (dichogamy) or space (hercogamy) or are incompatible (heterostyly and incompatibility). These examples are well known and explicit in existing literature, to which the reader is referred. Let us say that autogamy and self-fertilization are rare, and, aside from the classical examples of certain plants (some cereals, Leguminosae, and those with cleistogamous flowers), it exists only where pollinators are rare (subdeserts, deserts, and cold mountains) or absent. Autogamy is often caused by the visit of the insects themselves. There are extremely rare cases where the pollinator is extinct and the orchid has adopted autogamy.

When stamens of an hermaphroditic flower mature before the pistil, it is said to be protandrous. Conversely, if the pistil matures first, it is said to be protogynous. Examples of the pistils and anthers being spatially separated may be found in some orchids and milkweeds. A classic example of heterostyly — production of flowers with either a short style (brevistylous) and long stamens or with a long style (longistylous) and short stamens — is to be found among the primroses. The specialized nature of pollen grains practically excludes, among these plants, any self-fertilization.

Hercogamy, the condition in which self-fertilization is impossible, and dichogamy, the condition in which the maturing of the sexual elements takes place at different times, as mentioned by Jaeger (1976), favor cross-fertilization, but the only absolutely sure system for these are dioecy (plants with separate sexes) or autoincompatibility.

It is evident that many insects are capable of pollinating plants ranging from giant tropical butterflies to minute Thysanoptera, Thysanura, and even mites among the arachnids. However, Lepidoptera, Diptera, and Hymenoptera are the most efficient and the most common pollinators.

Magnolias, usually pollinated by beetles, have strong female organs. Some beetles may be satisfied only with pollen — for example, scarabs (cetonids) — but others often eat flower petals (some meloids and many galerucines). Rarely, other flowers are pollinated by beetles, such as *Welwitschia* sp. from the southwest African desert by Tenebrionidae. A famous species of *Amorphophallus*, a giant flower close to species of *Arum*, are full of small beetles which feed on the rich, oily, and starchy tissues of these plants. *Rafflesia* spp. flowers in Indonesia are also frequented by beetles, but the main pollinators of these gigantic flowers are Diptera.

The archaic cycads have male inflorescences with an unpleasant odor. The flowers are fertilized by numerous beetles, among which are the Languridae in tropical America and species of *Phleophagus* (Curculionidae) in austral Africa. Several entomophilous gymnosperms have inflorescences with colored bracts, such as some species of *Ephedra*, which are visited by tenebrionids. There are sometimes cases of insect-cycad symbiosis — for example, the case of the pollination of *Zamia purpuracea* (Zamiaceae) by a weevil in Mexico (Norstog and Fawcett, 1989). Crowson (1989, 1991) has cited many cases of beetle-cycad associations. This association probably dates from the Mesozoic when the Cycadales and Benettitales were dominant. Beetles are primitive and brutal pollinators and require strongly built flowers such as the Proteaceae, for instance.

Zamia pumila occurs abundantly in the pine forests of southeastern Florida. Two beetles are efficient pollinators of the plants. Adults and larvae are abundant on the male cones and are found occasionally on female cones. *Z. pumila* produces sugars and droplets rich in amino acids (Tang, 1987). Mimicry of male cones and use of female cones as refuges may account for beetle movement. Leaves of the *Zamia* are always the most visited by aulacosceline beetles in Central America.

Coleoptera, with the exception of a few genera with a proboscis such as some Meloids, are poorly specialized as honey-gathering insects. There seem to be some correlation between Coleoptera and primitive flowers, but as we will see later, this is uncertain. We must be careful to refrain from making generalizations.

Many beetles, flies, Hymenoptera, and Lepidoptera feed on pollen. Pollen starts to germinate in their gut or is crushed, as in the case of the beetle *Cyclocephala amazona* (Scarabaeidae), by plant trichomes (Rickson et al., 1990). The specialized plant cells function as gastroliths in the beetle's digestive tract. Beetles can be very efficient pollinators, such as *Elaeidobius kamerunicus* imported from west Africa to Malaysia for pollinating oil palm (*Elaeis guineensis*) flowers. The beetles are attracted to male flowers only during anthesis when a strong aniseed-like scent is emitted (Greathead, 1983). When the stigmas are receptive, the female flower emits short pulses of a similar odor which attracts the weevils away from the male flowers. Many beetles (Coccinellidae, Elateridae, Chrysomelidae, etc.), and other insects, including ants, feed on extrafloral nectaries (Jolivet, 1992; Pemberton and Vandenberg, 1993) and even on trophosomes (food bodies) of ant-plants.

Hymenopterans are the most efficient of all pollinators. The size and color of flowers are attractive to them; however, this is not equal for bees and humans. Their perceptions of pleasant or unpleasant odors may not agree. Flower bracts are colored in some plants, such as in species of *Poinsettia, Bougainvillea, Euphorbia,* and many others. The bracts surround small and inconspicuous flowers. In species of *Mussaenda*, one sepal is large, leafy, and brightly colored and helps to make the flowers conspicuous. Certain flowers have petals with spots or lines which are nectar guides. These guides are not always visible to the human eye, but can appear in the ultraviolet light range which is visible to bees. Humans see many contrasting colors, but bees apparently are unable to distinguish more than four colors: yellow, greenish blue, blue, and ultraviolet. This is a small number compared to the approximately fifty colors visible to humans.

Flower shape can also attract some insects. Some studies indicate that while bees show a preference for flowers with radial symmetry, bumble bees generally favor those that are flat, with vertical symmetry. Moeller (1995) has confirmed that bumble bees prefer flowers that are larger and more symmetrical. Nectar production seems to be greater in symmetrical flowers, which may explain bumble bee preferences for flower symmetry. Also, movement of the flowers by wind, as we have seen, is a strong source of attraction.

First discovered by observers at the beginning of the eighteenth century, visual landmarks present in certain flowers have recently received more scientific study. Flowers have groups of lines, dots, and colored spots visible or invisible (except in ultraviolet light), usually at the base of the petals. These are guides that direct bees toward hidden nectaries. Olfactory guides augment, at short distances, the optical guides and help insects to perceive the shape of objects (Meeuse, 1961). Extensive research on the subject has shown that nocturnal white flowers do not possess a visible nectar guide. Flowers of species of *Magnolia,* which are normally pollinated by beetles, do not have any guides. Hymenoptera which land on these flowers generally go back with an empty pollen basket because they do not succeed in finding the center of the flower. In addition to colored or odoriferous nectar guides, there are also guides composed of setae, fringes, and gutters. When touched by an insect, they help to direct it to the pollen. Heat also attracts insects. Often the temperature of the spathe of *Arum maculatum* rises above 40°C when the surrounding air is only 4°C.

The odor of flowers attracts pollinators. Some are sweet and pleasant to humans, while others are frankly unpleasant, as are the giant flowers of species of *Amorphophallus* or *Rafflesia. Amorphophallus* spp. smell of rotten fish and molasses. They attract beetles in large numbers until pollination is achieved. African species of *Stapelia* (Asclepiadaceae) attract flies by emitting a strong scent which mimics (with the same color) rotten flesh. So close is the resemblance that the flies lay eggs inside the flowers. Even the setae of the flowers resemble mold growing on rotten meat. *Rafflesia arnoldi,* an Indomalaysian plant, also is almost exclusively pollinated by flies attracted by its color and odor. Beetles, which also frequent those flowers, are poor pollinators. Blow flies are attracted to yellow or orange flowers even if they have no scent.

Generally, the scent of flowers is agreeable to both humans and insects, but these odors are not permanent. They can be discharged when the plant is shaken by an insect alighting or by the wind. They can be emitted at certain fixed times, perhaps only in the evening or throughout the night. Odor ceases after pollination is completed, or when the flower withers. Pollinators, whether insects, bats, or others, must be adapted to these rhythms. Birds, always diurnal pollinators, are attracted by bright colors instead of odor. Nectar production, too, follows certain rhythms. There exists a synchronism between the activity of the pollinators. Often the color changes after pollination is achieved (species of *Brunfelsia* and *Lantana*). The scent of the flower of *Calotropis gigantea* or *C. procera* (Asclepiadaceae) is not at all unpleasant. It is even candied by the Chinese in Java. The pollinators are species of *Xylocopa* bees and not, as for *Stapelia* spp., flies attracted by carrion.

Classifying flower odors is very difficult. Faegri and Van der Pijl (1966) distinguished between absolute odors (particularly smells from plants such as violet, rose, geranium, jasmine, and so on) and imitative odor, i.e., the odors which imitate ones to which the insects are conditioned by instinct and experience (sexual odors from species of *Ophrys*, odors of rotten meat, excreta, decayed fruit, and others). *Arum conophalloides*, for instance, imitates the odor of mammal skin. These plants attract small bloodsucking Diptera (Ceratopogonidae, Simuliidae, and Phlebotomidae). Mycetophilid flies are attracted by the fungus smell of certain inflorescences. Some odors seem to provoke mating between insects, namely odors from Apiaceae (Umbelliferae).

The question as to which odors repel and which are marker odors left by Hymenoptera is still being discussed. Not much debated now is Kullenberg's hypothesis, according to which the scents of some flowers imitate the body odor of the pollinators themselves or, better, their sexual pheromones. Saito and Harborne (1992) have found some correlation between anthocyanin type, pollinators, and flower color of the Lamiaceae; 49 species of cultivars of this family of plants were surveyed in this experiment. Knudsen and Stahl (1994) studied the floral odors of the Theophrastaceae (Primulales from the Neotropics) in a Swedish botanical garden. Floral scent was collected by head-space absorption, and by combining floral morphology and scent chemistry the authors concluded that two genera were sapromyophilous and two others mellittophilous, i.e., pollinated by flies or bees.

In some way or another, insects reach the entomogamous flowers and collect nectar, nutritive pollen tissues (hairs of certain orchids, staminodes, aroid tissues), waxes, resins, (orchids, Euphorbiaceae), and aromatic or nutritive oils. Some even find a breeding place inside the flower and probably often utilize other features, but they never use all of these elements at one time.

Beetle pollination in primitive angiosperms (Magnoliales) is regarded as a specialized interaction involving floral modifications: strong structures, nutritive tissues, sheltered location for breeding, and eventually a place for larval broods (Armstrong and Irvine, 1990). Several functional roles have been attributed to staminodes (abortive stamens) in flowers, including its roles as a physical barrier that separates fertile stamens and stigmas; as a food reward for pollinators, mostly beetles; and in the production of attractant odors. The pollination of *Eupomatia laurina* in Australia was reported to be done by weevils of the genus *Elleschodes*. The Eupomatiaceae are Magnoliales, beetle plants. The weevils feed on the staminodes, which are rich in starch, and produce a strong floral fragrance, providing a visual display. Those staminodes are sticky and repel ants but do not trap weevils. The beetles seem to be their only pollinators. Among Magnoliales, beetles mating within the flowers do not seem to be a very rare thing. They lay eggs inside, and the larvae feed on the remaining androecial tissue.

Nectar is composed of carbohydrates, but some flowers (the poppy, for instance) have no nectar. Instead, they offer the pollinator a quantity of pollen rich in proteins. The insect is attracted by the smell or color of the flower, and sometimes the flower is warmer inside, a feature that is rarely reported but which is combined with a fetid odor by many flowers. Some even temporarily capture their visitors, sometimes holding them for the entire night, time enough to complete pollination. The temperature of the spadix of *Arum orientale*, as for many others arums, does not go below 43°C, even when the outside temperature is as low as 15°C or even lower. A temperature higher than the ambient is often produced by arctic flowers in order to attract mosquitoes.

Orchid flowers, species of *Catasetum* (Catasetineae), are visited only by male euglossine bees, even though they produce no nectar and little pollen. Gongorinae, and some other orchids, are likewise pollinated only by males of these bees. The reason for this has been discovered only recently. Odor attracts these males which collect aromatic oil secreted by a special tissue of the flowers. This is stored by the bees in a specially modified groove of spongeous tissue of a chitinous nature on their hind tibiae. Often host specificity in tropical America is determined by an orchid species, its odor, and the euglossine bees which visit it. The use of this perfume is doubtful, but it seems that it may be a way of marking the bee's territory during its nuptial parade. This behavior may be compared to the ritual of color display among birds-of-paradise and other birds when vagabond males assemble to attract females. However, odor replaces colors for these bees. A similar case with *Gongora* spp. is considered later. This phenomenon is termed gamokinesis, or social pollination (lek behavior). Females of anthophorine bees in the same manner collect a nutritive oil from numerous plants, store it on the setae of the hind legs, and feed their larvae by mixing it with some pollen.

Orchid species of the genus *Maxillaria* also furnish a false pollen which clings to the flower labellum (or lower lip). Other plants produce waxes or resins in flowers for the same purpose.

Certain plants have transparent windows in the form of spots on their petals near the stigma. Insects are attracted to these false exits, and, because these windows are located near the stigma, they are drawn toward the stigmas so that pollination can take place. False windows are also known as traps among certain orchids and the Sarraceniaceae.

Depending on the type of pollinating insects, flowers have either a landing platform (the orchid labellum, which is the inferior lobe of the corolla) or nothing at all, if the insect hovers above the flower. Certain flowers "inform" their visitors that pollination has been completed by withering; closing, as in the case of peyote (a Mexican cactus); changing color (for example, magnolias); or ceasing production of the attracting odor or nectar. The African climbing lily, *Gloriosa superba*, changes its color from yellow to orange after fertilization, then to red when it stops producing nectar. *Lantana* bushes or the beautiful and fragrant species of *Brunfelsia*, those strange American solanaceous plants, also share that characteristic. Their old white flowers coexist with the recent purple ones on the same plant. Many

Boraginaceae also change color the same way. A red spot replaces the yellow nectar guide in the old flowers of horse chestnut trees and the effect is to ward off bees and to direct them toward young flowers.

Many insects and birds are in the habit of "stealing" nectar without pollinating by piercing holes in the base of the flower. Bees doing this often steal pollen.

Some anemophilous flowers, such as plantain, are also regularly pollinated by syrphids (Diptera). Sometimes they lay eggs in the flower and their maggots feed on the flower's pollen. This is true of cocoa, elephant's ear plant, and breadfruit, plants whose flowers and fruit feed their pollinators on the spot.

Some countries are devoid of nectariferous plants in any quantity. Because of this, a species of *Xylocopa* bees, usually the pollinators of *Calotropis procera*, which has a very short flowering season, must revert to other flowers for food. Or, during the dry winter of the Cape Verde islands, adult *Xylocopa* spp. pierce the base of the conical flower of *Tecoma stans* (Bignoniaceae) to get the nectar, as it cannot penetrate the yellow cone-shaped flowers, which are too narrow for its big body.

Some recent discoveries have proven the existence of amino acids in certain nectars. The concentration seems greater among more evolved plant families and especially those pollinated by Lepidoptera (except sphingids) or the ones that smell like carrion. Bees which receive their proteins from pollen generally visit flowers with glucose nectar. That plants supply proteins to insects via nectar is a rather recent hypothesis that had to be proved. Myrmecophilous plants are known to produce this kind of protein for ants. In the case of species of *Cecropia* trees, the trophosomes or Müllerian bodies contain glycogen instead of sugar, a good replacement of animal food for the *Azteca* ants.

According to Baker and Baker (1973), it seems that the chemical composition of nectar varies according to the plant and its pollinators. In addition to sugars and amino acids, some lipids, antioxidizers, alkaloids, and proteins are produced. Amino acids seem to be used when nectar is the only food source, as among adult butterflies or, for instance, when it is needed to attract blowflies. Hummingbirds, hawk moths (sphingids), bees, and flies particularly seem to benefit from lipids. Because bees receive their proteins from flower pollen and birds obtain theirs from the insects they eat inside the flowers, flowers visited by these two groups are relatively poor in amino acids.

A correspondence between the distance of the nectar inside the flower and the length of mouthparts of the corresponding insect exists. It is often a lock-and-key relationship. This is a strong indicator of coevolution between plants and insects even if the idea is somewhat contested now. For instance, the relationship between the Malagasy orchid *Angraecum sequipedale*, which has a nectar tube measuring 35 cm, and a sphingid moth with a long proboscis has been demonstrated. The same relationship has been found in tropical

America between the orchid *Habenaria* sp. and its moth, which also has a long proboscis used for sucking nectar. However, the relationships between some flowers and their pollinating insects are still to be determined, if, as is very probable, they do exist. Almost all the flowers of milkweeds (Asclepiadaceae) are able to close and hold a visiting insect inside for a time before releasing it. During the "incarceration" of the insect, the flower's pollen is released on the visitor in the form of pollinia and pollination is completed. As we have seen, pollinia are sacks of pollen with adhesive stalks. These sometimes resemble the palpi of visiting insects, especially when they are attached near the mouthparts. Some flowers have inwardly directed hairs which have an eelpot effect on foraging Diptera — for example, *Ceropegia* spp. (Asclepiadaceae) — or temporarily trap the bee *Xylocopa* spp., as for *Calotropis* spp. flowers and Asclepiadaceae. Pollination of *Calotropis procera* and the attachment of pollinia on the carpenter bee (*Xylocopa* spp.) legs has been studied in Dakar by Jaeger (1971). This also occurs in Vietnam and Thailand with the same species or *Calotropis gigantea*. The eumolpine genus *Platycorinus* has many species feeding on this plant but seems unconcerned with pollination in southeastern Asia. In Senegal, fertilization of the flowers takes place during the dry season (January) when the pollinia are mature, but frequently I have observed some species of *Xylocopa* collecting nectar long before the pollen is ripe. Other Hymenoptera, including bees, and some Lepidoptera also are involved. The flower may be permanently inhabited by small ants and Thysanoptera, which do not seem to play an important role in fertilization. It is when moving and trying to disengage their legs from the interstaminal groove that *Xylocopa* bees pull out the retinaculum, a small glandular mass, and its pair of pollinia. Species of *Xylocopa* pollinate many other flowers of species of *Acacia* when species of *Calotropis* are not flowering. These bees are not specific on flowers of species of *Calotropis*, unlike the specificity of certain insects to certain orchid flowers. In South America, bushes of *Calotropis procera* now grow subspontaneously, such as in Venezuela, where their relationship with bees has not been investigated. As with birds and bats, there must be a correspondence between the distribution of insects and the flowers to be pollinated. In species of the genus *Aconitum* (Ranunculaceae), for instance, pollination is effected by bumble bees with a long proboscis, but the species do not have overlapping distribution. Some species of *Xylocopa* replace bumble bees in tropical Africa, even though these insects coexist with the bumble bees in temperate areas.

Another case of adaptation of bumble bees to the life cycles of plants is exemplified by a composite from the Himalayas, *Saussurea sacra*. This plant is covered with hairs so dense that it resembles balls of fur. It often is covered with snow at these high altitudes. At the top of the plant is an opening through which the bumble bees enter. They often spend the night inside the flower where it is safe and warm, and at the same time they pollinate the plant. Flowers of this plant are odoriferous, which is rare for the flowers of Asteraceae (Compositae), but this seems to be a relationship with its entomophilous pollination. The remarkable species of the genus *Saussurea* are all extremely woolly. They are represented by a score of species growing above 4000 m and often as high as 5500 m in the Himalayas. Other species live in European mountains. *S. alpina* is sweet scented and must have a similar life cycle.

Less useful as pollinators are sometimes bumble bees and honey bees; even though they may succeed in penetrating the flowers of the red-hot poker, *Kniphofia* spp. (Aloeaceae), they may end up trapped and unable to leave. Normally the flowers of that plant are visited by sunbirds. Species of *Kniphofia* are abundant in the highlands of Ethiopia and east Africa, and are nicknamed torch lilies.

Each species of bumble bee *(Bombus* spp.) has its preferred flower, but occasionally they will also collect pollen from other flowers. The fact that they are able to use other flowers that exist in the same area, in case of rapid change in the composition of the vegetation, is a remarkable adaptation. Thus, these polyphagous species often have an advantage over monophagous species and their obvious lack of plasticity.

The enormous morphological and esthetic diversity of the orchids hides a sexual life so elaborate that it might be compared with a living Kuma Sutra. This sophistication depends on the very important role played by many species of insects. The orchids include 795 genera and perhaps as many as 25,000 species, many adapted in unique ways to species of insects; sometimes even snails are needed to pollinate some of the rare species. Certain orchids have no scent, while others produce a strong odor to attract insects, each varying according to the request of the "customer". For instance, some *Dendrobium* spp. change their perfume from that of a lily-of-the-valley during the day to that of a rose during the night, or from that of heliotrope in the morning to lilac during the night. Some orchids produce an odor similar to carrion, quite frankly unpleasant. They keep their flowers intact for months waiting for pollination, a feature useful to greenhouse orchid growers. Once pollinated, some orchids stop production of their attractant scent, while others lose their color and rapidly wither.

We actually know that orchids are a family in rapid evolutionary development as is shown by the many hybrids between species, or even genera, produced by growers. So far, 50 different specific odoriferous compounds have been isolated in orchids. Often they are mixed, which maintains the attraction of their specific pollinating insect and prevents natural cross-pollination.

The orchids that produce a rotten meat odor have flowers that even imitate the color of rotting flesh. These orchids are fertilized by flies, which either lick an agreeable secretion produced by the plant or are attracted by the deceptive odor, as the flower has nothing to offer other than this illusion. However, this fraud is not the rule, and many orchids offer a drink of nectar and food on hair-like projections to the insects which pollinate them. The food is rich in proteins and is eagerly chewed by the insects.

Certain orchid-eating chrysomelid beetles (*Petauristes* spp., for example) feed on parts of the flower and at the same time fertilize it, as is done by species of *Aulacophora* on the flowers of melons and pumpkins (Cucurbitaceae).

Orchids also attract insects by their color, shape, pilose fringes, trembling hairs, and many other "artificial" means. Insect guides are either morphological or nectarigenous areas and are a trap used to attract insects as they collect pollen or drop it on the stigma. Generally orchid flowers are hermaphroditic, but autogamy sometimes occurs. If the sexes are on separate flowers, these are dissimilar and sometimes have been described as separate species with different names. In the latter case, the plants are dioecious, but there are exceptions as some are known to be diclinous monoecious (different flowers on the same plant). As previously mentioned, orchid adaptations to attract insects are very numerous. Darwin was the first to outline these peculiarities. Since Darwin, many additional facts have been, and continue to be, discovered. Mainly, it is now known that very few species secrete "free" nectar. Almost always, the insect must find its way to the nectariferous tissue, although sometimes they may pierce the flower to do this.

Recently a paper by Nilsson (1992) summarized the biology of orchid pollination. Out of the 25,000 to 30,000 species of orchids, 8000 to 10,000 species do not provide a pollinator reward (nectar); instead, they act by deceit or fraud. Some orchids are self-pollinating, and a few are pollinated by birds, but most of them are pollinated by insects, flies, or Hymenoptera. To Nilsson, natural selection is the principal process behind orchid floral evolution and orchid-pollinator relationships. This is not coevolution but asymmetric evolution, most of it occurring on the plant side whether or not there is a floral reward. Nilsson (1988) confirms Darwin's hypothesis to explain flower depth primarily among orchids. To Darwin, the evolution of deep flowers is a response to a kind of "race" with pollinating insects.

The rostrellum of the stigma or several structures covering it often form obstacles to prevent self-fertilization, and often the pollen sacks are an agglomeration of ovoid masses forming pollinia which, as we have seen, are taken away to other flowers by the insect visitors. When they reach another flower these pollinia bend, in about 30 seconds, and meet the receiving surfaces where the pollen grains are deposited. As with rhododendrons, white orchids have a strong odor and attract moths at night. The flowers produce pollinia which attach to the head of the moth, sometimes even to its eyes.

The structures of the flowers of species of *Pterostylis* remind me of fly-catching plants. Insects land on the flower's labellum, a large petal which provides the landing platform in front of the flower. This is hinged and closes against the flower to force the pollinator to follow a path to the inside. In half an hour the labellum opens again. Many other structures, which remind us of the temporary trapping of *Xylocopa* species by the *Calotropis* flowers, exist among some orchids. Sometimes an insect is propelled to the inside of the flower by a shaking of the labellum or by rapid movement. The hinge system is sometimes even more complex and actually carries the insect to the nectar (its reward) and to the pollinia (its ransom).

Species of *Gongora*, orchids from tropical America, attract bees with odor from a scent gland, the osmophora. The slippery surface of the flower resembles a toboggan and causes the bee to slip on its back into the flower where it attracts the pollinia and fertilizes the stigma by the pollinia of another flower. Other orchids have different ways of capturing bees, including the transparent windows mentioned previously, which guide insects toward pollen or toward the stigma before they are released.

Other solutions are similar to the tactics used by carnivorous plants. Some insects literally become intoxicated by the odor of smelly tissue: they fall down into the liquid collected in a pail-shaped labellum, but they are able to escape through a gutter which directs them to the pollen or to the stigma. A similar process is found in a Central American orchid. The first hymenopteran to visit has difficulties coping with the rostrellum, but the later ones to arrive easily escape, soaked, drunk, and often having completed their pollinating function. It has been said that this orchid stops producing its odor if the insect takes too long to escape but starts it again the next day, probably as a method of avoiding self-pollination. Some reports indicate that certain orchids, such as species of *Gongora*, *Stanhopea*, *Catasetum*, and others, generally cause intoxication of insect visitors and that the anthropomorphic explanation given is that it "calms down" the hymenopteran. The intoxicating substances retained on the legs of the Hymenoptera might also play a role as markers or as sexual attractant (see Figure 5.2G,H).

Other orchids, among them species of *Catasetum* growing in tropical America, are provided with triggers (Arditti, 1966) — floral appendices, with a sticky viscidium in front, which release a catapult and attach the pollen mass to the insect. They are capable of projecting pollen a distance of one meter. Flower-eating insects chiefly visit *Catasetum* spp. They go directly to the pollen mass, although euglossine bees are also attracted. Most of these species mimic food to deceive the bees. They detain insects with a variety of dummy signals (Nilsson, 1992) and nectarless structures; often the insects do not find any pollen to feed on.

Insect traps occur in species of *Arum*, others among species of *Pterostylis* and *Masdevallia* plants. These are orchid plants with pitchers and a cover composed of a labellum which can stay closed for as long as 30 minutes. *Pterostylis* sp. usually has green flowers, and the *Masdevallia* sp. smells of bad meat but offers no reward to attract flies.

Certain orchids without nectar, oil, or edible tissue become prostitutes (a fact known since at least 1916 and restudied by Kullenberg, 1961, and others since). By mimicking a female insect, the flowers are fertilized by various Hymenoptera which are practically "masturbating" on the female insect-like flower. This is what Jaeger (1976) calls "flower sexual partners". This strange similarity between flower and insect was noted as early as 1831, but this fact was not yet connected to its role in fertilization. As stated by Huxley (1974), the flower offers its own flesh to satisfy her sexual desires.

Species of *Ophrys* orchids in Mediterranean Europe are distinguished by the form of their labellum. They resemble in shape, color, and velvety structure the body of an insect. Australian species of *Cryptostylis*, the South American species of *Paragymnomma*, and several other orchids also mimic insects to attract them. Numerous other examples certainly may be found wherever orchids grow. These orchids, devoid of the usual attractants, are however visited by Hymenoptera, flies, and other insects which pollinate them. The American orchid, *Trichoceros antennifera*, has evolved a similar relationship, not with a hymenopteran, but with certain flies.

Ophrys apifera, an English species, self-pollinates because it has lost its adopted insect pollinator as its sexual partner. Mating between insect and flower takes place only when the males are newly hatched and the females are still very rare. As we have already seen, the privilege of fertilizing the orchid by male insects exists among many species of the orchid family but nowhere else. Generally speaking, male aculeates (stinging Hymenoptera) emerge before the females, a trait generally believed to maximize mating success (Nilsson, 1992), but which also has an advantage for the orchids.

In north Africa, it is *Scolia ciliata*, a bee, that pollinates *Ophrys speculum* during false copulation. In northern Europe, *Ophrys muscifera* is pollinated by two species of wasps of the genus *Gorytes*. Numerous similar cases in the Mediterranean area involve other species of *Ophrys* and *Scolia* bees. Others are pollinated by species of *Eucera* and *Andrena* bees. Kullenberg (1961) has observed bigeneric attractiveness of *Ophrys scolopax* in Lebanon to males of species of *Eucera* and *Andrena* bees. To facilitate mating, the walls of epidermal cells of the labellum of *Ophrys* spp. is relatively strong. The primary stimulation is essentially olfactive, followed by a tactile and visual stimulation, the last being secondary and dependent first on tactile stimulation.

A layman imbued with a dose of finalism sees that the flower morphology of the species of *Ophrys* resembles an insect somewhat, mostly an hymenopteran. These "insects" have false eyes, antennae, and wings, and the "body" hairs (setae) are similar in shape, color, and texture to the hairs on the female insect. This observation is not to reflect Bernardin de Saint Pierre, who believed that the melon had rings to be eaten by a family, and his extreme finalism, but rather to suggest that there is at least a partial truth in the observation.

The true attraction for insects, however, comes from an odor similar to that produced by the female insect, a true pheromone mimic secreted by the plant itself. The volatiles released from flowers of species of *Ophrys* are similar to those produced by the insect glands in a number of pollination systems. This is a chemical mimesis (Borg-Karlson, 1990). So strong is that scent that males try to find flowers when the material is impregnated on paper, and the males of at least one species prefer the orchids to their own females. Flower morphology has only a secondary role in attracting insects, if such a role actually exists. To compensate for its immobility, the labellum on which the insect lands is dark purple or brownish with clear markings or is a mirror spot which reflects ultraviolet light. Once the insect lands on the orchid labellum, he finds the curved hairs, which persuade him that he has really found the female, and he starts a long and vigorous false copulation. The copulatory organ is extruded and is inserted into velvety areas of the labellum. Males, excited by smell, sometimes fight for the possession of a flower. A poetic picture isn't it? (See Figure 10.1E.)

However, the orchid is not enough for its loving partner; although the penis is erect and rubbed without interruption, there is no ejaculation, at least not for those visiting species of *Ophrys*. Sperm, however, has been produced by copulation with species of *Cryptostylis*, an Australian orchid pollinated by the ichneumon wasp *Lissopimpla semipunctata* (Erickson, 1965).

As Huxley (1974) wrote with humor, "Even so, the insect flies full of hope from flower to flower, practices again a simulacrum of copulation, and we hope finds its own pleasure, but, with great efficiency, it carries away pollen." Sometimes the labellum is violently bitten during these actions.

These prostitute orchids vary greatly in shape and pilosity, depending upon their "partners", their geographical origin, and the species of orchid. *Ophrys lutea*, for example, causes inverted mating because of its floral structure. Anyway, this resemblance between a flower and an insect is sufficient. It lacks any other source of attraction, and, because the insect does not receive any remuneration for the act of pollination, it is in all ways remarkable. It must be the fruit of a long evolution which began with a nectariferous plant. For these orchids, the erotic solution is their unique chance of survival, as cross-pollination is for them a necessity. Not only do male bees and flies pollinate orchids, but also some male ants, which according to Janzen (1977) are bad pollinators, use pseudocopulation to pollinate several Australian orchids, including *Leporella fimbriata* (Peakall and James, 1989; Peakall and Beattie, 1989, 1991; Peakall et al., 1990, 1991; Peakall and Handel, 1994). Ants, often repelled from flowers by sticky secretions (species of *Gaylussacia*, in Brazil, for instance) or extrafloral nectaries, pollinate several flowers and compete with mosquitoes on high mountains or in the Arctic (see Figure 5.1F).

Orchids are not the only example of production of a pheromone mimic by a plant. Truffles, subterranean fungi growing several inches to one meter below the soil surface, can be detected by their aroma, which resembles an androsterone of boars as well as of the human male. It is difficult to prove, but apparently this stimulant plays no role as a human sexual hormone, but for the sow it certainly does. In the south of France, female pigs have been trained to locate truffles by scent. If the sow eagerly digs in soil for a truffle, it is reacting to this scent with a mating behavior. Why, we ask? Perhaps coincidence, but also perhaps the fungus has "evolved" this solution in order to be disseminated. From wasps to pigs, there is a long taxonomic distance, but in both cases the plant has developed animal mimicking pheromones to effect its reproduction. There may be many other still unknown cases. A fly is also attracted to the truffle smell and can also be a precious indicator for the farmer. Dogs trained to locate the truffles seem to do so after training and not because of the sexual attraction.

Male tropical bees (species of the genus *Centris*) are attracted by the vibration of species of orchids of the genus *Oncidium*; believing they see a rival visiting their territory, they charge the flower and thus get the pollinia on their heads. In this case, odor, food, and color play no part in the attraction. It is only the territoriality of the insect that causes this behavior. How did such parallel evolution (or coevolution) occur? It is rather difficult to imagine.

Among the scolicid wasp species of the genus *Campsomeris*, females take the labellum of species of *Brassica* and *Calochilus* orchids to be a predatory insect. Such is at least one interpretation given for their behavior. They attack that false intruder, sting it, and hence pollinate the plant. This is what Van der Pijl and Dobson (see Gilbert and Raven, 1975) named pseudoparasitism (we could also use the name pseudocopulation for the *Ophrys* spp. and their pollinators). Also, with this deception the insect uses its energy, free of charge, in the service of the plant. The exchange is decidedly out of balance.

Other insects are imitated by orchids. Species of *Oberonia* in Sri Lanka produce a flat rachis with small greenish flowers imitating, to human eyes, honeydew-producing aphids. Is this the reason for the pollination of those flowers? This must be studied further.

Sapromyophily or egg laying by flies in flowers imitating the scent and color of rotten flesh is not always an unrewarded deception for the insect. In several cases, the larvae do not die but develop normally, producing a new generation of pollinators.

However, Dobson (see Gilbert and Raven, 1975) writes that most of the relationships between insects and orchids are based on deceptions for the pollinators. In many areas, flowers are separated by time and space, and no energetic equilibrium between pollination and pollinated can be maintained. There is no symbiosis in the strict sense, but instead parasitism of insects by plants. Pseudocopulation, pseudopredation, pseudoparasitism (for the parasitoids), pseudoterritoriality, pseudantagonism, pseudopollen on the labellum, pseudonectaries, pseudostamens, pseudoperfume, convergence in shape and color with the neighboring plants, and abundance of nectar are frequent phenomena among orchids, all of which are becoming better known and utilize, in some way, the insect's appetite. More than half of the known orchids do not provide food for their pollinators, instead, at the cost of a rather small energy expenditure, providing them only with illusions. Other species, apart from nectar, provide pollen, edible pseudopollen, or other food bodies. The edible pseudopollen is basically made of starch, as in species of *Maxillaria*, but more often it is empty and not edible as in species of *Polystachya*. Thus, the species of *Ophrys*, with their prostitute flowers, are not alone in selling illusions.

Fraud orchids sometimes go very far. Lepidoptera, for instance, pollinate nonrewarding *Epidendrum paniculatum* in South America in search of
pyrolizidin alkaloids that are used as precursors for pheromones for court-
ship and defense (Nilsson, 1992). That is very probably the reason why
males of the flea beetle species of *Gabonia* in Africa are attracted by wilted
leaves of species of *Heliotropium* (Boraginaceae), which can be related to some
kind of sexual (lek) behavior (Boppre and Scherer, 1981). Several species of
Epidendrum, such as *E. ibaguense*, are nectarless and mimic *Lantana camara* or
Asclepias curassavica in Panama to attract Monarch butterflies (Danainae).

Two sympatric species of *Oncidium*, cited previously, exist in the Baha-
mas. These are infertile between themselves but are exclusively fertilized by
the males of *Centris versicolor*, because of pseudoterritory inducement in one
species and by the females of the hymenopteran in the other. The latter
orchids are mimics of species of *Malpighia*, which are themselves nectar
producing. Hymenoptera are the dupes, but even so these complex associations have evolved and are maintained. The damage caused to the *Centris*
species must be slight, since it never ceases.

Pollination, with a long evolution of more than 200 million years, became
a problem for flowering plants only when they started to develop away from
water. The ones which returned to water, such as Zosteraceae, have found
new solutions completely independent of arthropods. Even though they are
terrestrial plants, the male gametes of species of *Cycas* and *Ginkgo* produce
antherozoa. These are mobile and swim in the moisture of their environment
before reaching the female nucleus. For higher plants, the microsporophyte
produces pollen, thanks to the development of pollen tubes, and joins and
fuses with the ovule to become the egg. The pollen of the zoogamous flowers
is sticky and its surface is covered with spines, crests, and rounded points,
unlike anemogamous pollen which is smooth, light, dry, and powdery.
Anemogamous flowers are small and greenish, without any bright colors or
nectar. The energy cost is less; wind works free of charge.

A flower with a deep corolla cannot survive if there are no insects with
a long proboscis or birds with a long beak that can suck out nectar and
pollinate it. The reverse is not true, as insects or birds with long appendages
can pollinate very well a short-corolla plant. However, flowers and pollinators seem for some people to evolve in parallel. It has been said also that
insects, and perhaps birds, have been the creators of the evolution of the
flowers in the primitive tropical forest. However, for Nilsson (1992), it is a
very asymmetric relationship, most of the evolution occurring generally on
the plant side.

The pollinating Hymenoptera have been mentioned previously, and
their model, so very well known, is the honey bee. However, numerous
dipterans and a few beetles also have a long, sucking proboscis. Many
beetles, however, are rough pollinators of primitive flowers, such as species
of magnolias and Calycanthaceae. These last plants do not provide nectar
and pollen, but rather food made of pseudostamens (staminodes).

As noted previously, insects, especially beetles, and some birds rob nectar and pollen from the flower by piercing the corolla; hence, they do not pollinate. Apparently, this was once true of most flowers, but now it is chiefly the very complicated flowers that are victim of this predation. It is now thought that angiosperm evolution has been rapid because of this. As early as the Carboniferous, insects must have fed on the spores of vascular cryptogams, followed by pollen of gymnosperms at the end of the Jurassic. Later, by selection, flowering plants produced nectar, odor, colors, guides, and other modifications until they reached the stage of "one insect for one flower" which, by reducing the possibilities of pollination, provokes a highly specialized adaptation of floral structure. Energy balances, or pollination costs, have been calculated by several authors in calories provided by nectar sugar of the flowers visited. Insects must rapidly visit the flower to procure an energy profit. Many flowers must, therefore, appear at the same time, or grow in colonies, to reduce to a minimum the distance traveled by the pollinators. These calculations are, however, hypothetical and no general statement applies to all pollinators. For instance, butterflies are less in need of energy than are bees. Butterflies do not feed their larvae, but bees do.

Hocking (see Gilbert and Raven, 1975) calculated the minimum distance covered by a nectariphagous female mosquito in the Canadian tundra. He estimated that sugar contained in the ventral diverticulum of a female mosquito's gut gives her a flight range of 25 km and that a single catkin of *Salix arctophila* produced enough energy for 950 mosquitoes/km per day. Further, it is known that flowers pollinated at low temperatures produce more calories (in quantity of nectar) than those which blossom at high temperatures. These flowers often grow in patches, a valuable asset in the Arctic and at high altitudes. The more the plants are grouped together, the shorter the distance to be covered by the pollinator and the smaller the energy expenditure. This is probably a genetic differentiation characteristic in local populations of plants.

There is certainly an equilibrium between plants and insects, writes Mesquin (1971), because in temperate regions insects compete for flowers, but elsewhere it seems to be the plants that compete for pollinators. To do this they must be "rivals" in developing special attractants. In American, African, and Papua New Guinean tropical mountains, birds are more efficient pollinators than insects, as they pollinate even during periods of cool, rainy weather, common in tropical mountains. This is something insects cannot do. This fact is well known for the species of *Lobelia* (Campanulaceae) that depend on sunbirds (nectarinians) in such places as Ruwenzori or Kenya.

In regard to an energy balance, there is the well-known example of the beautiful species of *Heliconius* tropical and subtropical butterflies. These pollinate species of *Anguria* and *Gurania* (Cucurbitaceae) and several other plants, but their caterpillars feed exclusively on Passifloraceae. Adults live for 6 months, an exceptionally long and active life for a butterfly. This is attributed to the fact that the butterfly feeds not only on nectar, but also on

pollen rich in proteins. However, exclusive pollen feeding does not lengthen
the lives of the very short-lived adult moths of the family Micropterygidae.
Gilbert (1972) saw the *Heliconius* spp. adults scraping the tip of their proboscis over floral anthers and carrying pollen loads on their tongue. Pollen-fed
butterflies produced, on average, five times as many eggs per day during
pollen-feeding time (Scoble, 1992). Benson et al. (1975) saw a "dynamic
coadaptative evolution" between *Heliconius* butterfly species and species of
Passiflora vines. For some other butterflies, pollen feeding is replaced by bird
droppings (Ithomiine butterflies), rotting fruit, fermenting sap, urine, dung,
and even, for certain moths, tears and blood. Some species of Ithomiine
adults can live 4 months which is rather long for a butterfly.

Just as we have classified the phytophagous insects into categories —
monophagous, oligophagous, and polyphagous — Hymenoptera, including
bees, have been divided into monotropic (flower visitors), monolectic (pollen
gathers), oligotropic or oligolectic, and polytropic and polylectic species.
However, most of the Hymenoptera species have the tendency to be generalists or no more than moderate specialists, and only in certain genera (such
as species of *Apis*, *Halictus*, and *Megachile*) is the percent of pure pollen
gatherers very high. Australian rain forests seem to be dominated by generalists pollinators (Williams and Adam, 1994), but this has to be verified
elsewhere.

As the plant seems to be able to induce the speciation of phytophagous
insects by selection to form biological races, the pollinators, such as bees, can
accelerate speciation of "their" plants when sympatric races exist. If verified,
this would be coevolution.

The flower being the seat of reproduction, it is normal that the plant
develops every way it can to protect it. It is the flower of *Cannabis sativa* that
has the highest concentration of the drug marijuana, or hypericin in the
flowers of St. John's wort, or pyrethrum in the flowers of species of *Chrysanthemum*, and so on. The flower is also morphologically protected from intruders by various structures in a multitude of shapes and strengths which attract
only the most common and hence the most useful pollinators. Their survival
and multiplication is at that price. Mammalian and avian pollinators most
likely preceded the more refined bee pollinators. Initially, flowers, or analogous structures, must have been robust, with protected ovules, mainly because beetles were probably the first important insect pollinators, preceding
many other insects (Hymenoptera, Lepidoptera, and Diptera) which treat
flowers with much more care. If we believe certain authors, the primitive
"flowers" of the Caytoniales were pollinated by small reptiles during the
Mesozoic and fossilized excreta have been found with the flower's print.
Extrafloral nectaries seem to have been produced to keep away ants and
other undesirable insects, as in the case of the glutinous traps of the stem of
many tropical and temperate plants.

In nature, pollinators vary according to climatic areas. For instance,
Lepidoptera tend to replace bees in temperate mountains; flies dominate (as

mosquitoes) in arctic areas, ants in semi-deserts, and birds in tropical mountains; where specially evolved pollinators are missing, "primitive" ones, such
as beetles, get their chance.

Among large insects, only adults pollinate flowers, although certain
small insect larvae can pollinate, but their role, in general, is negligible. A
balance must always exist between the energy expended by foragers, such as
bees, and the "caloric reward" provided by the nectar (Ehrlich and Raven,
1964).

It is possible that entomophily is primitive among the angiosperms and
that anemophily is a secondary phenomenon, but the question is complex
and far from being resolved. When some flowers are introduced into a new
geographical area, pollination is usually rapidly accomplished by insects and
birds, but they are never as efficient as those in the original country of the
plant. This can only be called pre-adaptation instead of coevolution.

On islands, the pollinator and plant assemblages are depauperate (Inoue,
1993). On Japanese islands, the long-tongued pollinators are absent, as are
bumble bees. Plants depending mostly on bumble bee pollination are absent.
When I visited Paquita island in Brazil a few years ago, I saw no butterflies,
ants, or bees; they were all killed by zealous insecticide sprayers, extermina-
tors fighting against mosquitoes. It is simple to eradicate insects so stupidly
from a small island. Repopulating can be achieved by wind from the continent, but it is a slow process and in the meantime the plants have to find a
way to survive or they perish.

In any case, when cross-pollination is obligatory, it contributes to gene
flow throughout the population, mostly thanks to insects. This gene flow
certainly contributes to the variation and diversification of the species and
varieties within a plant genus. It is, for instance, almost certain that some
pollinating Lepidoptera and Diptera show strong preferences for a color
variation (morph) of a given flower among polymorphic species. We have
seen previously the example illustrated by species of bumble bees. The
variations shown by plant species of the genera *Raphanus, Cirsium, Del-
phinium, Portulaca, Primula*, and many others are well known. For example,
in species of *Cirsium*, with white and red flowers, it is the white morphs that
are preferred by their pollinators. The abundance of white forms is interpreted as an adaptive answer from the plant to conditions of low pollination.
Without denying that ecological factors can influence certain variations, it is
a fact that the flowers often change color in mountainous areas. On the other
hand, pale or white flowers are more attractive during the night or at dusk,
perhaps because they are more visible, and therefore, their abundance in a
dense forest is explained.

The harmonic relations between pollinators and blossoms are shown in
Figure 12.2, where the blossoms are classified by shape. Although a variety
of pollinators are attracted; note that there is a remarkable selection both as
to shape of the flowers and their color. No doubt this strongly indicates
selection and adaptation between the plants and the animals involved.

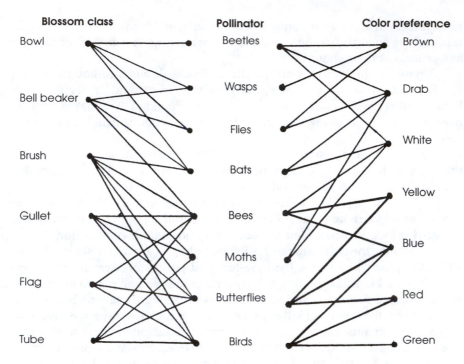

Figure 12.2 Interrelationships (harmony) between pollinators and various flower types. (From Faegri and van der Pilj, 1966.)

Even though the mechanism of pollination is not very well understood, we have seen that coevolution of insects and plants is probably due to the presence of phytophagous insects and pollinators, both vertebrates and invertebrates, and it is certainly insects that impose the most selective pressure on these plants.

chapter thirteen

Coevolution (or not) between insects and plants

To determine the origin of food selection is difficult. It is evidently much older than the Mesozoic period, as it started in the Paleozoic with the first plants that emerged from the water and the first arthropods that came out of the seas. The habit is very old indeed, and the first millipedes were saprophagous and fed on the spores of the first plants.

Hematophagy, or bloodsucking, appeared among insects in very different groups and probably independently from sap or nectar feeding or fruit piercing. It appeared among fleas; the dipterous families Culicidae, Tabanidae, and Glossinidae; the bugs Cimicidae and Triatomidae; and other families. Fleas seem to date from the Cretaceous period, even the lower Cretaceous, as one was found in Australia in Koonwarra silstone that probably lived in the fur of an early marsupial. *Saurophthirus longipes*, flea-like parasitic insects from Trans-baikalia, were probably parasitic on the fur of Pterosaurians (Ponomarenko, 1976). Fleas probably originated from Nematocera from the Triassic or lower Jurassic and these insects were nectar suckers (Riek, 1970).

Some groups considered to be purely phytophagous or aphagous (for instance, Chironomids) probably were derived from hematophagous ancestors, some of which are still exant in Australia and South Africa. To be hematophagous, insects should have had reptiles, small mammals, or at least Batracians to bite. The origin of that type of selection is very old, and parasites such as fleas (from Diptera) and lice (from Hemiptera) have evolved much later along with their bird or mammal hosts. However, it seems that at the beginning all of those hematophagous bloodsucking insects were phytophagous or nectariphagous, as males of mosquitoes still are. It should be noted that both male and female biting mosquitoes are also efficient pollinators and, in certain cases, specific pollinators of species of *Habenaria* (Orchidaceae) in Arctic regions. Biting bugs (Reduviidae) have certainly evolved by changing from sapsuckers to bloodsuckers and the rare ophthalmotrophic moths, the moths that feed on blood from the eyes of cattle, come from similar fruit-piercing species.

Phytophagy certainly started much before the appearance of flowering plants, which date from the Cretaceous, or at least from the Jurassic, with some primitive species in the Trias. It has been stated, for instance, that in the Gondwanian flora in South Africa, the leaves of species of *Glossopteris* were probably eaten by an insect during the Permian. In the Carboniferous, and probably much before, forest litter was occupied by detriticolous and phytophagous insects, and not only by primitive ones such as Collembola and Thysanura, but also by the more advanced cockroaches (Dictyoptera) and primitive millipedes, such as species of *Arthropleura*. They all fed on decaying vegetation and lycopod litter. Aphids probably began by living on gymnosperms. Even the more specialized groups, such as the primitive ancestors of the first Coleoptera, probably date from the end of the Paleozoic. These attacked the available plants, and (in the Permian period) the vascular cryptogams, by then much diversified since their Carboniferous origin; these were followed by the conifers, cycads, ginkgoes, and the Benettitales, a kind of prototype of the flowering plants. These are probably not the ancestors of the flowering plants, even if the tendency actually is to place them at the beginning of the genealogical tree of the angiosperms. The question is still being debated. The sexual parts of the Benettitales were enclosed by leaf-like organs, almost floral in appearance, with their ovules protected by scales. This group very much resembled Ranunculaceae, but the interpretation is difficult. It is probable that these ancient false flowers were brightly colored, bisexual, and attractive to their early pollinators, not all of which were insects, however. To some, these flowers were closed, at least some of them, and were penetrated only by beetles.

Coleoptera, such as species of Aulacoscelinae (primitive Chrysomelidae, closely associated to Sagrinae) still actually frequent the leaves of American cycads. Other families of Coleoptera do the same for some cycads, and tenebrionids frequent the flowers of the Gnetales (*Welwitschia mirabilis*) in southwest Africa. When they are imported from the tropics, *Cycas* flowers are often frequented by beetles.

It was not until the end of the Jurassic that the association of flowering plants and insects evolved and developed as they are now. At the beginning of the Cenozoic, most of the present-day adaptations must have existed if we take into account the state of evolution of insects now preserved as fossils, namely in amber.

The tendency of the "flowers" of the fossil Bennetitales to increase in size is very typical. They increased from about 5 mm to about 10 cm during the Cretaceous period. That was probably in conjunction with the appearance of large pollinating birds and mammals. Insects must have played a role in that pollination at the beginning of this evolution, but it is not clear the way in which those flowers were opened or closed to ordinary pollinators. Almost certainly specialized beetles, such as the chrysomelids, developed in the Cretaceous period parallel to the development of the angiosperms but mostly after the Eocene. Although ants are known from the Cretaceous, it does not seem that bees appeared as they are now, before the Oligocene, i.e., just after the appearance of the true Lepidoptera. Actually the primitive Lepidoptera were already diverse during the Cretaceous, even if evidence of their presence in the Jurassic and perhaps in the Triassic is known (Whalley, 1986). The most evolved of all pollinators, the bees, appeared last, along with the plants most adapted to that function, such as Asteraceae (Compositae), Fabaceae (Papilionaceae), Lamiaceae, Scrophulariaceae, and orchids. Together they have achieved this coevolution, a result perhaps of the perfection of their adaptations.

It also seems evident that before they began sucking nectar from flowering plants, lepidopterans ate only pollen. The present day Micropterygidae, one of the most primitive moth families closely related to Trichoptera, have chewing mouth parts used for this purpose. In Europe, for instance, these insects visit buttercups (Ranunculaceae) and eat their pollen, which is toxic to other insects, bees among them. The mouth parts of the Micropterygidae are used only for biting. Their long palpi gather pollen, scraping it from the anthers and bringing it to their mouth where it is crushed and swallowed. Theoretically, pollen eaters who previously could have been spore eaters are polyphagous. It is probably by foraging that the Benettitales with open flowers were pollinated before the appearance of nectar, though it is not sure at all that those "flowers" did not produce some.

To Roubik (1989), coevolution between flowering plants and bees occurs, but bee and flower coevolution seems diffuse most of the time. Bees sometimes can be specialists, and species of *Andrena* are known to concentrate on pollen toxic to other bees. Generalization, however, may help bees to dilute the specific toxin between many flower species to make the honey edible.

So, the nectariphagous insects, such as bees and other honey-gathering Hymenoptera, are theoretically polyphagous (with certain restrictions as already mentioned) as are the adults of Lepidoptera. They are not at all specifically associated with the entomogamous flowers, and some will collect, during the spring, the pollen of anemogamous flowers, such as those of poplars, Cyperaceae, and Poaceae (Gramineae). We know that the exception of the entomophilous Cyperaceae is limited to the tropics. Those plants have brightly colored bracts (yellow, white, red,) and attract many insects for pollen only. If beetles seem to have been the first true pollinators and still are often the pollinators of several gymnosperms and primitive angiosperms such as Magnoliaceae and Nymphaeaceae, Hymenoptera are generally associated with the most highly evolved flowers and probably the last ones to appear, such as the orchids and the Scrophulariaceae.

Leafcutting and fungus-growing ants in tropical America show the unique peculiarity of being both monophagous for the fungus they grow and polyphagous, as they cut leaves of many kinds of trees. This polyphagy, however, is only relative and varies according to whether the species attacks grasses or trees, or both. The choice also varies often among the trees of the forest and has a conservative effect on the vegetation. Man is totally responsible for the rapid change in the behavior of those ants and the destructive aspect of their attack of crops and prairies.

It could be that polyphagy was the primitive type, but that is not certain. Monophagy and oligophagy, including the "botanical sense", developed gradually by restrictive mutations and narrow adaptations to the food plant. This could explain, for instance, why parasitic fungi and aphids develop two generations on two unrelated plant families. It seems indeed, that if oligophagy (or monophagy) is primitive and not secondary, then fungi and aphids would have chosen related plants. It would be wrong, however, to consider polyphagous leaf miners as more primitive, since they belong to four different orders (Lepidoptera, Diptera, Hymenoptera, and Coleoptera), each highly evolved, even if some of them (Coleoptera) appeared during the Permian. Often polyphagous leaf miners, formerly monophagous, became polyphagous by secondary adaptation. The primitive polyphagy was forcibly some kind of monophagy since the diversity of the plant kingdom was very small and reduced to minimum. However, sapsuckers or saprophagous insects were not too choosy about their food plants when lycopods and early ferns dominated the planet.

Finally, primitive insects being polyphagous was a necessity during the Paleozoic and at the beginning of the Mesozoic. Reptiles and insects must have eaten, without much selectivity, vascular cryptogams and gymnosperms. Some primitive plants probably defended themselves by concentrating metabolites toxic or repulsive to insects and perhaps also to reptiles. That is one of the reasons invoked for the disappearance of reptiles at the end of the Cetaceous: the toxicity of the angiosperms. The majority of the insects avoided those plants, but some of them were able to accept them through selective mutations and pre-adaptation. These insects had an advantage over the others — the absence of competition. These toxic substances then became for these insects attractants or nutritional stimulants. What was probably at the beginning a chemical defense for the plant eventually became the reason for the specialized associations between the insect and the plant. Such is the theory developed by Fraenkel (1959) and by others. The scheme is perhaps a bit simple, but the example of St. John's wort, toxic because of the secretion of hypericin and repellent to many animals, is a classic. The plant attracts a whole group of leaf beetle species of the genus *Chrysolina* which have learned to detoxify the poison. For these insects, the repulsive became attractive. Hypericin-harboring plants are avoided by most of the large herbivorous mammals. They use experience instead of the genetic programming of insects for protection. However, cows never learned to avoid the *Cycas* flowers in New Guinea that rapidly kill them. How many millions of years would be necessary?

It is this parallel development of insects and plants, of biochemical barriers and the means to beat them, which has been called coevolution by Ehrlich and Raven (1964) as observed in the butterflies, but it was then applied to many associations between a host-specific phytophagous insect and a plant. The chemical products responsible for this process have been called allelochemicals. These play a great role in the way an insect behaves and also in host plant resistance.

Pollination by animals also represents a form, somewhat vague, of co-evolution which has lasted for more than 200 million years. During the Jurassic, some large dinosaurs probably fed on arborescent conifers. It seems that by the Cretaceous they moved over to angiosperms and possibly were poisoned by their alkaloids. This is rather a questionable theory, but it is certain that changes in fauna must have had a decisive influence on the diversification of the plants themselves and also on the evolution of phytophagous insects. Among the nearly one hundred hypotheses given for the death of the dinosaurs, this one is worth more than others, even including evidence of a meteorite landing in the Gulf of Mexico, such as the presence of iridium in the K/T boundary. Coevolutionary mechanisms, such as the production of a new alkaloid, provide a means to fight back, in turn causing food selection. Ehrlich and Raven (1964) also proposed that contact between plants and herbivores could be the main component of the interaction responsible for the appearance of organic diversity in a community of terrestrial organisms inhabiting a microhabitat.

We know that insects, except some butterflies, do not see red. For example, species of *Aquilegia* (Ranunculaceae), the columbines, are interesting because the red species (*A. formosa*) is pollinated by hummingbirds, and the sympatric white species (*A. pubescens*) by night-flying hawk moths. The color, flower position, and length of nectariferous spur vary according to the function of the pollinators. Hybrids produced by bumble bee cross-pollination probably exist and these also may be pollinated by birds or hawk moths. Variations are still occurring, resulting in secondary hybrids and further discrimination by pollinators. If so, insects can direct the evolution of species, but the contrary is equally true because plants have a real influence on the specialization of pollinators.

The concept of coevolution which was predominant in the 1970s and the 1980s soon went out of fashion (Bernays and Graham, 1988; Rausher, 1988). Already, Janzen (1980) was conscious of the relativity of the concept, namely in pollination. Some people now see only a sequential coevolution, and for them coevolution, as Ehrlich and Raven perceived it, is a rare phenomenon. "Sequential evolution by repeated colonization", despite the semantics, is still part of the coevolution paradigm.

According to Bernays and Graham (1988), insects seldom impose strong selection on plants, and as a consequence coevolution seldom occurs. Most of the reasoning is philosophical and not based much on experience. To Rausher (1988) and Bernays, natural enemies are responsible for the herbivore becoming specialized, and plant chemistry for it remaining specialized.

It is sure that feeding deterrents have often been found to be harmless, but that does not mean that plant chemistry is not the predominant element of the selection. What remains in the case of myrmecophytes or carnivorous plants to explain the extraordinary existing adaptations? For myrmecophytes, pre-adaptation of the plants to ants or coevolution seems to me a satisfying explanation even if some researchers, such as Hocking (1975), refer to the elusive Baldwin effect as being a semi-magical interpretation. Carnivorous plants alone seem to have developed an extraordinary and efficient mechanism to catch prey. Some regard this as mutualism, as the plants give nectar in exchange for some insects sacrified for their mutual interest (Joel, 1988). Let us agree that a clear explanation is difficult in this case since neither coevolution nor pre-adaptation fits easily. Plants of different families need nitrogen in a deficient soil. How do they develop such a sophisticated structure?

For Schultz (1988), plant chemistry is the predominant influence on the evolution of herbivore diets, and Berenbaum (1983) seems to accept a coevolutionary mechanism for insect-plants relationships. Miller (1987) proposes that insects have colonized their hosts subsequent to plant cladogenesis. No documented examples of parallel cladogenesis between insects and plants have yet been discovered. According to Nilsson (1992), orchid-pollinator interactions are asymmetrical, with most of evolution occurring on the plant's side. Futuyma (1994), who studied the phylogenetic aspects of host-plant affiliation in species of *Ophraella*, a galerucine leaf beetle, concludes that the phylogeny of *Ophraella* spp. is not congruent with that of its host plant. There is no "conspecification" of the beetle with the plant, and *Ophraella* species underwent host shifts among related and chemically similar plants.

Coevolution or not? It is difficult to decide, and further research is necessary. Whether diffuse coevolution (Janzen, 1980) or sequential evolution (Jermy, 1984), it is certain that phylogenies have not yet been constructed using cladistic methodologies and that no convincing evidence of parallel cladogenesis between insects and plants has been discovered. All attempts to correlate herbivore food preferences with the life history of these food plants have resulted in conflicting reports (Maiorana, 1978).

Practically, all plants, even *Ginkgo biloba* and *Azadirachta indica*, the neem tree, have enemies, at least in their native habitat. Ginkgoes, for instance, where they are still growing wild in the Chinese mountains, have their guild of herbivorous species. No plant, however, is attacked by all the phytophagous insects living in its area. If that were so, the plant probably would be condemned to immediate extinction. Neem trees (*Azadirachta indica* and *Melia azedarach*), endemic in tropical Asia, produce several compounds which seem to be toxic or at least to have an antifeeding effect on most insects. However, some insects that do feed on these trees locally are *Melolontha* spp., a beetle, scale insects, mealy bugs, bark-eating caterpillars (*Indarbela* spp.), and spider mites (Krantz et al., 1977). When imported into foreign countries, the "toxic" trees or grasses are attacked by herbivores, but only a few. Termites attack *Eucalyptus* spp. in Brazil, but nowhere outside of Australia and New Guinea are leaf miners found on those trees.

A question which is often raised is, "What is the relationship, if any, between primitive insects and plants?" In other words, would a primitive insect feed on primitive plants, and would this be indicative of the primitive nature of the plant and vice versa? It is evident that we are tempted sometimes to answer "yes". Caterpillar species of the genus *Micropteryx*, as previously mentioned, a somewhat aberrant genus between Trichoptera and Lepidoptera, live on *Hepatica* spp. (liverworts). These caterpillars, with very long antennae, are among the most primitive of the Lepidoptera. We have searched for some similar examples among the Chrysomelidae (Coleoptera) which, as for Lepidoptera, must have appeared with the angiosperms in late Jurassic. There are some archaic Sagrinae or related groups in Australia and America which frequent cycads, very primitive vascular plants, but in general those Chrysomelidae on conifers, vascular cryptogams (such as horsetails and ferns), and mosses are very highly evolved species which have abandoned the angiosperms through secondary adaptations. The same rule is verified by the fern Pyralidae, which belong to highly evolved species. Indeed there are very few insects on ferns, perhaps because of their hormone derivatives (phytohormones), certain repulsive compounds that many species produce.

We do not know of any butterfly (Papilionidae) which feeds on bryophytes or vascular cryptogams, and only certain moths, such as species of *Papaipema* (Noctuidae), do so. A few butterflies live on gymnosperms, including the Cycadaceae, and Lycaenidae also live on angiosperms. Here the antiquity of the insect group has nothing to offer in parallel with the antiquity of the plant. Lepidoptera on gymnosperms probably were derived from ancestors living on angiosperms. Lepidoptera on monocotyledons, except perhaps the Morphidae, are derived from ancestors living on dicotyledons. It remains to be demonstrated, however, whether the dicotyledons or the monocotyledons are the most primitive group. Both evolved during the Cretaceous, and very probably the monocots derive from the dicots. The leaf beetle *Aulacoscelis* sp. feeding on *Zamia* sp. (Cycadaceae) is primitive, while the langurid beetles feeding on these plants are not.

There is a possibility of coevolution between arthropods and pteridophytes (Gerson, 1973b), before and after the appearance of the angiosperms, which seems to be explained by the presence on ferns of both primitive and derived insect species. It may be, however, that ferns are actually underutilized by insects. Most of the ferns attacked are species of Polypodiaceae, which are sometimes associated with ants. Also, it must be noted, the older insect orders are better represented on ferns than the more recent ones.

The true significance of the previous example of cantharophily, or flower attraction for beetles to cycads and primitive angiosperms, is still doubtful. Certainly that may be an old relationship, but it is still to be proven. Finally, except for the pollinators, a rule linking the lower insects to the lower plants and vice versa is not evident. A primitive species may very well live on a derived plant species, and the reverse is true. This fact can be easily understood for leaf miners, as their adaptation to that diet came very late in their evolution when the flowering plants were already abundant and diversified and their conquest of the lower plants came about sporadically. Mines are very old, even older than the flowering plants, but they were rare at that time and not many were preserved.

Finally, we must assert that nonphytophagous species sometimes can be associated with plants for some unknown reason. For instance, in Phu Kae, Saraburi, Thailand, located between Bangkok and the national park of Khao Yai, we encountered an enormous aggregation (15 million or 170 kg) of three species of Tenebrionid beetles crawling over high trees including *Aegle marmelos, Dipterocarpus elatus, D. indicus, Schleichera oleosa, Stereospermum chelonoides,* and *Sterculia foetida.* Each tree was covered from the trunk to the top by kilograms of small insects all belonging to the genus *Mesomorphus.*

How can we explain this observation? In Jolivet (1971), I stated that there was no correlation with the rainy season. Was it an unknown meteorological phenomenon, a special trophic relationship? Being saprophagous, the beetles were feeding on the bark or lichens. I have seen enormous aggregations of species of *Calosoma* in Senegal, but in the latter case this was directly linked with the beginning of the rainy season. The aggregations of the *Mesomorphus* sp. remain mysteries and have nothing to do with fall hibernating insects such as ladybird beetle aggregations.

No one has yet given a satisfactory explanation for the seasonal aggregations of the endomychid beetle, *Stenotarsus rotundus,* on the same place and on the same plant, a palm tree (*Oenocarpus maposa*) on Barro Colorado island in Panama. New sites have been found elsewhere with other plants and other species, but the fundamental reason for this accumulation of beetles remains a complete mystery.

To come back to coevolution of insects-plants relationships, some cases of pollination are really difficult to explain without a coevolution hypothesis. A Ghanean plant, *Amorphophallus johnsonii* (Araceae), flowers during April, the main rainy season. Anthesis starts at dusk with fluid oozing from the upper spadix, producing a strong aminoid odor (Beath, 1996). Just after dark, several carrion beetles and dung fly species visit the inflorescences. The beetles become trapped in the lower spathe overnight and remain in the spadix until the following evening. In the evening of the following day, the anthers produce long threads of sticky pollen. The trapped beetles escape just after dark, past the dehisced anthers, and fly away from the spadix tip. Beetles transfer pollen from male-phase to female-phase inflorescences, generally the same night. The pollen is psilate and typical of beetle-pollinated Araceae.

chapter fourteen

A general review of insect-plant relationships; some thoughts about learning; the canopy

In the previous chapters, I have reviewed the main relationships between insects and plants from the Paleozoic until modern times. There is not much to say about arthropods other than insects, which are the most plentiful of the plant feeders. We have others, such as millipedes, terrestrial crustacea that are both polyphagous and scavengers. Of course, *Birgus latro*, the coconut crab, feeds on the endosperm of the coconut, but this is only part of its diet.

The notion of coevolution first raised by Ehrlich and Raven (1964) has since been enthusiastically adopted but not very closely critiqued. Really, though, this is the best explanation for the interpretation of the evolution of myrmecophytes, the other explanation being pre-adaptation of plants. For orchids, the relationships are sometimes so sophisticated that they seem to compel recognition of coevolution. Darwin himself was astonished by the adaptation between an orchid and its pollinator. Also, coevolution remains the "abominable mystery", more difficult to explain even than the origin of the flowering plants.

Plants, mostly green plants, are usually but not always at the base of the food chains. Bacteria and fungi can intervene in the ocean depths (vents) or in caves, but they are all exceptions. Exceptions can also appear in the feeding of a phytophagous insect which becomes carnivorous. Many leaf beetles and many other insects may feed on their own eggs and larvae; in the Amazon, one species of *Diabrotica* (Galerucinae) feeds exclusively on meloid beetles. This predatory beetle is potentially polyphagous on various plants and possibly has found in the meloids attractants able to trigger the feeding process (Mafra-Neto and Jolivet, 1994, 1996). Many cases of cannibalism are also listed in Chapter 1.

The chrysomelid beetle species of the subfamily Clytrinae are another exception. All the larvae (and some adults) are myrmecophilous, i.e., hosts of ants. The clytrine species live, to a great extent, at the expense of the eggs and larvae of ants, but they also eat plant debris found inside the ant nest. Adult clytrines are orange and black (probably an aposematic device), are polyphagous on bushes and grasses, and lay their eggs (which mimic seeds) in the vicinity of ant hills. These ants carry the eggs of the beetles inside the nest, probably mistaking them for seeds and more probably attracted by a pheromone. The hatched larvae live within cases made of excreta, the surface design of which is peculiar to each species. The front part is open to allow the head to emerge first. Often, within the nest, the larvae pull their heads back inside the case and the empty space at the end of the tube (the scatoshell) may be filled with eggs by the ant hosts. The eggs are immediately eaten by these strange "phytophagous-turned-carnivorous " insects. The same thing occurs with myrmecophilic adult clytrines and probably also with cryptocephalines, but almost no observations have been done inside the east African stipular thorns of species of *Acacia*.

Many lycaenid caterpillars are myrmecophilous, reared by ants and even brought within the nests by the ants themselves. Being phytophagous on dicotyledons, certain caterpillars of lycaenids become carnivorous, lichenivorous, or fungivorous, which results from a totally different selection process among closely related species. It is evident that the caterpillars of the moths living on sloths have food very different from free-living related species feeding on host's excreta. The adult moth eats desquamations of animal's skin. We could find numerous similar examples among the Tineidae and other Microlepidoptera. Some Latin American bees, species of *Trigona*, contain several species that are necrophagous.

Among phytophagous species, as we have seen previously, it is difficult to say which form, monophagy or polyphagy, is the most advantageous. Monophagy has an unquestioned monopoly, occupying a restricted ecological niche, both climatically and geographically. The insect's energy is saved by having to detoxify only a restricted number of toxins (Feeny, 1976). If, however, the plant is being wiped out, the specialist goes with it, as there is little chance that it could switch to some other food. The polyphagous or oligophagous species (i.e., the generalists) have a much greater margin of choice. These also have a wider distribution, hence more chance of survival and greater plasticity, but at the cost of a supplementary metabolic expense, as the toxins to be neutralized vary from one plant to another.

The first insects pierced plants for sap, as the aphids do now, but with much stronger beaks. When phytophagy appeared with the new chewing mouth parts, the plant's choice was not very great, and the surviving insects must have been polyphagous. Then when the angiosperms diversified during the Tertiary, food selection must have appeared, and oligophagy, then monophagy, developed from the primitive polyphagy.

True phytophagous insects represent about 50% of the living insects, perhaps more, and evidently during their evolution, phytophagy must have appeared first, followed by carnivores, which fed on the phytophages. We can see from the previously cited examples how different are the associations between insects and plants. However, as was said in the introduction, many problems have not been addressed: the role of plant pests in crop destruction; transmission of cryptogamic, mycoplasmic, bacterial, rickettsial, or viral diseases by insects; useful insects such as bees, cochineal, and silkworms which produce, from eating plants, useful substances; toxic plants which transmit their toxicity to insects and protect them indirectly (as in zygaenids, chrysomelids, and danaids); the symbiont problem necessary for certain phytophagous species (mostly those that are xylophagous) and for hematophagous species; vitamins; hormones; and certain other topics.

It must be stated, however, that toxicity of plants is not always responsible for toxicity of insects. Species of *Timarcha*, big black chrysomelids called bloodynose beetles, may feed on Rubiaceae or plantains in Europe. In either case, they produce the same toxic compounds during their autohemorrhaging. Pasteels and Daloze (1977) have demonstrated that cardiac glucosides effect protection against predators. These are not necessarily taken from their host plants by the leaf beetles, because they can be entirely synthesized by the beetle if the host plant does not provide it. Species of Danainae, on the contrary, obtain from their host plants (Asclepiadaceae, Apocynaceae, or Moraceae) not only the cardenolids, toxic to predators, but also the pheromone precursors.

A species of *Danaus* fed on cabbage loses its emetic agents and becomes edible, something that birds would learn rapidly if this happened in nature. Birds have recently learned that only the outer tissue of the larvae are poisonous, and, in the case of blue jays, they can scoop out and eat the inner parts of the larvae with safety. Some mimics imitate species of danaids and heliconiids, or both imitate each other (Müllerian and Batesian mimicry). It is the phenomenon of a "sheep in wolf's clothing" for the nontoxic species, while in the case of aggressive mimicry it is the wolf that hides in the sheep's skin to get its prey. Grasshoppers taken from species of *Calotropis* (Asclepiadaceae), when fed on lettuce or carrots, lose practically all of their toxic cardenolids, while still remaining aposematic and often capable of emission of blood and air (hemaphorrhea) to repel the eventual predator, generally a bird. However, we see that in this field we must not generalize.

In aggressive (Peckhamian) mimicry, a phytophagous insect, such as the poisonous species of Alticinae from the Kalahari, is imitated by a carnivorous species of carabid beetle (Lebiinae) which attacts and feeds on the alticine larva and adult. The "wolf" then carries the "sheep's clothing" to reach and capture its prey and thereby gets poisoned by the prey. So toxic are the pupae of these alticine victims and the lebiine predators that both are used as arrow poison by the Kalahari bushmen (Jolivet, 1967). Examples of Batesian and

Müllerian mimicry are common among Coleoptera in the tropics where, generally, phytophagous insects are involved, because they are toxic and aposematic and their toxicity is often due to their food plant.

The aposematic insects, i.e., those with a color indicating their toxicity (also called vexillar), such as the danaid butterflies, can bring some benefits to their host plant, such as a warning to eventual herbivores or by reinforcement of the protective odor of the plant. These advantages seem small, however, if we compare them with the ones the insects get from their toxic host plant. The quality of the food of silkworms also influences indirectly the quality and the secretion of their silk. This is so true that in Japan, during experiments in which caterpillars of *Bombyx mori* were fed on synthetic media instead of mulberry, the insects accepted the new food only if that medium contained at least 10% mulberry leaves. Also, in Japan they tried the culturing of mulberry tissue, and the results have been positive, provided the culture is produced in the light and if the cells contain a minimum of 1% of the normal chlorophyll content of the leaves. This experience shows us the complexity of the acceptance or refusal of specific plants by insects. Synthetic media are also a way to analyze insects' food preferences, the dosage of the attractants, and the necessary amount of various chemicals for rearing insects. In the previous examples, tissue culture with 1% chlorophyll has produced an excellent quality of silk. Many other caterpillars are less difficult to feed and will accept a 100% synthetic medium.

Outside of visual, olfactive, gustative, and chemical reasons for host-plant selection, the quality of the plant also influences the rapidity of the development and vigor of the insect. Choices of this nature are not so hidden as we might otherwise suppose.

All of this integration of plants and insects into adaptations, such as myrmecophily, carnivory, epizoic symbiosis, pollination, and its caprices, must have occurred during a relatively short evolutionary period, if we propose that the majority of these phenomena deal with the angiosperms, and that these plants appeared only at the end of the Jurassic and the beginning of the Cretaceous, which was even too early for the orchids. By the Cenozoic, all the present-day associations already existed and have not been greatly modified since then.

It is evident that food selection has a strong influence on a species' genesis, i.e., speciation. Currently, there exists in nature a whole series of biological races and developing species, not only among leaf miners, but also among externally phytophagous species, as well as many other species. Selection is often the function of existing pre-adaptations to a given diet. Polyphagous species have more chances to evolve quickly than monophagous species.

Roughly, plant diversity, as outlined by Hutchinson (1964), leads to increased diversity of phytophagous animals, but the contrary is equally true. The diversity of tropical species is due to climatic conditions, permanent

and optimal, which allowed development of the great diversity of phytophagous insects as well as the plants on which they feed. There should not be any
confusion between the superficial and reversible variations, i.e., somations in
relation to the biochemical and seasonal changes of the host plant (cassids,
nymphalids) and with the biological races which are ecologically and morphologically stable and genetically transmitted.

Finally, a great many insects are detriticolous or herbivorous. Among the
herbivores, some became selective. It is that passage from polyphagy to
oligophagy, then to monophagy, which has produced, by a series of restrictive mutations, the ecological potentialities of the insects. Some, such as leaf
miners, can sometimes partially recuperate their initial potentials. Nevertheless, the most striking adaptation (the result of a long selection process) is the
instinctive behavior that delimits the choice of host plant.

It is this mutual evolution (coevolution) during the past geological ages
of insects and plants, their interactions, reciprocal adaptations, symbioses,
defenses, nutrition, and protection that I have tried to review in this book.
The actions and reactions are so varied and so complex, and the subject so
vast that the previous chapters have only been an introduction to what surely
needs to be treated in as many different separate books.

A last word about learning. The matter has been dealt with in Section 3.2
as a theory which holds that the female of an insect breeding on two or more
hosts will prefer to lay eggs on the host on which said female was reared
(Torre-Bueno, 1989). The theory is debated, as it involves the memory of a
caterpillar being transmissible to the adult. It seems, however, that butterflies, as do bees, can rapidly learn to associate the color of a flower with the
presence of a reward. Butterflies also can display constancy to a single
rewarding plant within a multi-species rewarding patch (Weiss, 1995). Lewis
and Lipani (1990) assert that butterflies, as a group, have all the learning
abilities found in honey bees except those directly related to sociability.
Differences are due to variation in life history and in ecology. This is something of a rehabilitation for the butterflies, which were for a long time
considered "stupid". *Heliconius* spp. have a complex behavior, a long life,
and a variety of food, including pollen. Perhaps, their "intelligence" is greater
than that found among other genera.

To date, learning has been demonstrated in six orders of insects with
phytophagous habits (Papaj and Prokopy, 1989; Papaj and Lewis, 1992).
Food plant selection is not as passive as once was thought. The insect's
reaction is more complex than previously described. Social Hymenoptera
seem to remain the most advanced, but other groups, such as parasitic wasps,
also "learn" and remember past experience. Even leaf beetles, considered by
some as perfectly "stupid", can learn poison rejection. Food-aversion learning can be found in many insects, not only rats and slugs.

Learning is a change of behavior with experience which can be permanent (imprinting). In the absence of continued experience, an insect can,

however, forget what it has learned. It would be interesting to study phytophagous weevils, such as species of *Gymnopholus*, with a long life expect-
ancy (6 years) to see if they can learn something and if they remember for a
long time what they have learned. They are relatively polyphagous and do
not lead very active lives, being homochromous with their substrate (epizoic
symbiosis).

Larvae of some lepidopteran or coleopteran species will die of starvation
even if novel food is acceptable in insect standards. Colorado potato beetle
larvae when presented with a rejected *Solanum* species will not feed, a
phenomenon known as the so-called "starving to death at Lucullian ban-
quets". Learning is a difficult process sometimes encouraged by natural
selection; the ones that reject the new plant will die. Some will survive and
accept the plant in the future.

Ovipositing butterflies, such as several species of *Heliconius* (Heliconiinae)
or *Battus philenor* (Papilionidae) learn the shapes of the leaves of the host.
Bees fed at a target of a particular odor and color learn the association with
a sucrose reward. Modification of feeding preference of *Altica lythri*, an
alticine beetle, has been mentioned by Phillips (1977). Rausher (1983) could
not find the role of conditioning in the variation of oviposition of a cassid,
Deloyala guttata. If beetles can learn, then practically any insect is able to
learn, the social Hymenoptera being among the most "intelligent", not to
mention the spiders.

It is true that the degree of selectivity of an insect and the identity of
preferred plants can change during an individual's life time. These changes
sometimes can be due to previous experience, hence, probably some kind of
learning.

Imprinting, or programmed learning, known in vertebrates, is not very
clear among insects. Young ants seem to learn better than old ones but it is
very difficult to generalize, experiments having been conducted with few
insects. The number of host species that phytophagous insects can remember
is relatively limited, and the brain of an insect with a limited number of
neurons has its own limits.

It is evident that a bee or a butterfly will learn very quickly that certain
plants such as *Lantana camara*, *Brunfelsia americana*, *Gloriosa superba*, and so
many others will change color after fertilization. That means also that nectar
production has stopped. Such learning is very quick, and in certain cases is
genetically programmed.

For Papaj and Prokopy (1989), learning may be a force in plant-insect
coevolution. The authors add that we might be grateful that phytophagous
insects are capable of learning and equally grateful that plants are not. Is that
true? Under selective pressure, plants "learn" to avoid certain insects and
that is the way deterrents, antifeedants, and alkaloids have developed during
the geological past. The false eggs of certain *Passiflora* have been developed
through evolution to deter visually butterflies.

Depending upon their abilities, such as flight power, insects will choose herbs, shrubs, or big trees, even the canopy. It is evident that flightless insects will stay on small plants even if they are able to climb with their legs on some vines. Canopy species such as the leaf beetle *Megascelis* sp. will fly up and down and are frequently found on small shrubs in the beginning of the day. Some insects fly, some don't, and in some alate leaf beetles such as the *Chrysolina* spp. most of have lost their wing muscles and do not use their wings. The same plant families are often represented in the canopy and on the ground, but several families, such as the Cucurbitaceae are never (with one exception on Aldabra island) found growing as trees. The distribution of the plant-feeding species is a function of the availability of the host plant, of their flight aptitudes, and the climatic conditions. Many canopy dwellers are pollen or nectar feeders, but some feed on the leaves. This enchanted part of the forest requires a good observer and technical equipment. Even if the plants are known, their relationships with known and unknown insects still remain "a dark mystery". Research on the canopy fauna has been intensive; some are well founded, while others are inspired by some media exhibition-ism and are not very productive. The richness of the fauna depends mostly on the selectivity of the insects found on the trees. The tree diversity of the tropical forests is enormous, but is it the same with the insects? Moran et al. (1994) are not sure of the extent to which the insects associated with the trees in a South African forest are generalists or specialists. Erwin (1983a,b) esti-mated the total number of insects species at 30 million, linking plant diversity to animal diversity. Stork (1988) came back later with a more reasonable number and so did Basset (1990) in New Guinea, but no one has been able to give a reasonable estimate based on strict statistics. If most of the forest fauna is composed of specialists, then Erwin's estimate is low. Taxonomists and biologists are badly needed to revise this estimate of the fauna, although probably most of it will be extinct before the work is achieved.

Defoliation in tropical rain forests by insects has been estimated to be between 5 and 15%, but the overall damage to the canopy is just over 5%. In the tropics the leaves grow back very rapidly and the damage is not appar-ent, even that of leafcutter ants. Opinions vary as to whether vines were grazed less than the trees and ephemeral plants more than long-lived plants. Still a lot of research has to be done in the forest, what has been called "the last biotic frontier". The new cranes used to study canopies will certainly help in the search for more information about the biology of the treetop animals.

chapter fifteen

Epilogue

With present-day pollution and urbanization, it is certain that the biodiversity of plants and animals is diminishing very quickly. This destruction is impossible to evaluate correctly. However, there are still a great number of fungi, cryptogams, and vascular plants surviving and feeding arthropods. With about 300,000 living species of vascular plants, the diversity is really enormous and the number of insects has been estimated to be 100 times that figure. In a country such as Brazil, the Atlantic forest has been cut down more than 80%, and some of the original fauna has already disappeared. Some people, however, suggest that beetles and butterflies persist in refuge localities. This may be true for butterflies, but it is not true for apterous beetles — for instance, leaf beetles. The primary forest is cut, the host plant does not survive among the secondary new growth, and a dozen species of the genus *Elytrosphaera*, for example, are condemned to extinction rather rapidly. They feed on Asteraceae growing exclusively in forest openings (*Adenostemma* spp.) and on Solanaceae.

Apterous insects such as species of *Timarcha, Meloe,* and so many others are becoming more and more rare and will be, within a short time, only a memory in dusty collections. Pollution, urbanization, roads, buildings, cultivation, insecticides, and herbicides, together, are responsible for that sad future. As Burger (1981) wrote, we humans are just now beginning to understand the rich complexity of terrestrial ecosystems and we are part of the cultural forces that will decimate them.

It is not absolutely true to say, as did Schoonhoven (1990), that "no animal life exists in the absence of green plants, since they serve as the primary source of energy rich compounds for all heterotrophic organisms." There are other ways of synthesizing the proteins: in deep-sea vents, rocks, and caves, where some bacteria are purely chemotrophic. However, these are exceptions, and almost all life on Earth is based on green plants and on fungi which are themselves saprophytic.

It seems that the actual diversity of the plant kingdom is the result of the long-lasting cohabitation with animals and pollination. All the botanists do not agree on that, and there are endless discussions about the origin of the

217

vascular plant diversity (Burger, 1981; Stebbins, 1981); the number of plants has grown from 500 in the late Carboniferous to about 250,000 to 300,000 species today. Coevolution or no, in some cases such as orchid pollination pre-adaptation is not a sufficient explanation. Coevolution, whether diffuse or not, seems to impose itself exactly the same for myrmecophytes and carnivorous plants. I am not going to discuss philosophy here in this modest essay. Others have done it, or will do it, and much better than I could do.

As Futuyma (1983) said, "Behavioral specialization sets the stage for the evolution of physiological adaptations, such as detoxification, that are specific to the secondary compounds peculiar to the host." In simpler words, primitive insects were polyphagous and detritivorous, and monophagy came later as a sequential specialization. Aphids actually are selective, and there is no reason why the sapsucking Paleozoic Palaeodictyoptera were not selective, too, but the choice of plants was then small and we cannot speak of mono- or oligophagy. Actually, according to Kennedy (1986), 9 out of the 29 orders of insects contain species that feed on living plants. Some of them, however, have difficulties digesting cellulose. Among the mammals, pandas and gorillas digest only a small part of their bamboo food, while the ruminants and kangaroos, with their complex fermentation chambers, digest over 60% of the food they eat. Many phytophagous insects have bacterial or fungal symbionts that help in the digestion of cellulose, just as cows have their flagellates and termites their Hypermastigidae. For herbivorous insects, cellulose digestion is not the only problem. There also remains the detoxification of the toxins, specialization among attractants and deterrents, and many other problems linked with plant feeding. Perhaps if Hymenoptera, Coleoptera, Diptera, Lepidoptera, and Hemiptera are the most diverse orders, that perhaps is due to the close evolution with that of plants.

And what will happen to their diversity when perhaps the CO_2 level doubles and the fragile equilibrium between plants and insects is modified even more? Biosphere II cannot answer this question. Perhaps a new set of experiments will permit us to get a clearer answer. Whatever will be the result, the future of the fauna and flora for the next millennium is far from encouraging.

Glossary

Abaxial: side of leaves, petals, etc. facing away from the stem or main axis, i.e., lower (ventral) surface.

Acarodomatium (plur. acarodomatia): hollow protective structure formed by some mite-harboring plants; may reflect symbiosis.

Acarophytes: plants harboring mites.

Adaxial: side of lateral organs (leaf) facing towards the stem or main axis, i.e., upper side (or dorsal).

Allelochemicals: substances that affect individuals or populations of a species differently than they do at the source.

Allelopathic: see *Alleopathy*.

Allelopathy: harmful influence or effect of one plant or microorganism on another due to the release of secondary metabolic products into the environment.

Allomone: allochemical or allelochemical; of adaptive advantage to the organism producing it.

Allotrophy: see *Xenophagy*.

Ametabolus: without metamorphic stages.

Androecium: collective term for the stamens of a flower.

Anemogamy: see *Anemophily*; distribution of seeds by wind.

Anemophily: pollination by wind.

Anemotropism: a change in position of a plant in response to a wind current; anemotaxis

Angiospermae: plants with true flowers and seeds which are enclosed in a carpel.

Ant garden: association of various epiphytic plants harboring an ant nest in their roots.

Antennograms: see *Electroantennogram*.

Anthocyanin: plant pigments that contribute to the blue or pink coloration of flowers and fruit.

Anthoecium (plur. anthoecia): spikelet of such grasses as species of *Panicum*.

Anthophilous: fond of flowers, i.e., feeding on flowers (insects).

Antifeedant: a natural repellent for insects.

Antimicrobial: toxic to bacteria; bactericidal.

Antioxidizer, antioxidant: any substance inhibiting exidation; an inhibitor such as ascorbic acid effective in preventing oxidation by molecular oxygen.

Apetalous: a flower without petals.

Aphagous, aphagy: applies to those insects that do not feed in the adult stage.

Aposematic: with warning coloration, indicating that an insect is unpalatable (vexillar).

Appressed: closely flattened; compacted, but not joined.

Apterism, apterous: without wings, i.e., wingless adult insects

Arborescent: branched, as the branches of a tree.

Arenicolous: inhabiting a sandy area.

Aril: a fleshy, colored covering of the seed. It arises as an outgrowth of the funicle or base of the ovule and may be a tuft of setae.

Ascidium (plur. ascidia): a baglike or pitcher-shaped part.

Ascomycete: a fungus with spores borne in sac-like structures called asci.

Ascospore: haploid spore of ascomycete fungi, produced by an ascus.

Ascus: in Ascomycetes, large sac-like structure, usually the swollen tip of a hyphal branch within which ascospores are developed.

Autochory: self-dispersal of spores or seeds by an explosive mechanism.

Autogamy: self-fertilization.

Autohemorrhea: reflex bleeding in insects.

Auto-incompatibility: physiological differences preventing fertilization and seed development or reproduction in a plant.

Autosterile: see *Auto-incompatibility*.

Auxin: an accessory growth-promoting substance in the food of plants; a hormone produced in the growing tips of plants which travels through the plant from cell to cell.

Axil: dorsal angle between a leaf petiole and the stems.

Bakerian mimicry: plants having female flowers, without reward, and copying the nectariferous male flowers.

Baldwin effect: sequential process in which acquired characters under environmental factors are replaced by genetic characteristics, i.e., transmissible.

Basiconic sensilium: a thin-walled, peg-shaped sensilium with minute pores, functioning in chemoreception in insects.

Basidiomycete: a class of fungi with septate mycelia bearing spores on basidia.

Batesian mimicry: form of mimicry in which an edible species (the mimic) obtains security by counterfeiting the appearance of an inedible species (the model).

Beltian bodies: food bodies growing on certain Central American acacias; see *Trophosome*.

Biocenosis: community of organisms inhabiting a particular biotype.

Biodiversity: the diversity of life on Earth as represented by the species.

Bioinsecticide: substance produced by a living organism which is toxic to an insect.

Biomass: total weight of living organisms at any one time per unit area.

Biota: plants and animals of a given area.

Biotype: a strain of a species; a group of organisms having the same genotype.

Bisporangiate: with both microsporangia and megasporangia, e.g., both stamens and carpels on the same plant.

Bivoltine: having two generations a year.

Brachypterism: insects exhibiting short or abbreviated wings.

Bromatia: rounded swelling at the tips of the hyphae of fungi which are cultivated by Attini ants for food.

Buccal: pertaining to the mouth.

Bulbuous: basal swelling of the stem to form a bulb.

Bursa: a pouch.

Canopy: branches, leaves, flowers formed by woody plants above the ground.

Cantharidin: a chemical found mainly in the elytra of some beetles which produces blisters on the skin of humans and other mammals or, when eaten, causes severe irritation to the intestinal tract.

Cantharophily: flower attraction of beetles for pollination.

Capitate: head-shaped; with a globose head.

Cardenolids: compounds offering protection to plants against insect damage, often sequestered by insects.

Carnivorous: flesh-eating.

Carnivory: see *Carnivorous.*

Carotenoids: any of a group of orange-red pigments found in plants such as carrots.

Carpel: modified floral leaf in the seed-plants which forms the gynoecium of the flower part in which the seeds are formed.

Carpophore: the stipe, pileus, and lamellae of fungi.

Caruncle: outgrowth near the micropyle and hilum of the seed.

Capitulum: knob-like structure on some stick insect eggs, mimicking the seed elaiosomes.

Cecidium (plur. cecidia): plant gall caused by an insect, mite, or fungus.

Cecidogenous: producer of a plant gall.

Celandine: refers to a papaveraceous plant, *Chelidonium majus,* which has yellow sap and yellow flowers.

Cercus (plur. cerci): a paired terminal appendage on the abdomen of certain insects.

Cheirogamy: pollination of flowers by bats.

Chemoreceptor: sense organ composed of a group of cells sensitive to chemical properties of matter.

Chemosensory organ: see *Chemoreceptor.*

Chemosterilization: chemical sterilization of male insects.

Chemotaxis, chemotactic: reaction to a chemical stimulus.

Chemotrophic: organism able to synthesize food without the help of chlorophyll and which does not require light

Chemotropism: see *Chemotaxis.*

Chimeras: A plant having tissues of more than one genetic type. It can be
   produced by grafting different types or species of plant; for example:
   Crataegomespilus = a graft hybrid of *Crataegus* × *Mespilus* (Rosaceae).

Chitinase: enzyme secreted by a chitinivore to initiate the digestion of chitin.

Chitinous: composed of chitin.

Chloroplasts: minute flattened granules or plastids containing chlorophyll in
   plant cells.

Cibarial pump: modified cibarium (preoral cavity) and associated structures,
   forming a sucking pump.

Cibarium: mouth cavity of insects.

Ciliata: protoans with cilia.

Citral: a pale, yellow, water-insoluble liquid aldehyde with a strong lemon-
   like odor.

Cladistics: pertaining to cladistic analysis of characters of taxa for a cladistic
   classification.

Cladogenesis: diversification of a lineage of taxa by splitting, i.e., speciation.

Cleistogamous, cleistogamy: pollination and fertilization before the flower has
   opened.

Clonal: propagated from a bud.

Coadaptation: the correlated variation in two mutually dependent organisms.

Coelom: in arthropods, the hemocoel.

Coevolution: results of two or more populations interacting so closely that
   each serves as a strong selective force on the evolution of the other(s),
   resulting in reciprocal stepwise adjustments.

Commensal: one species benefiting by its relationship with another, nonpar-
   ticipating species

Conidium (plur. conidia): an asexual, thin-walled spore developed on mycelium or the hypha of fruiting bodies of a fungus.

Conspecific: belonging to the same species.

Coprolites: fossil fecal material.

Coprophagy: feeding on dung.

Corolla: the inner circle of petals which is usually the conspicuously colored
   part of a flower head.

Cryptogams: the large subdivision of the plant kingdom which contains those
   species that reproduce by spores.

Cultivar: a botanical variety that has originated under cultivation.

Cyanogenetic: producing cyanogen, several heterosides.

Cycloalexy: the ring defense behavior of certain insect larvae on plants.

Delpinian bodies: the food bodies of myrmecophilous species of the plant
   genus *Piper.*

Desquamation: the separation or shedding of the cuticle of epidermal cells in the form of flakes or scales.

Detriticolous: frequenting organic detritus.

Detritivores: heterotrophic animals that feed on organic detritus; see *Sapropha-gous.*

Diapause: a delay in development that is not the direct result of prevailing conditions.

Dichogamy: maturing of the anthers and ovules in the same flower at different times.

Diclinous: with flowers or fruiting organs of only one sex on a plant.

Dicotyledons: a plant with two seed leaves; a plant whose embryo has two cotyledons; one of the two great divisions of the angiosperms.

Dioecious: bearing male and female flowers on different plants.

Diterpenes: a group of terpenes with twice as many atoms as monoterpenes; contain isoprene units.

Diverticulum (plur. diverticula): any off-shoot from a vessel, usually blind or sac-like.

Domatium (plur. domatia): cavity formed by a plant in which ants or mites live apparently in symbiosis with the plant; see also *Myrmecodomatium; Acarodomatium.*

Dysplasy, dysplasia: abnormal tissue growth.

Ecdysone: steroid hormone secreted by insect prothoracic glands which stimu-lates the secretion of moulting fluid from glands in the epidermis.

Ectohormones: pheromones.

Ectoparasite: an external parasite of insects, vertebrates, and other organisms

Elaiosome: outgrowth from the surface of a seed, containing fats or oils; they are attractive to ants, thus resulting in seed dispersal; see *Aril, Myrmecochory.*

Electroantennogram (EAG): record of the sum of the potentials of a number of olfactory receptors on the antennae.

Electropalpogram (EPG): record of the sum of the potentials arising in the palpi.

Endoparasite: internal parasite; in insects, the parasite usually leaves the host before pupation.

Endosymbionts: internal symbionts generally on the gut wall or inside the Malpighian tubules of an insect.

Endozoic: living in an animal, frequently as a pathogen.

Endozoochory: the dispersal of seeds by animals after being passed through the digestive tract.

Entomocecidia: galls induced by insects.

Entomochory: depending on insects for dissemination.

Entomofauna: insect fauna.

Entomogamy: pollination of flowers by insects.

Entomophagous, entomophagy: eating of insects.

Entomophilous: pollination of a flower by the agency of an insect; insect pollinators.

Epicontinental: found or located in or on a continent, often used in reference to shallow inland seas of the past.

Epicuticular, epicuticle: thin outer layer of cutile of insects covering the exocuticle.

Epiphylls: algae, lichens, or moss prothalli growing on leaves in the humid tropical forest.

Epiphyte: plant growing on another plant but not deriving any food from it.

Epizoic symbiosis: external growth of cryptogam on the backs of insects in a humid environment, such as in tropical mountains of New Guinea.

Epizoochory: dispersion of plants whose seeds adhere to passing animals and thus are scattered.

Ergotism: disease caused by the fungus *Claviceps purpurea,* parasitizing rye seeds and other grains.

Extrafloral nectary: nectary formed in some place other than in the flower; not used in fertilization but utilized by many insects, especially ants, as food.

Fecundation: in plants or animals, fruitfulness or capacity of seed or young production of the female; productive capacity of the species; fertilization.

Filamentous: mycelium composed of loosely interwoven free hyphae.

Finalism: doctrine or belief that all events are determined by final causes.

Flavonoid: yellow pigments in plants, sometimes used by insects, absorbed unaltered from plants, and becoming a pigment in the coloration of the insect.

Floricolous: living on or frequenting flowers.

Folivore: leaf eaters.

Food bodies: production of several myrmecophytes on the stem or the leaf or the stipule feeding the ants which are generally rich in lipids, proteins, starch, or even glycogen, as in the case of *Cecropia* spp; see *Trophosome.*

Frass: solid excrement of insect larvae.

Fungivorous: fungus-feeding.

Funicle: that part of the insect antenna proximal to the antennal club.

Funiculus: stalk of an ovule of seed.

Gall: abnormal plant growth caused by the tissue penetration of an insect pest or a disease organism; see *Cecidium.*

Gallicolous: dwelling in galls either as gall producers or as inquilines.

Galligenous: producer or inducer of plant galls.

Gametogenesis: development of gametes; process of producing or forming gametes.

Gametophyte: phase of the life cycle of plants bearing the sex organs which produce gametes.

Gamokinesis: social pollination.

Gastroliths: hard or stony material used by certain animals to help crush food particles.

Genome: genetic material of an organism.

Geocarpic, geocarpy: ripening of fruits underground; growing fruits are pushed into the soil by a postfertilization curvature of the stalk.

Glabrous: without hair, bristles, setae, or scales.

Glanduliferous: bearing glands.

Glandulous: with the character or function of a gland.

Glucosides: any compound which, on hydrolysis, yields glucose together with one other substance, sometimes toxic.

Gondwanian: southern supercontinent of Gondwana during the Mesozoic era.

Gongylidium (plur. gongylidia): see *Bromatia.*

Gregarines: unicellular organisms living within the intestine of many insects and other invertebrates; Apicomplexa.

Guild: group of species that exploit the same class of environmental resources in a similar way; ecologically unified functional groups of organisms.

Gustative: relating to taste perception.

Gynoecium: single or compound carpel of a flowering plant; female organs of a flower.

Habitus: general form or appearance.

Hapaxanthic: flowering once, then dying.

Haustorium (plur. haustoria): an outgrowth of stem, root, or hyphae by certain parasitic plants or fungi, through which they obtain food from their host plant or insect host.

Hemaphorrhea: reflex bleeding from joints, the sera mixed with air, found in certain insects (some Orthoptera and Coleoptera).

Hematophagous, hematophagy, haematophagy: feeding on blood.

Hemimyrmecophyte: plant with extrafloral nectaries but not harboring ants in these structures.

Hemiparasitic: parasite which can also exist as a saprophyte; facultative saprophyte.

Hemorrhea: reflex bleeding among insects; same as *Autohemorrhea.*

Herbivory: feeding on vegetables.

Hercogamy: condition of a flower when the stamens and stigmas are so placed such that self-pollination is impossible.

Hermaphrodite: individual possessing both ovaries and testes or equilivant reproductive organs.

Heterosides: compound glysosides.

Heterostyly: division of a species into two or three kinds of individuals by the relative positions of the stigma and anthers.

Heterothallic: fungus with thalli separable into two or more morphologically similar strains.

Heterotrophic: obtaining nourishment as a parasite and a saprophyte; does not photosynthesize.

Hexanol: a water-, benzene-, and chloroform-soluble compound found in plants.

Homochromy: in U.S., change of color to resemble the environment; in Europe, of the same color as the surroundings (rocks, leaves, trunk, flowers).

Homoiothermic, homeothermic: in birds and mammals, warm-blooded.

Hopkins Host Selection Principle: theory which holds that the female of an insect breeding on two or more hosts will prefer to lay eggs on the host on which said female was reared.

Hydrogamy: aquatic plant submerged fertilization.

Hydrotropisms: growth response of plants stimulated by moisture.

Hyperparasite: parasite of a parasite.

Hyperparasitoid: parasitoid parasite of another parasitoid.

Hyperplasia, hyperplasy: excessive overgrowth of a part of a plant.

Hypha (plur. hyphae): individual filament of a fungus thallus.

Hypocotyle: short stem of an embryo seed plant; portion of the axis of the embryo seedling between the attachment of the cotyledons and the radicle of a seed.

Hypogaeic: subterranean.

Inflorescences: arrangements of flowers on a stem or axis.

Infrabuccal: under the mouth, as an infrabuccal chamber.

Inositol: aromatic substance occurring in many seeds and other parts of plants; part of the vitamin-B complex.

Inquilines: animal living in the home of another animal and sharing its food.

Internodes: stem between two successive nodes.

Interstaminal: between stamens.

Isothiocyanates: compound of the type R–N=C=S; may be an alkyl or aryl group, as for mustard oil.

Juglone: crystalline substance obtained from walnut trees, also called nucin.

Kairomone: allelochemical of adaptive value to the organism receiving it.

Kinesis: locomotory activity of an organism in response to a stimulus.

Kleptoparasitic: thievery by ants and some other insects.

Kohlrabies: see *Bromatia*.

Lamarckism: theory that the effects of use and disuse of the organs of an individual are transmitted to its offspring; known as the theory of inheritance of acquired characteristics.

Lamina: flattened part of the leaf, blade, or limb.

Lenticel: pore in the periderm of a woody stem (trees, shrubs), allowing the passage of air to internal tissues.

Lanceolate: spear-shaped.

Lek: specific site where males display and compete for the attention of females.

Lichenivorous: lichen feeders.

Lichenous: pertaining to or resembling lichens.

Lignins: substance deposited in plant cell walls forming woody tissue.

Limonene: liquid terpene in two optically different forms.

Lycophyte: Lycopodes, the club mosses.

Macrospore: large spore in heterosory; spore produced by female gameto-phyte.

Marcescent: withered, but remaining attached to the plant.

Maulik's Principle: if a phytophagous insect accepts as food a whole group of plants belonging to one or more genera, another insect that accepts one of those plant will accept all the plants of the group; only partially true.

Mechanoreceptors: sensillum or group of sensilla functioning in mechanore-ception.

Megasporophylls: leaf-like plant organ bearing or enclosing megasporangia; a carpel; female cone of gymnosperms.

Melanism: pigment darkening of structures.

Melittophilous: pollinated by bees.

Merostomata: subclass of arthropod class Arachnoidea, the king crabs and eurypterids.

Mesocarp: middle layer of the pericarp between the endocarp and the exo-carp; middle layer of the fleshy part of certain fruits.

Mesophyll: soft tissue (parenchyma) between the upper and lower epidermis of the leaf, the part entirely concerned with photosynthesis.

Metamorphorsis: the developmental stages of animals as they grow from the egg to the adult; in insects, the stages of development between moults.

Microclimatic: climatic factors in the microhabitat of an organism.

Microforest: groups of cryptogamic plants, such as mosses, lichens, and algae, growing on the dorsal surface of some insects, mainly in tropical rain forests.

Microhabitat: the immediate habitat of an organism comprising its environment.

Micropyle: a small opening in the integument at the apex of an ovule through which pollen enters.

Microspore: smaller spore in heterosporous plants, i.e., pollen grain; spore giving rise to the male gametophyte or prothallium, as in ferns.

Microsporophylls: small, leaf-like structures bearing microspores; a stamen.

Microsporophyte: male prothallus.

Mimetism: see *Mimicry.*

Mimicry: resemblance of one species (model) by another (mimic), living together in the same area, with protection from predators conferred upon the mimic.

Minimes, minima: in ants, minor workers.

Monoclinous: with stamens and carpel on the same flower; hermaphroditic.

Monocot, monocotyledonae: a division of the angiosperms with seeds with a single cotyledon.

Monoecious, monecious: with monosexual or bisexual flowers on the same plant.

Monogynous: plant flowers with only one pistil.

Monolectic: bees that use pollen of only a single plant species.

Monophagous, monophagy: species restricted to one species of food-plant.

Monophyletic: derived from a single ancestral form; a grouping recognized on the basis of synapomorphy.

Monosexual: with only one sex in the flower.

Monoterpenes: class of terpenes with two isoprene units.

Monotropic: bees that visit flowers of a single plant species; see *Monolectic.*

Monotypic: containing but one immediately subordinate taxon; as a genus, containing but one species.

Morphs: any of the genetic forms that account for polymorphism.

Müllerian mimicry: similarity of several species which are distasteful, poisonous, or otherwise harmful to predators.

Mutualism: same as symbiosis.

Mycangium (plur. mycangia): any one of a variety of special pocket-shaped receptacles used by insects to carry symbiotic fungi.

Myceplasmas: bacterioid elements living inside cells (belonging to a group of prokaryotic organisms, the mollicutes); nonspore-forming bacteria which lack a true cell wall.

Mycetocytes: in certain insects, cells containing symbiotic bacteria or fungi.

Mycetomes: a structure housing cells with intracellular symbiotes, or mycetocytes.

Mycocecidia: galls caused by fungi.

Mycophagous, mycetophagous: feeding on fungi.

Mycorrhiza: close plant association between fungi mycelia and the roots of various plants where both organisms benefit.

Mycotetes: bundles of thread emanating from the surface of the mycelium strands of the termite gardens.

Myrmecochory: ant dissemination of living seeds.

Myrmecodomatium (plur. myrmecodomatia): naturally hollow protective structure formed by some ant harboring plants.

Myrmecophily: symbiotic association between plants and ants or between insects and ants.

Myrmecophobic: antless species among a myrmecophilous genus of plants.

Myrmecophyte: plant harboring ants in its structures; these are sometimes used as food.

Myrmecotrophic: plant offering food to the ants.

Myrmecoxenic: plant offering both food and lodging to the ants.

Myrosin: enzyme found in seeds; in mustards it decomposes glucosid into allyl mustard oil, glucose, and potassium nitrate.

Necrophagous: feeding on dead and decaying animals.

Necrophilous: attracted to dead animals.

Nectary (plur. nectaries): nectar-secreting organ, nectarium, normally inside the flower; see *Extrafloral nectary.*

Nectariferous: producing nectar.

Nectarigenous: see *Nectariferous*.

Nectarinian: nectar-feeding birds, sunbirds in particular; brillantly colored passerine birds of warmer forested parts of the Old World, including Africa.

Nectariphagous, nectarivorous: nectar-sipping animals.

Nematocecidia: plant galls made by nematodes.

Nematophagy: feeders on nematodes.

Neotony: retention of juvenile characters in the adult stage.

Ochrea: cap-shaped structure around a stem formed by united stipules or united leaf bases.

Oligolectic: bees that use pollen of several plant species.

Oligophagous, oligophagy: with a restricted range of food plants of related genera, families, or orders.

Oligotropic: bees that visit few related plant species; see *Oligolectic*.

Ophthalmotrophic: frequenting the eye of animals.

Ornithochorous: seed distribution by birds.

Ornithogamy: pollinated by birds.

Paleomimetism: mimetism between ancient organisms and these organisms and plants.

Pantophagy: omnivory.

Pantropical: inhabitating both the Old World and New World tropics.

Parabiosis, parabiotic: living amicably in compound nest as ants of different species or genera.

Parasitoid: an internal or external parasite that slowly kills the host; parasite only at the larval stage (Hymenoptera, Diptera); protelian parasite.

Parthenogenetic: reproducing by parthenogenesis, i.e., egg development without fertilization.

Paurometabolous: insects with simple, direct, or gradual metamorphosis in which the series of immature forms are very much like the adult.

Pearl bodies: corpuscules growing on different parts of green plants and feeding ants; see *Food bodies, Trophosomes*.

Pedunculate: with a peduncle or stalk.

Perianth: calix and corolla collectively.

Pericarp: mature ovary wall; wall of fruit or seed vessel developed from the wall of the mature ovary.

Perithecium: globose or flask-shaped fruiting body characteristic of the pyrenomycetes fungi and of lichens.

Petaloid: shaped like the petal of a flower; colored like and resembling a petal.

Petiolar: growing on a small stalk; borne on or pertaining to a petiole.

Phagostimulant: chemical promoting feeding activity.

Phanerogams: flowering plants with stamens and pistil which develop seeds.

Phenolic: chemical containing phenol.

Phlorizin: bitter crystalline glucosid in the root bark of apple trees.

Phoretic, phoresy: a type of relationship in which one organism is carried on the body of a larger organism but does not feed on the latter.

Photoperiod: day length.

Phyllody: metamorphosis, under the influence of a mycoplasm, of an organ into a foliage leaf.

Phylogenetic, phylogeny: evolutionary history of a group of organisms.

Phytoecdysones: plant hormones mimicking insect moulting hormones.

Phytohormone: hormones produced by plants or one used to stimulate plant growth.

Phytojuvenile: plant chemicals mimicking insect juvenile hormones.

Phytoparasitic: parasitic on plants.

Phytophagy: feeding on plants.

Phytotelma (plur. phytotelmata): structures formed by terrestrial or epiphytic plants that impound water, such as modified leaves, leaf axes, stem holes, or depressions, flowers.

Pilosity: covering of fine hairs or setae.

Pinnule: secondary pinna of a dicompound or tripinnate leaf.

Pistillate: with pistils but not stamens; female flower with pistils only.

Placoid sensilla: plate-like sensilla.

Pleiotrophic, pleiotrophy: genes that affect several physical or physiological characteristics of the phenotype.

Pleometrosis: in Formicidae, primary polygyny; colony foundation by more than one female as in some social Hymenoptera.

Plumule: embryonic shoot or bud in the seed located between the cotyledons, from which develops the stem and leaves of the plant.

Poikilothermism, poikilothermic: cold-blooded animals.

Pollenivorous, pollenophagous: pollen-eating.

Pollinium: A mass of pollen grains held together by a sticky substance and transported as a whole during pollination.

Polygyny: coexistence in the same colony of ants of two or more egg-laying queens.

Polylectic: bees that use pollen of many plant species.

Polyphagy: eating many kinds of plants.

Polyphenols: phenol derivitives.

Polyphyletic: composite taxon whose members are derived from two or more ancestral sources.

Polytropic: bees that visit many plant species; see *Polylectic.*

Polytypic: containing two or more taxa in the immediate subordinate category, as a genus with more than one species.

Pre-adaptation: previously existing behavior patterns; physiological process or morphological structure already functional in some other context.

Predation: obtaining energy as food by comsuming prey organisms.

Presahelian: area between the Sahara Desert and the Sahelian Zone in Africa.

Prostoma (plur. prostomata): not prostomium; thinner area, generally devoid of vessels, becoming a stoma after being opened by ants.

Prothallus, prothallium: gametophyte in ferns and related plants.

Protocarnivorous: insect-eating plants that do not produce their own digesting enzymes.

Protogynous: flowering plants with receptive stigmas maturing before the anthers open.

Protoplast: protoplasm, as distinct from the cell wall.

Protothallus: hypothallus, the first formed stratum of a lichen.

Pseudo-antagonism: in orchids, antagonism of male bees against certain flowers.

Pseudobulb: one or several swollen internodes formed by some orchids often inhabited by ants.

Pseudocopulation: attempted copulation by some bees or flies with an orchid that resembles a bee or fly, which results in the pollination of the orchid from the pollen on the insect's body.

Pseudogall: swollen stipular thorns of species of the American and east African *Acacia* trees, which are preformed and not induced by the ants; the ants live inside.

Psilate: pollen with smooth walls and without conspicuous ornamentation.

Pyrethroids: synthetic chemicals similar in structure to pyrethrin.

Pyrolizidin alkaloids: substances present in wilting and dead plants, mostly of the family Boraginaceae, sought by certain distasteful or toxic insects.

Quadrilobate: with four lobes.

Quinones: group of chemical substances involved in the formation of some yellow or red pigments of insects and also in poisonous or repellent secretions.

Rachis: main flower stem to which the outer parts are attached; axis of a compound leaf.

Radicle: embryonic root; portion of the embryo of a seed below the cotyledons.

Repletes: among some ant species, an individual whose crop is greatly distended with liquid food.

Retinaculum: in plants, a small, viscid, gland-like body at the base of the stalk of a pollinium.

Rhizoids: rootlike structure or outgrowth of many mosses and thallophytes.

Ruderal: growing among rubbish or debris or in waste places.

Saccharose, sucrose: cane sugar.

Sacculine: isopod crustacea which is an internal parasite of crabs.

Saponine: any of a group of glucosides occurring in many plants.

Sapromyophilous: egg laying by flies in the flowers imitating the smell and color of rotten flesh.

Sapromyophily: insect-flower behavior involving carrion and dung insects.

Saprophagous: feeding on dead or decaying animal or vegetable matter.

Saprophyte: plant living on, or deriving its food from, dead organic matter.

Saproxylophagous: feeding on dead and rotten wood.

Saurophagy: feeding on reptiles.

Saxicolous: living on stones, rocks, and stone walls.

Sclerotized: hardening of the cuticle of an insect, involving the development of cross-links between protein chains.

Secondary substances: secondary metabolites, attractants, and repellents, responsible for host plant selection.

Secretory: concerned in the process of secretion.

Seismonastic: moving as a result of the stimulus of mechanical shock or vibration.

Seismonasty: response to a nondirectional mechanical stimulus.

Sensillum (plur. sensilla): small sensory setae in insects.

Sensillum basiconicum: basiconic sensory peg of insects.

Sensillum trichodeum: trichoid sensillum of insects.

Sesquiterpenes: any terpene with one and one-half times the terpene formula; moderately toxic, flammable, unsaturated liquid hydrocarbon found in essential oils and plant oleoresins.

Setal: of or pertaining to a seta.

Setiform: bristle- or seta-shaped.

Setose: furnished or covered with setae or still hair-like structures.

Silicaceous: showing a preference for flinty or sandy soils; impregnated with silicium, as in species of *Equisetum*, the horsetails.

Sinigrin: glucosid occuring in most Brassicaceae.

Skeletonize: refers here to a leaf eaten by an insect, leaving only the ribs and veins.

Solanaceous: of or pertaining to solanine containing plants.

Solanine: a poisonous, crystallizable alkaloid in many species of the plant genus *Solanum*.

Spadix: inflorescence consisting of a fleshy central column bearing many stamens below and many pistils above, surrounded by a large bract or spathe.

Sparteine: poisonous, colorless, oily alkaloid soluble in alcohol found in several species of Fabaceae.

Spathe: leaf-like colored bract surrounding the inflorescence of aroids and palms.

Spermatophore: capsule, sometimes with a stalk, formed by male colleterial glands, containing sperm to be transferred to female.

Spherules: small, spherical bodies.

Spinose: armed with thorny spines more elongate than echinate (bristly).

Spongeous: sponge-like; of a soft elastic tissue resembling a sponge.

Sporophore: of fungi; a general term for any structure producing and bearing spores.

Sporophyll: leaf that bears sporangia (ferns); stamens and carpels of flowering plants; see *Mega-* and *Microsporophylls*.

Sporulate, sporulation: act of forming spores.

Stamen: pollen-bearing organ of a flower.

Staminate: with, producing, or consisting of stamens.

Staminode: abortive stamen, sometimes eaten by pollinators.

Staminodia: imperfect organs occupying the position of and resembling stamens.

Staphyla (plur. staphylae): group of gongylidia; the swollen hyphae tips produced by fungi that live in symbiosis with Attini ants; see *Bromatia*

Stellate: with star-shaped hairs or setae.

Stigma (plur. stigmata): receptive part of the pistil.

Stipule: basal appendages of a leaf, a petiole; may be transformed into a swollen thorn (as in *Acacia* spp.); hence, stipular, meaning pertaining to the stipule.

Stolon: modified propagating underground stem.

Stoma (plur. stomata): opening in leaves of plants surrounded by guard cells which form an opening into internal air cavities; an opening in the stem of myrmecophilous plants to allow ants to penetrate inside when the prostoma has been pierced; pore in the epidermis of plants, present in large numbers.

Stomium: barbarism for stoma, but sometimes used.

Strobile, strobilus (plur. strobili): cone; a group of sporophylls with their sporangia more or less tightly packed around a central axis, forming a well-defined group.

Stylets: small style or stiff structure in the mouthparts of sucking insects.

Styloconic sensillum: olfactory cuticular sense organ consisting of one or more pegs.

Suffrutescent: describes a plant in which many of the branches die after flowering, leaving a persistent woody base.

Symbiont: organism living in symbiosis with another species.

Sympatric, sympatry: two or more populations of different species living in the same area.

Symphile, symphilic: symbiont; in particular, an insect or other kind of arthropod which is accepted to some extent by an insect colony (ants, termites).

Synapomorphy: shared derived features.

Synecthrans: animal living in the home of another and sharing its food, especially a guest in the nest of ants; treated generally with hostility or indifference.

Synergical, synergist: chemical substance used with an insecticide or other substance; will result in a greater total effect than the sum of their total individual effects.

Synoecete, synaecete: indifferently tolerated insect guests in an ant nest.

Systematics: study of classification, identification, evolution, genetics, ecology, and distribution of organisms; see *Taxonomy*.

Tagma: body section of segmented animals.

Tangoreceptors: trichoid sensilla, hair-like sensilla.

Tartrate: salt or ester of tartaric acid.

Taxon (plur. taxa): group of organisms irrespective of size or taxonomic rank; Latin barbarism.

Taxonomy: theory or practice of classifying organisms.

Termitophagous: eating termites.

Terpene: any hydrocarbon composed of two or more isoprene units; a moderately toxic, flammable, unsaturated liquid hydrocarbon found in essential oils and oleoresins.

Terpenoides: complex volatile compounds frequently employed as pheromones, similar to geraniol.

Terpineol: combustible, colorless liquid with a lilac scent derived from pine oil.

Thallus: plant body without differentiation into root, stem, and leaf; characteristic of thallophytes.

Therogamy: flower pollination by mammals such as rodents and marsupials.

Thiaminase: enzyme member of the vitamin-B complex.

Tomatine: glycosidal alkaloid obtained from leaves and stems of tomato plants.

Transgenic organisms: organisms into which genetic material from another organism has been experimentally transferred.

Transovarial transmission: transmission of microorganisms from one generation to the next by way of the egg.

Trichilium (plur. trichilia): setose cushion at the base of the petiole among which Müllerian bodies are formed (*Cecropia* spp.).

Trichomes: botanical, an outgrowth of plant epidermis which can absorb nutrients; also plant setae covering the domatia; zoological, tufts of setae carried by certain myrmecophilous insects which diffuse aromatic secretions, sometimes sweet and often licked by ants.

Tritrophic: interrelationships between a plant, a phytophagous insect, and a parasitoid; often the parasitoid is attached to the food plant of its host.

Trophic: of or pertaining to food or eating.

Trophosomes: food-bodies of myrmecophilous plants; there are Beltian (*Acacia* spp.), Müllerian (*Cecropia* spp.), Beccarian (*Macaranqa* spp.), and Delpinian (*Piper* spp.) food bodies.

Tubiform: in the shape of a tube.

Umbel: inflorescence with the pedicels arising at a common point but with the main axis bearing a flower with outer flowers blooming later than the inner flowers.

Univoltine: having one generation a year.

Urticating: nettling, causing a stinging or burning sensation of the skin when touched.

Utricule: more or less inflated, membranous, bladder-like envelope surrounding some fruits and fructification of fungi; bladder-like, inflated plant structure.

Vernalization: process of shortening the vegetative period of plants by treating the seeds.

Vesicant: blister raising or blistering substance found in the bodies of beetles of the families Meloidae, Staphylinidae, Cantharidae, and Oedemeridae.

Vexillar: see *Aposematic.*

Volatiles: readily vaporized.

Whorled: with more than two leaves at a node in a circle around it; arranged in whorls.

Xanthophyll: constituent of chlorophyll, yellow in color and insoluble in water.

Xenophagy: drastic change of the food habits of an animal.

Xenophily: opposite of, and rarer than, xenophoby; the acceptance by an insect of a foreign plant related to its usual host.

Xenophoby: rejection by many insects, mono- or oligophagous, of a foreign plant closely related to the usual host; it seems that the rejection is more or less proportional to the distance of the plant from its place of origin.

Xenotrophic: eating foreign food plants.

Xerophyte: plant or other organism adapted to dry conditions or environment.

Xylophagous: feeding on xylem or wood.

Zoidogamous, zoidogamy: fertilization of a flower by antherozoids, water-motile pollen.

Zoogamous, zoogamy: sexual reproduction in animals and plant with motile sex cells.

Zoophagous: consuming animals or animal products.

References

Abisgold, J. D. and Simpson, S. J. (1987). The physiology of compensation by locusts for changes in dietary problems. *J. Exp. Biol.* 129: 329–346.

Abrahamson, W. G., Ed. (1989). *Plant-Animal Interactions*. McGraw-Hill, New York: 481 pp.

Ahmad, S. (1983). *Herbivorous Insects: Host-Seeking Behaviour and Mechanisms*. Academic Press, New York: 257 pp.

Aiello, A. and Jolivet, P. (1996). Myrmecophily in Keroplatidae (Dipt. Sciaroidea). *J. N.Y. Entomol. Soc.* 104(3-4): 226–230.

Albert, V. A., Williams, S. E., and Chase, M. W. (1992). Carnivorous plant phylogeny and structural evolution. *Science*, 257: 1491–1495.

Alluaud, E. and Jeannel, E. (1922). Notes sur les galles d'*Acacia drepanolobium* in Santschi, *Voyage Alluaud et Jeannel en Afrique Orientale*: 97–98, 147–148.

Almeda, F. (1989). Five new berry-fruited species of American Melasteomataceae. *Proc. Calif. Acad. Sci.* 46(5): 137–150.

Ananthakrishnan, T. N. (1984). *Biology of Gall Insects*. Edward Arnold, London: 362 pp.

Ananthakrishnan, T. N. (1994). *Functional Dynamics of Phytophagous Insects*. Science Publishers, Lebanon, NM: 304pp.

Ananthakrishnan, T. N. and Gopichandran, R. (1993). *Chemical Ecology in Thrips Host-Plant Interactions*. Backhuys Publishers, Leiden: 125 pp.

Ananthakrishnan, T. N. and Raman, A. (1993). *Chemical Ecology of Phytophagous Insects*. Backhuys Publishers, Leiden: 332 pp.

Antor, R. J. and Garcia, M. B. (1995). A new mite-plant association: mites living amidst the adhesive traps of a carnivorous plant. *Oecologia*, 101(1): 51–54.

Aoxi, J. I. (1966). Epizoic symbiosis: an orbatid mite, *Symbioribates papuensis*, representing a new family, from cryptogamic plants growing on back of Papuan weevils. *Pacific Insects*, 8(1): 281–289.

Arditti, J. (1966). Orchids. *Sci. Am.* 214(1): 70–78.

Arditti, J., Ed. (1994). *Orchid Biology. VI. Reviews and Perspectives*. John Wiley & Sons, New York, 610 pp.

Armbruster, W. S. (1992). Phylogeny and the evolution of plant-animal interactions. *BioScience*, 42(1): 12–20.

Armstrong, J. A. (1979). Biotic pollination mechanisms in the Australian flora. *N.Z. J. Botany*, 17: 467–508.

Armstrong, J. E. and Irvine, A. K. (1990). Functions of staminods in the beetle-pollinated flowers of *Eupomatia laurina*. *Biotropica*, 22(4): 429–431.

Arnett, R. H., Jr. (1963). *The Beetles of the United States: Chrysomelidae*, The Catholic University of America Press, Washington, D.C.: 899–947.

Arnett, R. H., Jr. (1968). Pollen feeding by Oedemeridae. *Bull. Entomol. Soc. Am.* 14: 184.

Arnett, R. H., Jr. (1985). *American Insects*. Van Nostrand-Reinhold, New York: 850 pp.

Arnett, R. H., Jr. and Jacques, R. L., Jr. (1985). *Insect Life*. Prentice Hall, Englewood Cliffs, NJ: 354 pp.

Arnol'di, L. U., Zherikhin, V. V., Nikritin, L. M., and Ponomarenko, A. G. (1992). *Mesozoic Era Coleoptera*. Smithsonian Institution, Washington: 285 pp.

Atkins, M. D. (1980). *Introduction to Insect Behavior*. Macmillan, New York: 237 pp.

Atsatt, P. R. and O'Dowd, D. J. (1976). Plant defense guilds. *Science*, 193: 14–19.

Augner, M. (1994). Should a plant always signal its defense against herbivores? *Oikos*, 70(3): 323–332.

Ayal, Y. (1994). Time-lags in insect response to plant productivity significance for insect-plant interactions in deserts. *Ecol. Entomol.* 19(3): 207–214.

Baffray, M., Brice, F., and Danton, P. (1985). *Les Plantes Carnivores de France*. Sequences Publishers, Aigre: 154 pp.

Baffray, M., Brice, F., Danton, P., Tournier, J. P. (1989). *Nature et Culture des Plantes Carnivores*. Edisud, Aix: 178 pp.

Baker, H. G. and Baker, I. (1973). Amino acids in nectar and their evolutionary significance. *Nature*, 24: 543–545.

Balachowsky, A. S. (1951). *La Lutte Contre les Insectes*. Payot, Paris: 380 pp.

Balazuc, J. (1988). Laboulbeniales (*Ascomycetes*) parasitic on Chrysomelidae, in *Biology of Chrysomelidae*, Jolivet, P., Petitpierre, E., and Hsiao, T. H., Eds., Kluwer Academic, Dordrecht: 389–398.

Balazuc, J. (1989). Quelques insectes fossiles des diatomites de Saint Bauzile (Ardeche). *Bull. Soc. Linn. Lyon*, 58(8): 240–245.

Baldwin, I T. and Schultz, J. C. (1983). Rapid changes in tree leaf chemistry. Evidence for communication between plants. *Science*, 221: 277–279.

Balick, M. J. and Furth, D. G. (1978). Biochemical and evolutionary aspects of arthropod predation on ferns. *Oecologia*, 35(1): 55–89.

Bänziger, H. (1971). Bloodsucking moths of Malaya. *Fauna*, 1: 14–18.

Barbosa, P. and Letourneau, D. K., Eds. (1988). *Novel Aspects of Insect-Plant Interactions*. John Wiley & Sons, New York: 362 pp.

Barrett, S. C. (1987). Mimicry in plants. *Sci. Am.* 257(3): 76–83.

Barron, G. L. (1977). *The Nematode-Destroying Fungi*. Canadian Biological Publications, Guelph, Ontario: 140 pp.

Barth, F. B. (1985). *Insects and Flowers: The Biology of a Partnership*. Princeton University Press, NJ: 297 pp.

Bass, M. and Cherrett, J. M. (1994). The role of leaf-cutting ant workers (Hym. Form.) in fungus gardens maintenance. *Ecol. Entomol.* 19: 215–220.

Bass, M. and Cherrett, J. M. (1995). Fungal hyphae as a source of nutrients for the leaf-cutting ant *Atta sexdens*. *Physiol. Entomol.* 20(1): 1–6.

Bass, M. and Cherrett, J. M. (1996). Leaf-cutting ants (Form. Attini) prune their fungus to increase and direct its productivity. *Funct. Ecol.* 10: 55–61.

Basset, Y. (1990). The arboreal fauna of the rainforest tree *Argyrodendron actinophyllum* as sampled with restricted canopy fogging: composition of the fauna. *Entomologist,* 109: 173–183.

Basset, Y. (1992). Host specificity of arboreal and free-living insect herbivores in rain forests. *Biol. J. Linn. Soc.* 47: 115–133.

Basset, Y. (1994). Palatability of tree foliage to chewing insects: a comparison between a temperate and a tropical site. *Acta Oecologica,* 15(2): 181–191.

Basset, Y. and Samuelson, G. A. (1996). Ecological characteristics of an arboreal community of Chrysomelidae in Papua New Guinea, in *Chrysomelidae Biology,* vol. II, Jolivet, P. and Cox, M., Eds., SPB Academic Publishers, Amsterdam: 21 pp.

Basset, Y., Springate, N. D., and Samuelson, G. A. (1994). Feeding habits and range of body size: a case study in Papua New Guinea using arboreal leaf-beetles and weevils (Col. Chrys. Curcul.). *Mitt. Schweiz. Entomol. Ges.* 67(3–4): 347–361.

Batra, L. R. (1979). *Insect Fungus Symbiosis in Mutualism and Symbiosis.* John Wiley & Sons, New York: 276 pp.

Batra, S. W. and Batra, L. R. (1967). Fungus gardens of insects. *Sci. Am.* 217(5): 112–120.

Batra, S. W. and Batra, L. R. (1977). The fungus gardens of insects, in *The Insects,* Eisner, P. R. and Wilson, C. L., Eds., W.H. Freeman, San Francisco: pp. 203–210.

Beaman, J. H. and Adam, J. (1983). Observations on *Rafflesia* in Sabah. *Sabah Soc. J.* 7(3): 208–212.

Beaman, R. S., Decker, P. J., and Beaman, J. H. (1988). Pollination of *Rafflesia* (Rafflesiaceae). *Am. J. Bot.* 75(8): 1148–1182.

Beath, D. D. (1996). Pollination of *Amorphophallus johnsonii* (Araceae) by carrion beetles (*Phaeochrous amplus*) in a Ghanaian rain forest. *J. Trop. Ecol.* 12: 409–418.

Beattie, A. J. (1985). *The Evolutionary Ecology of Ant-Plant Mutualism.* Cambridge University Press, London: 182 pp.

Beccari, O. (1884). Piante ospitatrici ossia piante formicarie della Malesia e delle Papuasia. *Malesia,* vol. 2, 340 pp.

Beccari, O. (1885). Plantes a fourmis de l'archipel Indo-Malais et de la Nouvelle Guinée. *Arch. Ital. Biol.* 6(3): 41 pp.

Becerra, J. K. and Vanable, D. L. (1990). Rapid-terpene-bath and "squit gun" defense in *Bursera schlechtendalii* and the counterploy of Chrysomelid beetles. *Biotropica,* 22(3): 320–323.

Beck, S. D. (1965). Resistance of plants to insects. *Ann. Rev. Entomol.* 10: 207–232.

Becker, M. (1994). The female organs of symbiont transmission in the Eumolpinae, in *Novel Aspects of the Biology of Chrysomelidae,* Jolivet, P., Cox, M. L., and Petitpierre, E., Eds., Kluwer Academic, Dordrecht: pp. 363–370.

Behan, M. and Schoonhoven, L. M. (1978). Chemoreception of an oviposition deterrent associated with eggs in *Pieris brassicae. Ent. Exp. Appl.* 24: 163–179.

Belt, Th. (1874). *The Naturalist in Nicaragua.* The University of Chicago Press, Chicago: 403 pp.

Benedict, J. H., Wolfenbarger, D. A., Bryant, V. M., and George, D. M. (1991). Pollens ingested by boll weevils (Col. Curc.) in southern Texas and northeastern Mexico. *J. Ecol. Entomol.* 84(1): 128–131.

Benson, W. W. (1985). Amazon ant-plants, in *Amazonia*, Prance, G. T. and Lovejoy, T. E., Eds., Pergamon Press, Oxford: pp. 239–266.

Benson, W. W., Brown, K. S., and Gilbert, L. E. (1975). Coevolution of plants and herbivores: passion flower butterflies. *Evolution*, 29: 659–680.

Bentley, B. L. and Elias, T. (1983). *The Biology of Nectaries*. Columbia University Press, New York: 259 pp.

Benzing, D. H. (1970). An investigation of two bromeliad myrmecophytes, *Tillandsia butzii* Mez. and *T. caputmedusae* E. Morron, and their ants. *Bull. Torrey Bot. Club*. 97(2): 109–115.

Benzing, D. H. (1980). *The Biology of Bromeliads*, Mad River Press, Eureka: 305 pp.

Benzing, D. H. (1986). Foliar specialization for animal-assisted nutrition in Bromeliaceae, in *Insects and the Plant Surface*, Juniper, B. E. and Southwood, R., Eds., Edward Arnold, London: pp. 235–256.

Benzing, D. H. (1987). The origin and rarity of botanical carnivory. *Trends Ecol. Evol. (TREE)*, 2: 364–369.

Bequaert, J. (1922). Ants in their diverse relations to the plant world. *Bull. Am. Mus. Nat. Hist.* 45: 333–348.

Berenbaum, M. R. (1983). Coumarins and caterpillars: a case for coevolution. *Evolution*, 37: 163–179.

Berenbaum, M. R. (1995). The chemistry of defense: theory and practice. *Proc. Natl. Acad. Sci.* 92(1): 2–8.

Berg, C. C. (1990). Reproduction and evolution in *Ficus* (Moraceae). Traits connected with the adequate rearing of pollinators. *Mem. N.Y. Botan. Garden*, 55: 169–185.

Bernays, E. A. (1988). Host specificity in phytophagous insects; selection pressure from generalist predators. *Entom. Exp. Appl.* 49: 131–140.

Bernays, E. A., Ed. (1989–1994). *Insect-Plant Relationships*. I–V. CRC Press, Boca Raton, FL: (I) 164 pp., (II) 199 pp., (III) 258 pp., (IV) 240 pp., (V) 240 pp.

Bernays, E. A. (1992). Plant sterols and host-plant affiliations of herbivores, in *Insect-Plant Interactions*, Bernays, E. A., Ed., CRC Press, Boca Raton, FL: pp. 45–57.

Bernays, E. A. and Woodhead, S. (1982). Plant phenols utilized as nutrients by a phytophagous insect. *Science*, 216: 201–203.

Bernays, E. A. and Chapman, R. F. (1987). The evolution of deterrent responses in plant feeding insects, in *Perspectives in Chemoreception and Behavior*, Chapman, R. F., Bernays, E. A., and Stoffolano, J. G., Eds., Springer-Verlag, New York: pp. 159–173.

Bernays, E. A. and Chapman, R. F. (1994). *Host-Plant Selection by Phytophagous Insects*. Chapman and Hall, New York: 312 pp.

Bernays, E. A. and Graham, M. (1988). On the evolution of host specificity in phytophagous arthropods. *Ecology*, 69(4): 886–892.

Bernays, E. A. and Janzen, D. H. (1988). Saturnid and sphingid caterpillars: two ways to eat leaves. *Ecology*, 69 (4): 1153–1160.

Bernays, E. A. and Lee, J. C. (1988). Food aversion learning in the polyphagous grasshopper *Schistocerca americana*. *Physiol. Entomol.* 13: 131–137.

Blaney, W. M., Winstanley, C., and Simmonds, M. S. (1985). Food selection by locusts: an analysis of rejection behaviour. *Entomol. Exp. Appl.* 38: 35–40.

Blaney, W. M., Schoonhoven, L. M., and Simmonds, M. S. (1986). Sensitivity variations in insect chemoreceptors: a review. *Experientia*, 42: 13–19.

Blatter, E. (1928). Myrmecosymbiosis in the Indo-malayan flora. *J. India Bot. Soc.* 7: 176–185.

Bloch, P. (1965). Abnormal development in plants: a survey. *Handbuch der Plant Zenphysiologie.* 15(2): 145–183.

Blum, M. S. (1981). *Chemical Defenses of Arthropods.* Academic Press, New York: 562 pp.

Boardman, M., Askew, R. R., and Cook, L. M. (1974). Experiments on resting site selection by nocturnal moths. *J. Zool. Lond.* 172: 343–355.

Boer, G. de et al. (1977). Chemoreceptors in the preoral cavity of the tobacco hornworm, *Manduca sexta*, and their possible function in feeding behavior. *Entomol. Exp. Appl.* 21: 287–298.

Boer, G. de and Hanson, F. E. (1984). Foodplant selection and induction of feeding preference among host and non-host plants in larvae of the tobacco hornworm, *Manduca sexta. Entomol. Exp. Appl.* 35: 177–193.

Bogacheva, I. A. (1994). Leaf size selection by insects: a phenomenon created by random sampling. *Oikos*, 69(1): 119–124.

Boivin, G. and Vincent, Ch., Eds. (1986). Les relations insectes-plantes. *Rev. Entomol. Quebec*, 31(1–2): 86 pp.

Bolton, B. (1994). *Identification Guide to the Ant Genera of the World.* Harvard University Press, Boston: 222 pp.

Boppre, M. and Scherer, G. (1981). A new species of flea-beetle (Alticinae) showing male-biased feeding at withered *Heliotropium* plants. *Syst. Entomol.* 6: 347–354.

Borg-Karlson, A. K. (1990). Chemical and ethological studies of pollination in the genus *Ophrys* (Orchidaceae). *Phytochemistry*, 29(5): 1359–1387.

Borg-Karlson, A. K. and Tengo, J. (1986). Odor mimetism? Key substances in *Ophrys lutea, Andrena* pollination relationship Orchidaceae-Andrenidae. *J. Chemical Ecol.* 12(9): 1927–1941.

Boucher, D. H., Ed. (1985). *The Biology of Mutualism.* Croom Helm, London: 388 pp.

Bourdouxhe, L. and Jolivet, P. (1981). Nouvelles observations sur le complexe mimétique de *Mesoplatys cincta* Olivier (Col. Chrys.) au Sénégal. *Bull. Soc. Linn. Lyon*, 50(2): 46–48.

Bowers, W. S., Ohta, T., Cleere, J. S., and Marsella, P. A. (1976). Discovery of insect anti-juvenile hormones in plants. *Science*, 193: 542–547.

Brackenbury, J. (1995). *Insects and Flowers: A Biological Partnership.* Cassell Blandford, London, 160 pp.

Bradshall, W. E. and Creelman, R. E. (1984). Mutualism between the carnivorous purple pitcher plant and its inhabitants. *Am. Midl. Nat.* 112: 294–304.

Brener, A. G. and Silva, J. F. (1996). Leaf-cutter ants (*Atta laevigata*) aid the establishment success of *Tapirira velutinifolia* (Anacardiaceae) seedlings in a parkland savanna. *J. Trop. Ecol.* 12: 163–168.

Bronstein, J. L. (1994). Our current understanding of mutualism. *Q. Rev. Biol.* 69(1): 31–51.

Brooks, D. R. (1979). Testing the context and extent of host-parasite evolution. *Syst. Zool.* 28: 299–307.

Brooks, D. R. and McLennan, D. A. (1991). *Phylogeny, Ecology, and Behavior.* The University of Chicago Press: 434 pp.

Brooks, H. K. (1955). Healed wounds and galls on fossil leaves from the Wilcox deposits (Eocene) of western Tennessee. *Psyche*, 62: 1–255.

Brower, L. (1969). Ecological chemistry. *Sci. Am.* 220: 22–29.

Brues, C. T. (1924). The specificity of foodplants in the evolution of phytophagous insects. *Am. Nat.* 58: 127–144.

Brues, C. T. (1946). *Insect Dietary.* Harvard University Press, Boston: 466 pp.

Brues, C. T. (1951). Insects in amber. *Sci. Am.* 838 (November): 6 pp.

Bruin, J., Sabelis, M. W., and Dicke, M. (1995). Do plants tap SOS signals from their infested neighbours? *Trends Ecol. Evol. (TREE)*, 10(4): 167–170.

Buchmann, S. L. and Cane, J. H. (1989). Bees assess pollen returns while sonicating *Solanum* flowers. *Oecologia*, 81(4): 292–294.

Buckley, R. C. (1982). *Ant-Plant Interactions in Australia.* Junk, The Hague: 162 pp.

Buhr, H. (1964). *Bestimmungstabellan der Gallen (Zoo-und Phytocecidien) an Pflanzen Mittel-und Nordeuropas* (2 vols.). Gustav Fisher-Verlag, Iena.

Bultman, T. L. and White, J. F. (1988). Pollination of a fungus by a fly. *Oecologia*, 75: 317–319.

Burger, W. C. (1981). Why are there so many kinds of flowering plants? *Bioscience*, 31(8): 577–581.

Buscaloni, L. and Huber, J. (1900). Eine neue theorie des Amerisenpflanzen. *Beih. Bot. Centralbl.* 9(2): 85–88.

Buttiker, W. (1967). Biological notes on eye-frequenting moths from northern Thailand. *Mitt. Schweit Entomol. Gesell.* 39: 151–178.

Buttiker, W. (1973). Vorlaufige Beabachtungen an augenbesuchenden Schmetterlingen in der Elfenbeinkustes. *Rev. Suisse Zool.* 80: 1–43.

Campbell, B. C. (1989). On the role of microbial symbiotes in herbivorous insects, in *Insect-Plant Interactions*, Bernays, E. A., Ed., CRC Press, Boca Raton, FL: p. 3.

Campbell, R. K. and Eikenbary, R. D. (1990). *Aphid-Plant Genotype Interactions.* Elsevier, Amsterdam: 378 pp.

Cano, R. J. and Borucki, M. K. (1995). Revival and identification of bacterial spores in 25- to 40-million-year-old Dominican Amber. *Science*, 268(5213): 1060–1064.

Cano, R. J., Poinar, H. N., Pieniazek, N. J., Acra, A., and Poinar, G. (1993). Amplification and sequencing of DNA from a 120–135 million-year-old weevil. *Nature*, 363: 536–538.

Carpenter, F. M. (1956). Personal communication.

Carpenter, F. M. (1971). Adaptations among Paleozoic insects. *Proc. North Am. Paleontol. Conv.* 1: 1236–1251.

Carpenter, F. M. (1992). *Treatise on Invertebrate Paleontology:* vol. 3, *Arthropods;* vol. 4, *Superclass Hexapoda.* The Geological Society of America, Boulder, CO: 695 pp.

Carter, W. (1962). *Insects in Relation to Plant Disease.* Wiley Interscience, New York: 705 pp.

Cartier, J. J. (1968). Factors of host-plant specificity and artificial diets. *Bull. Ent. Soc. Am.* 14: 18–21.

Castro, N. M. (1986). Estudos morfologicos dos orgaos vegetativos de especies de *Paepalanthus* Kunth (Eriocaulaceae) de Serra do Cipo. (Minas Gerais). Masters Thesis, University of São Paulo, Brazil.

Cates, R. G. and Orians, G. H. (1975). Successional status and the palatability of plants to generalized herbivores. *Ecology*, 56: 410–418.

Cates, R. G. and Rhoades, D. F. (1977). *Prosopis Leaves as a Resource for Insects*, in *Mesquite*, Simpson, S. J., Ed., Dowden, Hutchinson and Ross, Stroudsberg, PA.

Chaloner, W. G. and MacDonald, R. (1980). *Plants Invade the Land*. The Royal Scottish Museum, Edinburgh: 17 pp.

Chapman, R. F. and Bernays, E. A., Eds. (1978). Insects and host plants. *Proc. Int. Symp. Entomol. Exp. Appl.* 24: 201–766.

Chapman, R. F. and Boer, G. de, Eds. (1995). *Regulatory Mechanisms in Insect Feeding*. Chapman and Hall, New York: 398 pp.

Chase, M. W. and Hills, H. G. (1992). Orchid phylogeny, flower sexuality, and fragrance-seeking. *BioScience*, 42(1): 43–49.

Chattopadhyay, A. K. and Sukul, N. (1994). Anti-predator strategy of larval aggregation pattern in *Aspidomorpha miliaris* (Chrys. Col.). *Entomology*, 9(3–4): 125–130.

Cheers, G. (1983). *Carnivorous Plants*. Globe Press, Melbourne: 95 pp.

Cherrett, J. M. (1972). Some factors involved in the selection of vegetable substrate of *Atta cephalotes* (L.) (Hym. Form.) in tropical rain forest. *J. Animal Ecol.* 41(3): 647–660.

Cherrett, J. M. (1986). The biology, pest status, and control of leafcutting ants, in *Agricultural and Zoological Reviews I*: 1–37.

Cherrett, J. M. and Cherrett, F. J. (1989). A bibliography of the leaf-cutting ants, *Atta* and *Acromyrmex* spp. up to 1975. *Odnri Bull.* 14: 1–58.

Cherrett, J. M., Powell, R. J., and Stradling, D. J. (1988). The mutualism between leaf-cutting ants and their fungus, in *Insect-Fungus Interactions*, Wilding, N. et al., Eds., Academic Press, New York: 93–120.

Chown, S. L. and Scholtz, C. H. (1989). Cryptogam herbivory in Curculionidae (Col.) from the sub-antarctic prince Edward Islands. *Col. Bull.* 43(2): 165–169.

Chrispeels, M. and Sadavu, D. (1977). *Plants, Food, and People*. W.H. Freeman, San Francisco: 278 pp.

Claridge, M. F., Reynolds, W. J., and Wilson, M. R. (1977). Oviposition behavior and food plant discrimination in leaf hoppers of the genus *Oncopsis*. *Ecol. Entomol.* 2: 19–25.

Claridge, M. F. and Wilson, M. R. (1978). British insects and trees: a study in island biogeography or insect/plant coevolution. *Am. Nat.* 112: 451–456.

Coleman, E. (1928). Pollination of an Australian orchid by the male ichneumonid *Lissopimpla semipunctata* Kirby. *Trans. Entomol. Soc. Lond.* 76: 533–539.

Coleman, E. (1929). Pollination of *Cryptostylis subulata* (Labill.) Reichb. *Victorian Nat.* 46: 62–66.

Coleman, E. (1930). Pollination of some west Australian orchids. *Victorian Nat.* 46: 203–206.

Coleman, J. S. and Leonard, A. S. (1995). Why it matters where on a leaf a folivore feeds. *Oecologia*, 101(3): 324–328.

Coley, P. D. (1987). Patronas en las defensas de las plantas: porque los herbivores prefieren ciertas especies? *Rev. Biol. Tron.* 35(suppl. 1): 151–164.

Collinge, S. K. and Louda, S. M. (1989). Influence of plant phenology on the insect herbivore bittercress interactions. *Oecologia*, 79: 111–116.

Collinson, M. E. (1986). Catastrophic vegetation changes. *Nature*, 324: 112.

Compton, S. G. (1987). *Aganais speciosa* and *Danaus chrysippus* (Lep.) sabotage the latex defences of their host plants. *Ecol. Entomol. Appl.* 12: 115–118.

Compton, S. G. (1989). Sabotage of latex defenses by caterpillars feeding on fig trees. *Suid-Afrikaanse Tydskrif vir Wetenskap*, 85(9): 605–606.

Conn, E. E. (1981). Secondary plant products, in *The Biochemistry of Plants*, vol. 7, Conn, E. E., Ed., Academic Press, New York: 798 pp.

Cooper, E. (1965). *Insects and Plants*. Butterworth Press, London: 142 pp.

Corbet, S. A., Chapman, H., and Saville, N. (1988). Vibratory pollen collection and flower form: bumble bees on *Actinidia*, *Symphytum*, *Borago*, and *Polygonatum*. *Functional Ecol.* 2: 147–155.

Cornet, B. (1993). Dicot-like leaf and flowers from the late triassic tropical Neward supergroup rift zone, USA. *Mod. Geol.* 19: 81–99.

Cota, J. H. (1993). Pollination syndromes in the genus *Echinocereus*: a review. *Cactus Succulent J.* 65(1): 19–26.

Cottrell, C. B. (1984). Aphytophagy in butterflies: its relationship to myrmecophily. *Zool. J. Linn. Soc.* 79: 1–57.

Craighead, F. C. (1921). Hopkins Host Selection Principle as related to certain cerambycid beetles. *J. Agric. Res.* 22(4): 189–220.

Crane, P. R. (1993). Time for the angiosperms. *Nature*, 366: 631–632.

Crawley, M. J. (1983). *The Herbivory: The Dynamics of Animal-Plant Interactions*. University of California Press, Berkeley: 437 pp.

Crepet, W. L. (1979a). Insect pollination: a paleontological perspective. *BioScience*, 29: 102–108.

Crepet, W. L. (1979b). Some aspects of the pollination biology of Middle Eocene angiosperms. *Rev. Palaeobot. Palynol.* 27: 213–238.

Crepet, W. L. and Friis, E. M. (1987). The evolution of insect pollination in angiosperms, in *The Origins of Angiosperms and Their Consequences*, Friis, E. M., Chaloner, W. G., and Crane, P. R., Eds., Cambridge University Press, London: pp. 181–201.

Crepet, W. L., Dilcher, D. L., and Potter, F. W. (1974). Early angiosperm flowers. *Science*, 185(4153): 781–784.

Crepet, W. L., Friis, E. M., and Nixon, K. C. (1991). Fossil evidence for the evolution of biotic pollination. *Phil. Trans. R. Soc. Lond.* 333: 187–195.

Cresswell, J. E. (1991). Capture rates and composition of insect prey of the pitcher plant, *Sarracenia purpurea*. *Am. Midl. Nat.* 125(1): 1–9.

Cribb, J. W. (1969). Pollination, with special reference to Queensland plants. *Queensland Nat.* 19(4–6): 70–75.

Croizat, L. (1960). *Principia Botanica*. Caracas: pp. 25–256.

Cronquist, A. (1981). *The Integrated System of Classification of Flowering Plants*. Columbia University Press, New York: 1262 pp.

Cronquist, A. (1988). *The Evolution and Classification of Flowering Plants*. The New York Botanical Garden, NY: 555 pp.

Crowson, R. A. (1981a). A Russian palaeoentomological view of insect phylogeny. *Entomol. Gener.* 7(1): 105–108.

Crowson, R. A. (1981b). *The Biology of Coleoptera*. Academic Press, London: 802 pp.

Crowson, R. A. (1989). *Relations of Beetles to Cycads*. International Congress of Coleopterists, EAC, Barcelona: 13–15.

Crowson, R. A. (1991). The relations of Coleoptera to Cycadales, in *Advances in Coleopterology*, Zunino, M., Belles, X., and Blas, M., Eds., University of Barcelona: pp. 13–28.

Daccordi, M. (1994). Notes for phylogenetic study of Chrysomelidae, with description of new taxa and a list of all the known genera, in *Proc. 3rd Int. Symp. Chrysom. Beijing*, Furth, D., Ed., Backhuys Publishers, Leiden: pp. 60–84.

Dafni, A. (1984). Mimicry and deception in pollination. *Ann. Rev. Ecol. Syst.* 15: 259–278.

Dafni, A. (1985). Pollination in *Orchis* and related genera: evolution from reward to deception, in *Orchid Biology Reviews and Perspectives*, Arditti, J., Ed. Comstock Publishers, Ithaca, NY: pp. 81–104.

Dafni, A. and Eisikowitch, D., Eds. (1990). Advances in pollination and ecology. *Israel J. Bot. Spec. Issue*, 39(1–2).

Daly, H. V., Doyen, J. T., and Ehrlich, P. R. (1978). *Introduction to Insect Biology and Diversity.* McGraw-Hill, New York, 564 pp.

Darlington, A. (1975). *The Pocket Encyclopedia of Plant Galls in Colour.* Blandford Press, Poole: 191 pp.

Darwin, C. (1875). *Insectivorous Plants.* Appleton and Co., New York: 462 pp.

Darwin, F. (1877). On the glandular bodies of *Acacia spaherocephala* and *Cecropia peltata*, serving as food bodies. *J. Linn. Soc. Bot.* 15: 398–409.

Davidson, D. W. (1988). Ecological studies of Neotropical ant gardens. *Ecology,* 69(4): 1138–1152.

Davidson, D. W. and Epstein, W. W. (1990). Epiphytic associations with ants, in *Phylogeny and Physiology of Epiphytes*, Lüttge, V., Ed., Springer-Verlag, Berlin: pp. 200–233.

Davidson, D. W. and McKey, D. (1993a). The evolutionary ecology of symbiotic ant-plant relationships. *J. Hym. Res.* 2(1): 13–83.

Davidson, D. W. and McKey, D. (1993b). Ant-plant symbioses: stalking the Chuyachaqui. *Trends Ecol. Evol. (TREE)*, 8(9): 326–332.

de Abbott, H. C. (1987). Comparative chemistry of higher and lower plants, *Am. Nat.* 21: 800–810.

de Kesel, A. (1995). Relative importance of direct and indirect infection in the transmission of *Laboulbenia slackensis* (Ascom. Laboulb.). *Belq. J. Bot.* 128(2): 124–130.

de Wilde, J, Ed. (1983). *Proc. 5th Int. Symp. on Insect-Plant Relationships.* Pudoc, Wageningen: pp. 464pp.

de Wilde, J. and Schoonhoven, L. M. (1969). *Insect and Host Plants.* North-Holland, Amsterdam: 610 pp.

Dejean, A., Olmsted, I., and Snelling, R. R. (1995). Tree-epiphyte-ant relationships in the low inundated forest of Sian Katan Biosphere Reserve, Quintana Roo, Mexico. *Biotropica*, 27(1): 57–70.

Delevoryas, T. (1971). Biotic provinces and the Jurassic-Cretaceous period floral transition, in *Proc. North Am. Paleontol. Conv. Part I*: pp. 1660–1674.

Delpino, F. (1886). Funzione myrmecophila del regno vegetale. Prodroma d'una monografia delle piante formicarie. *Mem. R. Acc. Scienze Ist. Bologna.* 7: 215–323.

Denlinger, D. L. (1994). The beetle tree. *Am. Entomol.* 4: 168–171.

Denno, R. F. and Dingle, H., Eds. (1981). *Insect Life History Patterns Habitat and Geographic Variation.* Springer-Verlag, Berlin: 225 pp.

Denno, R. F. and McClure, M. S. (1983). *Variable Plants and Herbivores in Natural and Managed Systems.* Academic Press, New York: 717 pp.

Dethier, V. G. (1941). The function of the antennal receptors in lepidopterous larvae. *Biol. Bull. Woods Hole, MA.* 80: 403–414.

Dethier, V. G. (1954). Evolution of feeding preferences in phytophagous insects. *Evolution*, 8(1): 33–54.

Dethier, V. G. (1960). The designation of chemicals in terms of responses they elicit from insects. *J. Econ. Entomol.* 53: 134–136.

Dethier, V. G. (1970). Chemical interactions between plants and insects, in *Chemical Ecology*, Sendheimer, E. and Simeone, J. B., Eds., Academic Press, New York: pp. 83–102.

Dethier, V. G. (1976). *Man's Plague.* Dawin Press, Princeton: 237 pp.

Dethier, V. G. (1977a). The role of chemosensory patterns in the discrimination of food plants. *Colloques CNRS, Paris*, 265: 103–114.

Dethier, V. G. (1977b). *Gustatory Sense of Complex Mixed Stimuli by Insects.* Olfaction and Taste Symposium, Paris: pp. 323–331.

Dethier, V. G. (1978a). Other tastes, other worlds. *Science*, 201: 224.

Dethier, V. G. (1978b). Studies on insect/plant relations, past and future. *Entomol. Exp. Appl.* 24: 559–566.

Dethier, V. G. (1980). Food-aversion learning in two polyphagous caterpillars, *Diacrisia virginica* and *Estigmene congrua*. *Physiol. Entomol.* 5: 321–325.

Dethier, V. G. (1988). Induction and aversion-learning in polyphagous arctiid larvae (Lep.) in an ecological setting. *Can. Entomol.* 120: 125–131.

Dethier, V. G. and Schoonhoven, L. M. (1968). Evaluation of evaporation by cold and humidity receptors in caterpillars. *J. Insect Physiol.* 14: 1049–1054.

Dethier, V. G. and Schoonhoven, L. M. (1969). Olfactory coding by lepidopterous larvae. *Entomol. Exp. Appl.* 12: 535–543.

Dethier, V. G. and Kuch, J. H. (1971). Electrophysiological studies of gustation in lepidopterous larvae. I. Comparative sensitivity to sugars, amino acids, and glycosides. *Z. Vergl. Physiol.* 72: 343–363.

Dethier, V. G. and Yost, M. T. (1979). Oligophagy and absence of food aversion learning in tobacco hornworms, *Manduca sexta*. *Physiol. Entomol.* 4: 125–130.

Dethier, V. G. and Canjar, R. M. (1982). Candidate codes in the gustatory system of caterpillars. *J. Gen. Physiol.* 79: 543–569.

Dieleman, F. L. (1969). Effects of gall midge infestation on plant growth and growth regulating substances. *Entomol. Exp. Appl.* 12(12): 745–749.

Dillon, P. M., Lowrie, S., and McKey, D. (1983). Disarming the "evil woman": petiole constricting by a sphingid larva circumvents mechanical defenses of its host plant. *Biotropica*, 15(2): 112–116.

Dixon, A. F. G. (1973). Biology of aphids, in *Studies of Biology*, 44, Edward Arnold, London: 150 pp.

Dixon, A. F. G. (1985). *Aphid Ecology.* Chapman and Hall, New York: 157 pp.

Dixon, A. F. G. and Kindlmann, P. (1994). Optimum body size in aphids. *Ecol. Entomol.* 19(2): 121–126.

Dixon, C. A., Erickson, J. M., Kellett, D. N., and Rothschild, M. (1978). Some adaptations between *Danaus plexippus* and its food plant with notes on *Danaus chrysippus* and *Euploea core* (Ins. Lep.). *J. Zool. Lond.* 185: 437–467.

Dobson, C. H., Dressler, R. L., Hills, H. G., Adams, R. M., and Williams, N. H. (1969). Biologically active compounds in orchid fragrance. *Science*, 164: 1243–1248.

Doss, R. P., Shanks, C. H., Chamberlain, J. D., and Garth, J. K. L. (1987). Role of leaf hairs in resistance of a clone of beach strawberry, *Fragaria chiloensis*, to feeding by adult black vine weevil, *Otiorhynchus sulcatus* (Col., Curc., O.). *Environ. Entomol.* 16(3): 764–768.

Doss, R. P., Shanks, C. H., and Chamberlain, J. D. (1988). The influence of leaf pubescence on the resistance of selected clones of beach strawberry (*Fragaria chiloensis* (L.) Duchesne) to adult black vine weevils (*Otiorhynchus sulcatus* F.). *Scientia Horticulturae*, 34: 47–54.

Downes, J. A. (1955). The food habits and description of *Atrichopogon pollinivorus* sp. n. (Dipt. Ceratop.). *Trans. R. Entomol. Soc. Lond.* 106: 439–453.

Dreger-Jauffret, F. and Shorthouse, J. D. (1992). Diversity of gall-inducing insects and their galls, in *Biology of Insect-Induced Falls,* Shorthouse, J. and Rohfritsch, O., Eds., Oxford UniversityPress, London: pp. 8–33.

Dumpert, K. (1978). *The Special Biology of Ants.* Pitman Advanced Publishers, London: 298 pp.

Dumpert, K. (1981). *The Social Biology of Ants.* Pitman Advanced Publishers, Boston: 298 pp.

Dussourd, D. E. (1990). The vein drain: or how insects outsmart plants. *Nat. Hist.* 2: 44–48.

Dussourd, D. E. and Eisner, T. (1987). Vein-cutting behavior: insect counterploy to the latex defense of plants. *Science*, 237(4817): 898–901.

Dussourd, D. E. and Denno, R. F. (1991). Deactivation of plant defense: correspondence between insect behavior and secretory canal architecture. *Ecology*, 72: 1383–1396.

Dwyer, J. D. (1980). Rubiaceae, part I, in *Flora of Panama*, Woodson, R. E. and Schery, R. W., Eds., Missouri Botanical Garden, St. Louis, MO: pp. 222–245.

Eastop, V. F. (1981). *Coevolution of Plants and Insects*, in *The Evolving Biosphere*, Forey, P. L., Ed., Cambridge University Press, London: pp. 179–190.

Edmunds, G. F. and Alstad, D. N. (1978). Coevolution in insect herbivores and conifers. *Science*, 19(9): 941–945.

Edwards, P. B. and Wanjura, W. J. (1989). Eucalyptus-feeding insects bite off more than they can chew: sabotage of induced defences. *Oikos*, 54: 246–248.

Edwards, P. J. and Wratten, D. S. (1980). *Ecology of Insect Plant Interactions*, Edward Arnold, London: 60 pp.

Ehrlich, P. R. and Raven, P. H. (1964). Butterflies and plants: a study in coevolution. *Evolution*, 18: 586–608.

Ehrlich, P. R. and Raven, P. H. (1977). Butterflies and plants, in *The Insects*, Eisner, T. and Wilson, C. L., Eds., W.H. Freeman, San Francisco: pp. 195–202.

Ehrlich, P. R. and Murphy, D. D. (1988). Plant chemistry and host range in insect herbivores. *Ecology*, 69(4): 908–909.

Eisner, T. (1972). Chemical ecology: on arthropods and how they live as chemists. *Verh. Deutsch. Zool. Gesellsch.* 65: 123–137.

Eisner, T. and Aneshansley, D. J. (1983). Adhesive strength of the insect trapping glue of a plant (*Befaria racemosa*). *Ann. Entomol. Soc. Am.* 76: 295–298.

Eisner, T. and Meinwald, J. (1995). Chemical ecology. *Proc. Natl. Acad. Sci.* 92: 1.

Elias, S. A. (1991). Insects and climate change. Fossil evidence from Rocky mountains. *BioScience*, 41(8): 552–559.

Elias, S. A. (1994). *Quaternary Insects and Their Environments*, Smithsonian Institution Press, Washington, D.C.: 256 pp.

Ellis, A. G. and Midgley, J. J. (1996). A new plant-animal mutualism involving a plant with sticky leaves and a resident hemipteran insect. *Oecologia*, 106: 478–481.

Elton, C. S. (1958). *The Ecology of Invasions by Animals and Plants*. Methuen and Co., London: 181 pp.

Emberger, L. (1960). *Traité de Botanique. II. Les Végétaux Vasculaires*. Masson et Cie, Paris: 1540 pp.

Emberger, L. (1968). *Les Plantes Fossiles Dans Leurs Rapports Avec les Végétaux Vivants*. Masson et Cie, Paris: 758 pp.

Emmons, L. H., Nias, J., and Briun, A. (1991). The fruit and consumers of *Rafflesia keithii* (Rafflesiaceae). *Biotropica*, 23(2): 197–199.

Engler, A. and Prantl, C. (1930). *Die Naturlichen Pflanzenfamilien*. Engelmann, Leipzig: 492 pp.

Erickson, P. R. (1965). Insects and Australian orchids. *Australian Orchid Rev.* 30: 172–174.

Erickson, P. R. (1978). *Plants of Prey*. University of Western Australia Press, Osborne Park: 300 pp.

Erickson, P. R. and Feeny, P. (1974). Sinigrin: a chemical barrier to the black swallowtail butterfly, *Papilio polyxenes*. *Ecology*, 55: 103–111.

Erwin, T. L. (1982). Tropical forests: their richness in Coleoptera and other arthropod species. *Coleop. Bull.* 36(1): 74–75.

Erwin, T. L. (1983a). Beetles and other insects of tropical forest canopies at Manaus, Brasil, sampled by insecticidal fogging, in *Tropical Rain Forest Ecology and Management*, Sutton, S. L. et al., Eds., Blackwell Scientific, Oxford: pp. 59–75.

Erwin, T. L. (1983b). Tropical forest canopies: the last biotic frontier. *Bull. Entomol. Soc. Am.* 29: 14–19.

Fabre, J. H. (1989). *Souvenirs Entomologiques* (2 vols.), Robert Laffont, Paris: 1138 and 1187 pp. (original edition, 1925).

Faegri, K. and van der Pijl, L. (1966). *The Principles of Pollination Ecology*. Pergamon Press, New York: 291 pp.

Farrell, B. D., Mitter, C., and Futuyma, D. J. (1992). Diversification at the insect-plant interface. *BioScience*, 42(1): 34–42.

Feeny, P. (1970). Seasonal changes in oak leaf tannins and nutrients as a cause of spring feeding by winter moth caterpillars. *Ecology*, 51: 565–581.

Feeny, P. (1976). Plant apparency and chemical defense, in *Biochemical Interrelation Between Plants and Insects*, Wallace, J. M. and Mansell, R. L., Eds., Plenum Press, New York: pp. 1–40.

Feeny, P. (1977). Defensive ecology of the Cruciferae. *Ann. Miss. Bot. Garden*, 64: 221–234.

Feeny, P. (1992). The evolution of chemical ecology: contributions from the study of herbivorous insects, in *Herbivores: Their Interactions With Secondary Metabolites*, Rosenthal, G. A. and Berenbaum, M. R., Eds., Academic Press, New York: pp. 1–44.

Feeny, P., Rosenberry, L. B., and Carter, M. (1983). Chemical aspects of oviposition behavior in butterflies, in *Herbivorous Insects: Host-Seeking Behaviour and Mechanisms*, Ahmad, S., Ed., Academic Press, New York: pp. 27–47.

Fessler, A. (1982). *Fleischfressende Pflanzen für Haus und Garten.* Kosmos, Stüttgart: 112 pp.

Fiala, B., Maschwitz, U., and Pong, T. Y. (1991). The association between *Macaranga* trees and ants in southeast Asia, in *Ant-Plant Interactions*, Huxley, E. R. and Cutler, D. F., Eds., Oxford University Press, London: pp. 263–270.

Fiala, B., Grunsky, H., Maschwitz, U., and Linsenmair, K. E. (1994). Diversity of ant-plant interactions: protective efficacy in *Macaranga* species with different degrees of ant association. *Oecologia*, 97: 186–192.

Fiedler, K. (1996) Lycaenid butterflies and plants: is myrmecophily associated with amplified host/plant diversity? *Ecol. Entomol.* 19(1): 79–82.

Fiedler, K. and Saam, C. (1995). Ants benefit from attending facultatively myrmecophilous Lycaenidae caterpillars: evidence from a survival study. *Oecologia*, 104: 316–322.

Figueira, J. E. C. (1989). Associacao entre *Paepalanthus bromeloides* Silv. (Eriocaulaceae) Aranhas e Termites. Masters thesis, University of Campinas, Brazil.

Figueira, J. E. C. and Vasconcellos-Neto, J. (1993). Reproductive success of *Latrodectus geometricus* (Theridiidae) on *Paepalanthus bromeloides* (Eriocaulaceae): rosette size, microclimate and prey capture. *Ecotropicos*, 5: 1–10.

Figueira, J. E. C., Vasconcellos-Neto, J., and Jolivet, P. (1994). Une Nouvelle plante protocarnivore, *Paepalanthus bromelioides* Simv. (Eriocaulaceae) du Brésil. *Rev. Ecol.* 49: 3–9.

Filip, V., Dirzo, R., Maass, J. M., and Sarukhan, J. (1995). Within- and among-year variation in the levels of herbivory on the foliage of trees from a Mexican tropical deciduous forest. *Biotropica*, 27(1): 78–86.

Finch, S. (1978). Volatile plant chemicals and their effect on host plant finding by the cabbage root fly (*Delia brassicae*). *Entomol. Exp. Appl.* 24: 150–159.

Fink, L. S. (1995). Food-plant effects on colour morphs of *Eumorpha fasciata* caterpillars (Lep. Sphingidae). *Biol. J. Linn. Soc.* 56: 423–437.

Fish, D. (1976). Structure and Composition of the Aquatic Invertebrate Community Inhabiting Epiphytic Bromeliaceae in South Florida and the Discovery of an Insectivorous Bromeliad. Ph.D. thesis, University of Florida, Gainesville: p. ix.

Flanders, S. E. (1962). Did the caterpillar exterminate the giant reptile? *J. Res. Lep.* 1(1): 85–88.

Flenley, J. (1993). The origins of diversity in tropical rain forests. *Trends Ecol. Evol. (TREE)*, 8(4): 119–120.

Folgaralt, P. J. and Davidson, D. W. (1994). Antiberbivore defenses of myrmecophytic cecropia under different light regimes. *Oikos*, 71(2): 305–320.

Folliot, R. (1964). *Contribution a l'étude de la Biologie des Cynipides Gallicoles (Hym. Cynip.).* Masson et Cie, Paris: 564 pp.

Fonseca, C. R. Herbivory and the long-lived leaves of an Amazonian ant-tree. *J. Ecol.* 82(4): 833–842.

Forbes, W. T. M. (1924). The occurrence of nygmata in the wings of Insecta Holometabola. *Entomol. News*, 35: 230–232.

Forey, P. L. (1980). *The Evolving Biosphere.* British Museum of Natural History, London: 305 pp.

Foster, W. A. (1996). Duelling aphids: intraspecific fighting in *Astacopteryx minute* (Homopt. Hormaphid.). *Animal Behav.* 51: 645–655.

Fowler, S. V. and Lawton, J. H. (1984). Trees don't talk: do they even murmur? *Antenna*, 8(2): 69–71.

Fox, L. R. (1988). Diffuse convolution within complex communities. *Ecology*, 69(4): 906–907.

Fraenkel, G. (1959). The *raison d'être* of secondary plant substances. *Science*, 129: 1466–1470.

Fraenkel, G. (1969) Evaluation of our thoughts on secondary plant substances. *Entomol. Exp. Appl.* 12: 473–486.

Frank, J. H. (1983). Bromeliad phytotelmata and their biota, especially mosquitoes, in *Phytotelmata: Terrestrial Plants as Hosts for Aquatic Insect Communities*, Frank, J. H. and Lounibos, L. P., Eds., Plexus Publishers, Medford: pp. 101–128.

Frank, J. H. and Lounibos, L. P., Eds. (1983). *Phytotelmata: Terrestrial Plants as Hosts for Aquatic Insect Communities*, Plexus Publishers, Medford: 293 pp.

Frank, J. H. and O'Meara, G. F. (1984). The bromeliad *Catopsis berteroniana* traps terrestrial arthropods but harbors *Wyeomyia* larvae (Dipt. Cul.), *Florida Entomol.* 67(3): 418–424.

Frazier, J. L. and Hanson, F. E. (1986). Electrophysiological recording and analysis of insect chemosensory responses, in *Insect-Plant Interactions*, Miller, J. R. and Miller, T. A., Eds., Springer-Verlag, New York: pp. 285–330.

Free, J. B. (1993). *Insect Pollination of Crops*, 2nd. ed., Academic Press, London: 768 pp.

Freitas, A. V. L. and Oliveira, P. S. (1996). Ants as selective agents on herbivore biology: effects on the behavior of a non-myrmecophilous butterfly. *J. Animal Ecol.* 65: 205–210.

Friis, E. M., Chaloner, W. G., and Crane, P. R., Eds. (1987) *The Origins of Angiosperms and Their Biological Consequences*. Cambridge University Press, London: 358 pp.

Fry, G. L. A. and Wratten, S. D. (1979). Insect-plant relationships in ecological teaching. *J. Biol. Educ.* 13: 267–274.

Fukuyama, K., Maeto, K., and Kirton, L. (1994). Field tests of a balloon suspended trap system for studying insects in the canopy of tropical rain forests. *Ecol. Res.* 9(3): 357–360.

Funk, D. J., Futuyma, D. J., Orti, G., and Meyer, A. (1995). Mitochondrial DNA sequences and multiple data sets: a phylogenetic study of phytophagous beetles (Chrysomelidae: *Ophraella*). *Mol. Biol. Evol.* 12(4): 627–640.

Fushun, Y., Evans, K. A., Stevens, L. H., van Beek, T. A., and Schoonhoven, L. M. (1990). Deterrents extracted from the leaves of *Ginkgo biloba*: effects on feeding and contact chemoreceptors. *Entomol. Exp. Appl.* 54: 57–64.

Futuyma, D. J. (1992). Genetics and the phylogeny of insect-plant interactions, in *Proc. 8th Int. Symp. Insect-Plant Relationships*, Menken, S. B. J., Visser, J. H. E., and Harrewigen, P., Eds., Kluwer Academic, Dordrecht: pp. 191–200.

Futuyma, D. J. (1983a). Selective factors in the evolution of host choice by phytophagous insects, in *Herbivorous Insects: Host-Seeking Behaviour and Mechanisms*, Ahmad, S., Ed., Academic Press, New York: pp. 227–245.

Futuyma, D. J. (1983b). Evolutionary interactions among herbivorous insects and plants, in *Coevolution*, Futuyma, D. J. and Slatkin, M., Eds., Sinauer Associates, Sunderland, MA: pp. 207–231.

Futuyma, D. J. (1994). Genetic and phylogenetic aspects of host plant affiliation in *Ophraella* (Chrys. Gal.), in *Novel Aspects of the Biology of Chrysomelidae*, Jolivet, P., Cox, M. L., and Petitpierre, E., Eds., Kluwer Academic, Dordrecht: pp. 249–258.

Futuyma, D. J. and Slatkin, M., Eds. (1983). *Coevolution*. Sinauer Associates, Sunderland, MA: 555 pp.

Futuyma, D. J. and Peterson, S. C. (1985). Genetic variation in the use of resources by insects. *Ann. Rev. Entomol.* 30: 217–238.

Futuyma, D. J. and McCafferty, S. S. (1990). Phylogeny and the evolution of host plant associations in the leaf beetle genus *Ophraella*. *Evolution*, 44(8): 1885–1913.

Futuyma, D. J. and Keese, M. C. (1992). Evolution and coevolution of plants and phytophagous arthropods, in *Herbivores: Their Interactions With Secondary Plant Metabolites*, 2nd ed., Rosenthal, G. A. and Berenbaum, M. R., Eds., Academic Press, New York: pp. 439–475.

Futuyma, D. J., Herrmann, C., Milstein, S., and Keese, M. C. (1993). Apparent transgenerational effects of host plant in the leaf beetle *Ophraella notulata* (Col. Chrys.). *Oecologia*, 96: 365–372.

Gall, J. C. (1995). *Paléoécologie: Paysages et Environnements Disparus*. Masson et Cie, Paris: 239 pp.

Gallun, R. L. (1972). Genetic relationships between host plants and insects. *J. Environ. Qual.* 1: 259–265.

Garber, P. A. (1988). Foraging decisions during nectar feeding by tamarin monkeys in Amazonian Peru. *Biotropica*, 20(2): 100–106.

Gaston, K. J. (1993). Herbivory at the limits. *Trends Ecol. Evol. (TREE)*, 8(6): 193–194.

Gerson, U. (1969). Moss-arthropod associations. *Bryologist*, 72: 495–500.

Gerson, U. (1973a). Lichen-arthropod associations. *Lichenologist*, 5: 434–443.

Gerson, U. (1973b). The associations between pteridophytes and arthropods. *Fern. Gaz.* 12(1): 29–45.

Gerson, U. (1974). The associations of algae and arthropods. *Rev. Algologique*, 2: 18–41, 213–247.

Gerson, U. (1982). Bryophytes and invertebrates, in *Bryophyte Ecology*, Smith, P. M., Ed., Chapman and Hall, London: pp. 291–332.

Gerson, U. and Seaward, M. R. D. (1977). Lichen-invertebrate associations, in *Lichen Ecology*, Seaward, M. R. D., Ed., Academic Press, London: pp. 69–119.

Gibson, R. W. (1971). Glandular hairs providing resistance to aphids in certain wild potato species. *Ann. Appl. Biol.* 68: 113–119.

Gibson, R. W. (1974). Aphid-trapping glandular hairs in hybrids of *Solanum tuberosum* and *S. berthaultii* (Solanaceae). *Potato Res.* 17: 152–154.

Gibson, R. W. and Turner, R. H. (1977). Insect-trapping hairs on potato plants. *PANS*, 22(3): 272–277.

Gilbert, L. E. (1972). Pollen feeding and reproductive biology of *Heliconius* butterflies. *Proc. Nat. Acad. Sci. USA*, 69(6): 1403–1407.

Gilbert, L. E. (1980). Ecological consequences of a coevolved mutualism between butterflies and plants, in *Coevolution of Animals and Plants*, Gilbert, L. E. and Raven, P. M., Eds., University of Texas Press, Austin: pp. 210–240.

Gilbert, L. E. and Raven, P. M., Eds. (1980). *Coevolution of Animals and Plants*. University of Texas Press, Austin: 246 pp.

Giridhar, G., Santosh, G., and Vasudevan, P. (1988). Antitermite properties of *Calotropis* latex. *Pesticides*: pp. 31–33.

Givnish, T. J. (1989). Ecology and evolution of carnivorous plants, in *Plant-Animal Interactions*, Abrahamson, W. G., Ed., McGraw-Hill, New York: 481 pp.

Givnish, T. J., Burkhardt, E. L., Rappel, R. E., and Weintraub, J. D. (1984). Carnivory in the bromeliad *Brocchinia reducta*, with a cost/benefit model for the general restriction of carnivorous plants to sunny, moist, nutrient-poor habitats. *Am. Nat.* 124: 479–497.

Godfray, H. C. J. (1994). *Parasitoids: Behavioral and Evolutionary Ecology.* Princeton University Press, NJ: 473 pp.

Godfray, H. C. J., Agassiz, J. L., Nash, D. R., and Layton, J. H. (1995). The recruitment of parasitoid species to two invading herbivores. *J. Animal Ecol.* 64(3): 393–402.

Godman, M. (1979). *Island Ecology.* Chapman and Hall, London: 79 pp.

Graham, L. E. (1993). *Origin of Land Plants.* John Wiley & Sons, New York: 287 pp.

Grassé, P. P. (1949). Insectes, in *Traité de Zoologie*, Grassé, P. P., Ed., Masson et Cie, Paris: 1117 pp.

Grassé, P. P. (1951). Insectes Supérieurs et Hémiptéroides, in *Traité de Zoologie* (2 vols.), Grassé, P. P., Ed., Masson et Cie, Paris: 1948 pp.

Grassé, P. P. (1982). *Termitologia.* Masson et Cie, Paris: 676 pp.

Greathead, D. J. (1983). The multi-million dollar weevil that pollinates oil palms. *Antenna*, 7(8): 105–107.

Greigh-Smith, P. (1986). The trees bite back. *New Scientist*, (1 May): 33–35.

Gressitt, J. L. et al. (1965). Flora and fauna on backs of large Papuan moss-forest weevils. *Science*, 150(3705): 1833–1835.

Gressitt, J. L. (1966a). The weevil genus *Panthorytes* (Coleopt.) involving cacao pests and epizoic symbiosis with cryptogamic plants and microfauna. *Pacific Insects*, 8(4): 15–65.

Gressitt, J. L. (1966b). Epizoic symbiosis: the Papuan weevil genus *Gymnopholus* (Leptopiinae) symbiotic with cryptogamic plants, oribatid mites, rotifers and nematodes. *Pacific Insects*, 8(1): 221–280.

Gressitt, J. L. (1966c). Epizoic symbiosis: cryptogamic plants growing on various weevils and on a colydiid beetle in New Guinea. *Pacific Insects*, 8(1): 294–297.

Gressitt, J. L. (1969). Epizoic symbiosis. *Entomol. News*, 80(1): 15.

Gressitt, J. L. (1970). Papuan weevil genus *Gymnopholus*: second supplement with studies in epizoic symbiosis. *Pacific Insects*, 12(4): 753–762.

Gressitt, J. L. (1977). Papuan weevil genus *Gymnopholus*: a third supplement with studies in epizoic symbiosis. *Pacific Insects*, 17(23): 179–185.

Gressitt, J. L. and Sedlacek, J. (1967). Papuan weevil genus *Gymnopholus*: supplement and further studies in epizoic symbiosis. *Pacific Insects*, 481–500.

Grinfeld, E. K. (1975). Anthophily in Coleoptera and a critique of the cantharophily hypothesis. *Entomol. Oborz.* 54(3): 507–514.

Grounds, R. (1980). *The Private Life of Plants.* Davis-Poynter, London: 205 pp.

Grzimek, B., Ed. (1976). *Encyclopedia of Ecology.* Van Nostrand-Reinhold, New York, 705 pp.

Gutierrez-Ochoa, M., Camino-Lavin, M., Castreion-Ayala, F., and Jimenez-Perez, A. (1993). Arthropods associated with *Bromelia hemisphaerica* (Bromeliaceae) in Morelos, Mexico. *Florida Entomol.* 76(4): 616–621.

Hainsworth, F. R. (1988). Criteria for efficient energy use: using exceptions to prove rules, in *Proc. Int. 100 Do-G Meeting, Bonn*: pp. 311–316.

Hainsworth, F. R. (1989). "Fast food" vs. "haute cuisine": painted ladies, *Vanessa cardui* (L.), select food to maximize net meal energy. *Funct. Ecol.* 3: 701–707.

Halle, F. and Blanc, P. (1990). *Mission Radeau des Cimes.* Dept. Xylochimie, Fondation Elf, Paris: 231 pp.

Halle, F. and Pascal, D., Eds. (1991). *Biologie d'Une Canopée de Foret Equatoriale.* II. Fondation Elf, Paris: 288 pp.

Hanski, I. and Cambefort, Y., Eds. (1991). *Dung Beetles Ecology.* Princeton University Press, NJ: 481 pp.

Hanson, F. E. (1983). The behavioral and neurophysiological basis of food plant selection by lepidopterous larvae, in *Herbivorous Insects: Host-Seeking Behaviour and Mechanisms,* Ahmad, S., Ed., Academic Press, New York: pp. 3–23.

Harborne, J. B. (1972). *Phytochemical Ecology.* Academic Press, New York: 272 pp.

Harborne, J. B. (1977). *Introduction to Ecological Biochemistry.* Academic Press, London: 243 pp.

Harborne, J. B. (1978). *Biochemical Aspects of Plant and Animal Coevolution.* Academic Press, New York: 425 pp.

Hartzell, A. (1967). Insect ectosymbiosis, in *Symbiosis,* Mark-Henry, S., Ed., Academic Press, New York: pp. 107–140.

Haslett, J. R. (1989). Adult feeding by holometabolous insects: pollen and nectar as complementary nutrient sources for *Rhingia campestris* (Dipt. Syrph.). *Oecologia,* 81: 361–363.

Hatcher, P. E., Paul, N. D., Ayres, P. G., and Whittaker, J. B. (1994). Interactions between *Rumex* spp., herbivores and a rust fungus: *Gastrophysa viridula* grazing reduces subsequent infection by *Uromyces rumicis. Funct. Ecol.* 8(2): 265–272.

Haukioja, E. (1982). Inducible defences of white birch to a geometric defoliator, *Epirrita autumnata,* in *Proc. 5th Int. Symp. Insect-Plant Relations,* Visser, J. H. and Minks, A. K., Eds., PUDOC, Wageningen: pp. 199–203.

Haukioja, E. and Niemala, P. (1977). Retarded growth of a geometric larva after mechanical damage to leaves of its host tree. *Ann. Zool. Fern.* 14: 48–52.

Haukioja, E. and Niemela, P. (1979). Birch leaves as a resource for herbivores. Seasonal occurence of increased resistance in foliage after mechanical damage of adjacent leaves. *Oecologia,* 39: 151–159.

Hawkes, C., Patton, S., and Coaker, T. H. (1978). Mechanisms of host plant finding in adult cabbage root fly, *Delia brassicae. Entomol. Exp. Appl.* 24: 219–227.

Hawkins, C. P. and MacMahon, J. A. (1989). Guilds, the multiple meanings of a concept. *Ann. Rev. Entomol.* 34: 423–451.

Heads, M., Craw, R. C., and Gibbs, G. W., Eds. (1984). Croizat's pangeography and principia botanica. Search for a novel biological synthesis. *Tuatara,* 27: 1–75.

Heard, T. A. (1994). Behaviour and pollinator efficiency of stingless bees on macadamia flowers. *J. Apic. Res.* 33(4): 191–198.

Hedin, P. A., Ed. (1983). Plant resistance to insects. *Am. Chem. Soc. Symp.* 208: 375 pp.

Hedstrom, I. and Thulin, M. (1986). Pollination by a hugging mechanism in *Vigna vexillata* (Leg. Papil.). *Plant Syst. Ecol.* 154: 275–283.

Hegarty, M. P. (1992). Defence systems in plants: variants on don't bug me. *Queensland Nat.* 31(3–4): 58–62.

Heim, R. (1977). *Termites et Champignons.* Boubée, Paris: 190 pp.

Heinrich, B. and Raven, P. H. (1972). Energetics and pollination ecology. *Science,* 176: 597–602.

Heinrich, O. (1979). *Bumble Bee Economics.* Harvard University Press, Boston: 245 pp.

Hendry, L. B., Kostelc, J. G., Hindenlang, D. M., Wichmann, J. K., Fix, C. J., and Korzeniowski, S. H. (1976). Chemical messengers in insects and plants. *Recent Adv. Phytochem.* 10: 351–384.

Hering, M. (1935–37). *Die Blattminen Mittel-und Nord-Europas Einschliesslich Englands.* Junk, Den Haag: 631 pp.

Hering, M. (1950a). Die oligophagie phytophager Insekten als Hinweiss auf eine verwandtschaft der Rosaceae mit den Familien der Amentiferae, in *Trans. 9th Int. Congr. Entomol. Stockholm, 1949*: 74–79.

Hering, M. (1950b). Monophagie und Xenophagie. Die Nahrungswahl phytophager Insekten and die geographische Herkunft ihrer wirtspflanzen. *Naturwissenschaften,* 37(23): 531–536.

Hering, M. (1951). *Biology of the Leaf Miners.* Junk, Den Haag: 420 pp.

Hering, M. (1957). *Bestimmungstabellen der Blattminen van Europa Einschliesslich des Mittelmeerbeckens und der Kanarischen Inseln* (3 vols.). Junk, Den Haag: 1406 pp.

Heslop-Harrison, Y. (1976). Carnivorous plants a century after Darwin. *Endeavour,* 35(126): 114–122.

Heslop-Harrison, Y. (1978). Carnivorous plants. *Sci. Am.* 238(2): 104–115.

Heywood, V. M. (1975). *Taxonomy and Biology.* Academic Press, New York: 370 pp.

Hill, D. (1994). *Agricultural Entomology.* Timber Press, London: 635 pp.

Hill, J. K., Hamer, K. C., Lace, L. A., and Banham, W. M. T. (1995). Effects of selective logging on tropical forest butterflies on Buru, Indonesia. *J. Appl. Ecol.* 32: 754–760.

Hill, T. A. (1977). *The Biology of Weeds.* Edward Arnold, London: 64 pp.

Hocking, B. (1970). Insect associations with swollen thorn acacias. *Trans. Roy. Entomol. Soc. Lond.* 122(7): 211–255.

Hocking, B. (1975). Ant-plant mutualism: evolution and energy, in *Coevolution of Insects and Plants,* Gilbert, L. E. and Raven, P. M., Eds., University of Texas Press, Austin: pp. 78–89.

Hodkinson, I. D. and Hughes, M. K. (1982). *Insect Herbivory.* Chapman and Hall, New York: 77 pp.

Holloway, J. D. and Herbert, P. D. N. (1979). Ecological and taxonomic trends in macrolepidopteran host plant selection. *Biol. J. Linn. Soc.* 112: 229–251.

Holttum, R. E. (1977). *Plant Life in Malaya.* Longman, London: 254 pp.

Horn, J. M., Lees, D. C., Smith, N. G., Nash, R. J., Fellows, L.E., and Bell, E. A. (1987). The *Urania, Omphalea* interaction: host-plant secondary chemistry, in *Insects-Plants,* Labeyrie, V., Fabres, C., and Lachaise, D., Eds., Junk, Dordrecht: 394 pp.

Hosozawa, S., Kato, N., Munakata, K., and Chen, Y. L. (1974). Antifeeding active substances for insects in plants. *Agric. Biol. Chem.* 38(4): 1045–1048.

Hovanitz, W. (1959). Insects and plant galls. *Sci. Am.* (Nov.): 151–162.

Howe, H. F. and Westley, L. C. (1988). *Ecological Relationships of Plants and Animals.* Oxford University Press, London: 273 pp.

Hsiao, T. H. (1969). Chemical basis of host selection and plant resistance in oligophagous insects. *Entomol. Exp. Appl.* 12: 777–788.

Hsiao, T. H. (1978). Host plant adaptations among geographic populations of the Colorado potato beetle. *Entomol. Exp. Appl.* 24: 437–447.

Hsiao, T. H. (1986). Specificity of certain Chrysomelids for Solanaceae, in *Solanaceae: Biology and Systematics*, d'Arcy, W. G., Ed., Columbia University Press, New York: pp. 345–377.

Hsiao, T. H. (1988). Host specificity, seasonality and bionomics of *Leptinotarsa* beetles, in *Biology of Chrysomelidae*, Jolivet, P, Petitpierre, E., and Hsiao, T. H., Eds., Kluwer Academic, Dordrecht: pp. 581–599.

Hsiao, T. H. (1989). Host-plant affinity in relation to phylogeny of *Leptinotarsa* beetles. *Entomography*, 6: 413–422.

Hsiao, T. H. and Fraenkel, G. (1968). Selection and specificity of the Colorado potato beetle for solanaceous and non-solanaceous plants. *Ann. Entomol. Soc. Am.* 61: 493–502.

Hughes, L. and Westoby, M. (1992). Capitula on stick insect eggs and elaiosomes on seeds: convergent adaptations. *Funct. Ecol.* 6: 642–648.

Hughes, N. F. (1976). *Palaeobiology of Angiosperm Origin*. Cambridge University Press, New York: 242 pp.

Hulley, P. E. (1968). Caterpillar attacks plant mechanical defence by moving trichomes before feeding. *Ecol. Entomol.* 13: 239–241.

Humber, F. A. (1996). Fungal pathogens of the Chrysomelidae and prospects for their use in biological control, in *Chrysomelidae Biology* (2 vols.), Jolivet P. and Cox, M. L., Eds., SPB Academic Publishers, Amsterdam: pp. 93–115.

Hutchinson, J. (1959). *The Families of Flowering Plants*, vols. I and II. Clarendon Press, Oxford: 510 + 282 pp.

Hutchinson, J. (1964). *Evolution and Phylogeny of Flowering Plants*. Academic Press, New York, 717 pp.

Huxley, A. (1974). *Plant and Planet*. Allen Lane, London: 428 pp.

Huxley, C. R. (1991). Ants and plants: a diversity of interactions, in *Ant-Plant Interactions*, Huxley, C. R. and Cutler, D. F., Eds., Oxford University Press, London: pp. 1–11.

Huxley, C. R. and Cutler, D. F., Eds. (1991). *Ant-Plant Interactions*. Oxford University Press, London: 601 pp.

Huxley, C. R. and Jebb, M. H. P. (1991). The tuberous epiphytes of the Rubiaceae. 1A. New subtribe: the Hydnophytinae. *Blumea* 36: 1–20.

Huxley, C. R. and Jebb, M. H. P. (1993). The tuberous epiphytes of the Rubiaceae. 5. A revision of *Myrmecodia*. *Blumea* 37: 271–334.

Inoue, K. (1993). Evolution of mutualism in plant-pollinator interactions on islands. *J. Biosci.* 18(4): 525–536.

Ishikawa, S. (1966). Electrical response and function of a bitter substance receptor associated with the maxillary sensilla of the silkworm, *Bombyx mori* L. *J. Cell. Physiol.* 67: 1–11.

Ishikawa, S., Tazima, Y., and Hirao, Y. (1963). Responses of the chemoreceptors of maxillary sensory hairs in a "non-preference" mutant of the silkworm. *J. Sericult. Sci. Jpn.* 32 125–129.

Ishikawa, S., Hirao, T., and Arai, N. (1969). Chemosensory basis of host plant selection in the silkworm. *Entomol. Exp. Appl.* 12: 544–554.

Jacobs, M. (1966). On domatia, the view points and some facts. *Kon. Meded. Akad. Wet.*, Ser. C, 69: 275–289.

Jacques, R. L. (1988). *The Potato Beetles*. St. Lucie Press, Boca Raton, FL: 144 pp.

Jaeger, P. (1971). Contribution a l'étude de la biologie florale des Asclépiadacées, le *Calotropis procera*. *Bull. IFAN, Dakar*, 33(A1): 32–43.

Jaeger, P. (1976). Relations entre insectes et plantes, in *Traite de Zoologie: Insectes,* Grassé, P. P., Ed., Masson et Cie, Paris: pp. 799–842.

Jaenike, J. (1982). Environmental modification of oviposition behavior in *Drosophila. Am. Nat.* 119(6): 784–802.

Jaffe, K., Michelangeli, F., Gonzalez, J. M., Miras, B., and Ruiz, M. C. (1992). Carnivory in pitcher plants of the genus *Heliamphora* (Sarraceniaceae). *New Phytologist*, 122: 733–744.

Jaffe, K., Pavis, C., Vansuyt, G., and Kermarrec, A. (1989). Ants visit extrafloral nectaries of the orchid *Spathonlotis plicata* Blume. *Biotropica*, 21(3): 278–279.

Jaffe, K., Blum, M. S., Fales, H. M., Mason, N. T., and Cabrera, A. (1995). On insect attractants from pitcher plants of the genus *Heliamphora* (Sarraceniaceae). *J. Chem. Ecol.* 21(3): 379–384.

Jaffe, K., Ramos, C., and Issa, S. (1995). Trophic interactions between ants and termites that share common nests. *Ann. Entomol. Soc. Am.* 88(3): 328–333.

Janzen, D. H. (1966). Coevolution of mutualism between ants and acacias in Central America. *Evolution* 20(3): 249–275.

Janzen, D. H. (1967). Interaction of the bulls horn (*Acacia cornigera*) with ant inhabitant (*Pseudomyrmex ferruginea*) in eastern Mexico. *Kans. Univ. Sci. Bull.* 47: 315–558.

Janzen, D. H. (1969). Allelopathy by myrmecophytes: the ant *Azteca* as an allelopathic agent of cecropia. *Ecology* 50(1): 147–153.

Janzen, D. H. (1972). Protection of *Barteria* (Passifloraceae) by *Pachysima* ants (Pseudomyrmecinae) in a Nigeria rain forest. *Ecology* 53(5): 885–892.

Janzen, D. H. (1974). Epiphytic myrmecophytes in Sarawak: mutualism through the feeding of plants by ants. *Biotropica* 6: 237–259.

Janzen, D. H. (1975). *The Ecology of Plants in the Tropics.* Edward Arnold, London: 66 pp.

Janzen, D. H. (1977). Why don't ants visit flowers? *Biotropica* 9: 252.

Janzen, D. H. (1980). When is it coevolution? *Evolution*, 34: 611–612.

Janzen, D. H., Ed. (1983). *Costa Rican Natural History.* The University of Chicago Press: 816 pp.

Janzen, D. H. (1988). On the broadening of insect-plant research. *Ecology* 69(4): 905.

Jarzembowski, E. A. (1985). On the track of giant dragonflies. *Antenna,* 9(3): 126–127.

Janzembowski, E. A. and Ross, A. (1994). Progressive palaeontology. *Antenna,* 18(3): 123–126.

Jeannel, R. (1979). *Paléontologie et Peuplement de la Terre,* 2nd ed., Boubée, Paris: 101 pp.

Jennings, D. M. (1975). *Symbiosis.* Cambridge University Press, London: 633 pp.

Jepson, P. C., and Healy, T. P. (1988). The location of floral nectar sources by mosquitoes: an advanced bioassay for volatile plant odours and initial studies with *Aedes aegypti. Bull. Entomol. Res.* 78: 641–650.

Jermy, T. (1958). Untersuchungen uber Auffinden und Wahl der Nahrung beim Kartoffelk afer (*Leptinotarsa decemlineata* Say). *Entomol. Exp. Appl.* 1: 197–208.

Jermy, T., Ed. (1976a). *The Host-Plant in Relation to Insect Behaviour and Reproduction,* Plenum Press, New York: 322 pp.

Jermy, T. (1976b).Insect-host plant relationships: coevolution or sequential evolution? *Symp. Biol. Hung.* 16: 109–113.

Jermy, T. (1983). Multiplicity of insect antifeedants in plants, in *Natural Products for Innovative Pest Management*, Whitehead, A. T. and Bowers, W. S., Eds., Pergamon Press, Oxford: pp. 223–236.

Jermy, T. (1984). Evolution of insect/host plant relationships. *Am. Nat.* 124: 609–630.

Jermy, T. (1987). The role of experience in the host selection of phytophagous insects, in *Perspectives in Chemoreception and Behavior*, Chapman, R. F., Bernays, E. R., and Stoffolano, J. G., Eds., Springer Verlag, New York: pp. 143–157.

Jermy, T. (1988). Can predation lead to narrow food specialization in phytophagous insects? *Ecology*, 68(4): 902–504.

Jermy, T., Hanson, F. E., and Dethier, V. G. (1969). Induction of specific food preference in lepidopterous larvae. *Entomol. Exp. Appl.* 11: 211–230.

Joel, D. M. (1985). Leaf anatomy of *Caltha dionaeifolia* Hooker (Ranunculaceae). Is this species carnivorous? *Botanical J. Linn. Soc.* 90: 243–252.

Joel, D. M. (1988). Mimicry and mutualism in carnivorous pitcher plants Sarraceniaceae, Nepenthaceae, Cephalotaceae, and Bromeliaceae. *Biol. J. Linn. Soc.* 35: 185–197.

Joel, D. M., Juniper, B. E., and Dafni, A. (1985). Ultraviolet patterns in the traps of Carnivorous plants. *New Phytol.* 101: 585–593.

Johnson, D. K. and Lanza, J. (1991). Effects of amino acids in floral nectars on the behavior and fecundity of *Pieris rapae*. *Bull. Ecol. Soc. Am.* 72(suppl. 2): 154–155.

Johnson, K. S. and Scriber, J. M. (1994). Geographic variation in plant allelochemicals of signifiance to insect herbivores, in *Functional Dynamics of Phytophagous Insects*, Ananthakrishnan, T. N., Ed., Science Publishers, Lebanon, NM: pp. 7–31.

Johnson, W. T. and Lyon, H. H. (1976). *Insects that Feed on Trees and Shrubs*, Cornell University Press, Ithaca: 464 pp.

Jolivet, P. (1952). Quelques données sur la myrmécophilie des Clytrides (Col. Chrys.). *Bull. Inst. R. Soc. Nat. Belg.* 28(8): 1–12 pp.

Jolivet, P. (1954). *Phytophagie et Sélection Trophique*. Livre Jub. V. Van Straelen, Brussels: pp. 1101–1134.

Jolivet, P. (1955). Recherches sur les organes facettiques des ailes des Insectes. *Bull. Inst. R. Soc. Nat. Belg.* 31(17): 1–23.

Jolivet, P. (1967). Les Alticides vénéneux de l'Afrique du Sud. *L'Entomol.* 23(4): 100–111.

Jolivet, P. (1971a). Une extraordinaire concentration de Ténébrionides dans é arboretum de Phu Kae, pres Sara Buri (Thailande). *Bull. Ann. Soc. Entomol. Belg.* 106(10–11): 323–326.

Jolivet, P. (1971b). La Nouvelle Guinée Australienne: Introduction écologique et entomologique. *Cah. Pacifique* 15: 41–70.

Jolivet, P. (1973). Les plantes myrmecophiles du sud est Asiatique. *Cah. Pacifique* 17: 41–65.

Jolivet, P. (1979). Les Chrysomelidae (Col.) des Citrus et apparentes (Rutaceae) en zone tempérée et tropicale. *Bull. Soc. Linn. Lyon*, 43(4): 197–200, 245–256.

Jolivet, P. (1980a). *Les Insectes et l'Homme*. Presses Universitaires de France: 126 pp.

Jolivet, P. (1980b). Les mannes: entomologie et botanique. *Bull. Soc. Linn. Lyon*, 49(9): 17–22.

Jolivet, P. (1983). Un hémimyrmécophyte à Chrysomélides (Col.) du sud est Asiatique, *Clerodendrum fragrans* (Vent.) Wild. (Verbenaceae). *Bull. Soc. Linn. Lyon*, 52(8): 242–261.

Jolivet, P. (1984). *Phaedon fulvescens* (Weise Col. Chrys.).Un auxiliaire possible dans le contrôle des rubus aux tropiques. *Bull. Soc. Linn. Lyon*, 53(7): 235–246.

Jolivet, P. (1986). *Les Fourmis et les Plantes: Un Exemple de Coevolution.* Boubée, Paris: 254 pp.

Jolivet, P. (1987). *Les Plantes Carnivores.* Le Rocher, Paris: 127 pp.

Jolivet, P. (1991). Ants, plants and beetles: a triangular relationshop, in *Ant-Plant Interactions*, Huxley, C. R. and Cutler, D. F., Eds., Oxford University Press, London: 601 pp.

Jolivet, P. (1992). *Insects and Plants: Parallel Evolution and Adaptations*, 2nd ed., St. Lucie Press, Boca Raton, FL: 190 pp.

Jolivet, P. (1993a). La Serra do Cipo au Brésil. *Bull. ACOREP*, 17: 7–12.

Jolivet, P. (1993b). Mimétisme comportemental sous les tropiques. *Bull. ACOREP*, 18: 29–36.

Jolivet, P. (1994). Physiological colour changes in tortoise beetles, in *Novel Aspects of the Biology of Chrysomelidae*, Jolivet, P., Cox, M. L., and Petitpierre, E., Eds., Kluwer Academic, Dordrecht: pp. 331–335.

Jolivet, P. (1995). A status report of the species of *Timarcha* (Col. Chrys.). *Insecta Mundi*, 9(1–2): 153–154.

Jolivet, P. (1996a). *Ant Plants: An Example of Coevolution* (enlarged edition). Backhuys Publishers, Leiden: 303 pp.

Jolivet, P. (1996b). *Biologie des Coleopteres Chrysomélides*, Boubée, Paris: 279 pp.

Jolivet, P. and Petitpierre, E. (1973). Plantes-hôtes connues des *Timarcha* Latr. Quelques considerations sur les raisons possibles du trophisme collectif. *Bull. Soc. Entomol. Fr.* 78(1–2): 9–25.

Jolivet, P. and Petitpierre, E. (1976). Les Plantes-hôtes connues des *Chrysolina* (Col. Chrys.). Essai sur les types de sélection trophique. *Ann. Soc. Entomol. Fr. (N.S.)*, 12(1): 123–149.

Jolivet, P. and van Parys, E. (1977). Un cas inédit de mimétisme agressif entre un Chrysomélide (*Mesoplatys cincta* Olivier) et un carabique *Cyaneodinodes ammon* (F.) (Col.). *Bull. Soc. Linn. Lyon*, 46(6): 168–180.

Jolivet, P. and Vasconcellos-Neto, J. (1993). Un genre aptère de Coléopteres Chrysomélides: *Elytrosphaera* en voie d extinction et sa distribution dans le SE brésilien. *Nouv. Rev. Entomol. (N.S.)*, 10(4):321–325.

Jolivet, P. and Hawkeswood, T. J. (1995). *Host Plants of the Chrysomelidae of the World: An Essay About the Relationships Between the Leaf-Beetles and Their Food Plants.* Backhuys Publishers, Leiden: 261 pp.

Jolivet, P. and Cox, M. L., Eds. (1996). *Chrysomelidae Biology* (3 vols.). SPB Academic Publishers, Amsterdam: 444 + 465 + 365 pp.

Jolivet, P., Petitpierre, E., and Hsiao, T. H., Eds. (1988). *Biology of Chrysomelidae.* Kluwer Academic, Dordrecht: 615 pp.

Jolivet, P., Cox, M. L., and Petitpierre, E., Eds. (1994). *Novel Aspects of the Biology of Chrysomelidae.* Kluwer Academic, Dordrecht: 582 pp.

Jones, C. G., Hopper, R. F., Coleman, J. S., and Krischik, V. A. (1993). Control of systematically induced herbivore resistance by plant vascular architecture. *Oecologia*, 93(3): 452–456.

Jones, D. A. (1972). Cyanogenic glycosides and their function, in *Phytochemical Ecology*, Harborne, J. B., Ed., Academic Press, London: 272 pp.

Jones, D. L. (1970). The pollination of *Corybas diemenicus*. *Victorian Nat.* 87: 372–374.

Jones, O. T. and Coaker, T. H. (1978). A basis for host plant finding in phytophagous insects. *Entomol. Exp. Appl.* 24: 272–284.

Jones, R. A. (1994). Ants feeding directly on plant sap. *Br. J. Entomol. Nat. Hist.* 7(4) 139–140.

Jong, R. and de Visser, J. H. (1988a). Specificity-related suppression of responses to binary mixtures in olfactory receptors of the Colorado potato beetle. *Brain Res.* 447: 18–24.

Jong, R. and de Visser, J. H. (1988b). Integration of olfactory information in the Colorado potato beetle brain. *Brain Res.* 447: 10–17.

Juniper, B. E., Robins, R. J., and Joel, D. M. (1989). *The Carnivorous Plants*. Academic Press, London: 353 pp.

Juniper, B. E. and Southwood, R., Eds. (1986). *Insects and the Plant Surface*. Edward Arnold, London: 368 pp.

Kahn, D. M. and Cornell, H. V. (1989). Leaf-miners, early abcission and parasitoids: a tritrophic interaction. *Ecology*, 70(5): 1219–1226.

Kasasian, L. (1971). *Weed Control in the Tropics*. Leonard Hill, London: 307 pp.

Kato, M. (1993). Floral biology of *Nepenthes gracilis* (Nepentaceae) in Sumatra. *Am. J. Botany*, 80(8): 924–927.

Kennedy, C. (1986). Tiptoeing through the trichomes. *Antenna*, 10(2): 75–78.

Kennedy, J. S. (1953). Host plant selection in Aphididae. *Trans. IXth Int. Congr. Entomol.* 2: 106–110.

Kennedy, J. S. (1965). Mechanisms of host-plant selection. *Ann. Appl. Biol.* 56: 317–322.

Kennedy, J. S. and Booth, C. O. (1951). Host alternation in *Aphis fabae* Scop. I. Feeding preferences and fecundity in relation to the age and kind of leaves. *Ann. Appl. Biol.* 38: 25–64.

Kennedy, J. S. and Stroyna, H. L. G. (1959). Biology of aphids. *Ann. Rev. Entomol.* 4: 139–160.

Kennedy, J. S. and Brooke, I. H. (1972). The plant in the life of an aphid: insect-plant relationships, in *Symp. Roy. Entomol. Soc. London 6*, Van Emden, H. F., Ed., Blackwell Scientific, Oxford: 215 pp.

Kennedy, J. S., Ibbotson, A., and Booth, C.D. (1950). The distribution of aphid infestation in relation to leaf age. *Ann. Appl. Biol.* 37: 651–679.

Kevan, P. G. (1973). Flowers, insects, and pollination ecology in the Canadian high Arctic. *Polar Rec.* 16(104): 667–674.

Kevan, P. G., Chaloner, W. G., and Savile, D. B. O. (1975). Interrelationships of early terrestrial arthropods and plants. *Palaeontology*, 18(2): 391–417.

Keys, R. N., Buchmann, S. L., and Smith, S. E. (1995). Pollination effectiveness and pollination efficiency of insects foraging *Prosopis velutina* in southeastern Arizona. *Appt. Ecol.* 32: 519–527.

Kingsolver, J. M. (1995). On the family Bruchidae. *Chrysomela*, 30: 3.

Kirk, H. M. (1988). Cannibalism in a chrysomelid beetle, *Gastrophysa viridula*. Ph.D. dissertation, Liverpool University: pp. 1–289.

Kirk, W. D. J. (1987). How much pollen can thrips destroy? *Ecol. Entomol.* 12: 31–40.

Kjellsson, G., Rasmussen, F. N., and Dupuy, D. (1985). Pollination of *Dendrobium infundibulum*, *Cymbidium insigne* (Orchidaceae) and *Rhododendron lyi* (Ericaceae) by *Bombus eximius* (Apidae) in Thailand: a possible case of floral mimicry. *J. Trop. Ecol.* 1:289–302.

Knudsen, J. T. and Stahl, B. (1994). Floral odours in the Theophrastaceae. *Biochem. Syst. Ecol.* 22(3): 259–268.

Kogan, M. (1977). The role of chemical factors in insect/plant relationships. *Proc. Int. Congr. Entomol.* XV: 211–227.

Koptur, S. (1989). Mimicry of flowers by parasitoid wasp pupae. *Biotropica,* 21(1): 93–95.

Krantz, J. et al. (1977). *Diseases, Pests, and Weeds in Tropical Crops.* Paul Parey, Berlin: 666 pp.

Krieger, K. L., Feeny, P. P., and Wilkinson, C. F. (1971). Detoxication in the guts of caterpillars: an evolutionary answer to plant defenses? *Science,* 172: 579–581.

Krzeminska, E., Krzeminski, W., Haenni, J. P., and Dufour, C. (1992). *Les Fantomes de l'Ambre. Insectes Fossiles Dans l'Ambre de la Baltique.* Musée d'Histoire Naturelle de Neuchatel: 142 pp.

Kullenberg, B. (1961). *Studies on* Ophrys *pollination.* Almguist Wiksell, Uppsala: 34 pp.

Kullenberg, B. and Bergstrom, G. (1975). Chemical communication between living organisms. *Endeavour,* 122: 58–60.

Kullenberg, B. and Bergstrom, G. (1976). The pollination of *Ophrys* orchids. *Bot. Not.* 129(1): 11–19.

Kursar, T. A. and Coley, P. O. (1992). Delayed greening in tropical leaves: an anti-herbivore defense? *Biotropica,* 24(2): 256–262.

Kuschel, G. and May, B. (1990). Palophaginae, a new subfamily for leaf beetles, feeding as adult and larva on *Araucaria* pollen in Australia. *Invertebr. Taxon.* 3: 697–719.

Kuschel, G. and May, B. (1996). Discovery of Palophaginae (Col. Megal.) on *Araucaria araucana* in Chile and Argentina. *N.Z. Entomol.* 19: 1–13.

Labandeira, C. C. and Sepkoski, J. J. (1993). Insect diversity in the fossil record. *Science,* 261: 310–315.

Labandeira, C. C., Beall, B. S., and Hueber, F. M. (1988). Early insect diversification: evidence from a Lower Devonian bristletail, *Science,* 242: 913–916.

Labandeira, C. C., Dilcher, D. L., Davis, D. R., and Wagner, D. L. (1994). Ninety-seven million years of angiosperm insect association: paleobiological insights into the meaning of convolution. *Proc. Natl. Acad. Sci. USA,* 91(25): 12278–12282.

Labeyrie, V. (1977). *Comportement des Insectes en Milieu Trophique.* Centre National de la Rechereche Scientifique, Paris: 493 pp.

Labeyrie, V., Ed. (1981). *The Ecology of Bruchids Attacking Legumes (Pulses).* Junk, The Hague, 233 pp.

Labeyrie, V., Fabres, C., and Lachaise, D., Eds. (1987). *Insects-Plants,* Junk, Dordrecht: 459 pp.

Laboissière, V. (1934). Galerucinae de la Faune Française (Col.). *Ann. Soc. Ent. Fr.* 103: 1–108.

Larew, H. G. (1992). Fossil galls, in *Biology of Insect-Induced Galls,* Shorthouse, R. and Rohfritsch, O., Eds., Oxford University Press, London: pp. 50–59.

Larson, P. P. and Larson, M. W. (1965). *Ants Observed.* Scientific Book Club, London: 192 pp.

Larsson, S. G. (1978). *Baltic Amber: A Palaeobiological Study.* Scandinavian Society Press, Klampenborgh: 192 pp.

Larsson, S. G. and Ekbom, B. (1995). Oviposition mistakes in herbivorous insects: confusion or a step towards a new host plant? *Oikos*, 72(1): 155–160.

Lawrence, J. F. and Britton, E. B. (1994). *Australian Beetles*. Melbourne University Press: 192 pp.

Lawton, J. H. and Schroder, D. (1977). Effects of plant type, size of geographical range and taxonomic isolation on number of insect species associated with British plants. *Nature*, 265: 137–140.

Lawton, J. H. and Price, P. W. (1979). Species richness of parasites on hosts: agromyzid flies on the British Umbelliferae. *J. Animal Ecol.* 48: 619–638.

Le Gall, Ph. (1989). Le choix des plantes nourricieres et la spécialisation trophique chez les Acridoidea. *Bull. Ecol.* 20(3): 245–261.

Lecoufle, M. (1989). *Plantes Carnivores*. Bordas, Paris: 144 pp.

Letourneau, D. K. (1990). Code of ant-plant mutualism broken by parasite. *Science*, 24B: 215–217.

Letourneau, D., Feynner-Arias, G. J., and Jebb, M. (1991). Interactions among *Endospermum* ants, and herbivores in Papua New Guinea. *Bull. Ecol. Soc. Am.* 72(2): 173.

Levin, D. A. (1973). The role of trichomes in plant defense. *Q. Rev. Biol.* 48: 3–15.

Levin, D. A. (1976). Alkaloid-bearing plants: an ecogeographic perspective. *Am. Nat.* 110: 261–284.

Lewis, A. C. and Lipani, G. A. (1990). Learning and flower use in butterflies: hypotheses from honey bees, in Bernays, E. A., Ed. *Insect-Plant Relationships*, vol. 2, CRC Press, Boca Raton, FL: pp. 95–110.

Lewis, S. E. and Carroll, M. A. (1991). Coleopterous egg deposition on alder leaves from the Klondike mountain formation (Middle Eocene), northeastern Washington. *J. Paleontol.* 65(2): 334–335.

Lewis, T. (1987). Thrips. A microcosm of the insects or beyond the fringes? *Antenna,* 11(3): 92–99.

Lichtwardt, P. W. (1986). *The Trichomycetes: Fungal Associates of Arthropods*. Springer-Verlag, Berlin: 343 pp.

Lincoln, D. E., Fajer, E. D., and Johnson, R. H. (1993). Plant-insect herbivore interactions in elevated CO_2 environments. *Trends Ecol. Evol. (TREE)*, 8(2): 64–68.

Linsenmaier, W. (1972). *Knaurs grosses Insektenbuch*. Th. Knaur Nachf.: 380 pp.

Lloyd, F. E. (1942). *The Carnivorous Plants*. Ronald Press, New York: 352 pp.

Lonsdale, D. (1995). Some observations on the pros and cons of being a bark-feeding insect. *Br. J. Entomol. Nat. Hist.* 8: 129–137.

Lonsdale, W. M., Farrell, G., and Wilson, C. G. (1995). Biological control of a tropical weed: a population model and experiment for *Sida acuta*. *J. Appl. Ecol.* 32(2):391–399.

Louda, S. M. and Rodman, J. E. (1996). Insect herbivory as a major factor in the shade distribution of a native crucifer (*Cardamine cordifolia* A. Gray, bittercress). *J. Ecol.* 84: 229–237.

Louveaux, J. (1965). *Plantes Carnivores et Végétaux Hostiles*, Hachette, Paris: 108 pp.

Lowman, M. D. and Moffett, M. (1993). The ecology of tropical rain forest canopies. *Trends Ecol. Evol. (TREE)*, 8(3): 104–107.

Lowrie, A. (1987). *Carnivorous Plants of Australia*. University of Western Australia Press, Nedlands: 200 pp.

Lundstroem, A. N. (1887). Von Domatien. *Nova Acta Reg. Soc. Upsal.* 13(3): 1–88.

Mabberley, D. J. (1987). *The Plant Book.* Cambridge University Press, London: 706 pp.

Macior, L. W. (1971). Coevolution of plants and animals: systematic insights from plant-insect interactions. *Taxon,* 20(1): 17–28.

Madelin, M. F. (1960). Internal fungal parasites of insects. *Endeavour,* 76: 181–190.

Maeyama, T., Terayama, M., and Matsumoto, T. (1994). The abnormal behavior of *Colobopsis* sp. (Hym. Form.) parasitized by *Mermis* (Nematode) in Papua New Guinea. *Sociobiology,* 24(2): 115–119.

Mafra-Neto, A. and Jolivet, P. (1994). Entomophagy in Chrysomelidae: adult *Aristobrotica angulicollis* (Erichson) feeding on adult meloids (Col.), in *Novel Aspects of the Biology of Chrysomelidae,* Jolivet, P., Cox, M. L., and Petitpierre, E., Eds., Kluwer Academic, Dordrecht: pp. 171–178.

Mafra-Neto, A. and Jolivet, P. (1996). Cannibalism in leaf beetles, in Jolivet, P. and Cox, M. L., Eds. (1996). *Chrysomelidae, Biology 2,* SPB Academic Publishers, Amsterdam: pp. 195–211.

Maguire, B. (1978). Sarraceniaceae in the botany of Guyana highlands. *Mem. N.Y. Bot. Garden. Bronx,* 29: 36–62.

Maiorana, U. C. (1978). What kinds of plants do herbivores rearly prefer? *Am. Nat.* 112(985): 631–635.

Mamay, S. H. (1976). Paleozoic origin of the cycads. *U.S. Geol. Surv. Prof. Paper.* 934: 48 pp.

Mani, N. S. (1964). *The Ecology of Plant Galls.* Junk, The Hague: 434 pp.

Manier, J. F. (1950). *Recherches sur les Trichomycetes.* These, Paris: 162 pp.

Manning, A. (1956). The effect of honey guides. *Behavior,* 9: 114–139.

Mark, G.A. (1982). Induced oviposition preference, periodic environments and demographic cycles in the Bruchid beetle *Callosobruchus maculatus.* *Entomol. Exp. Appl.* 32: 155–160.

Mark-Henry, S., Ed. (1966–1967). *Symbiosis* (2 vols.). Academic Press, New York: 478 + 443 pp.

Marohasy, J. (1994). Biology and host-specificity of *Weiseana barkeri* (Col. Chrys.): a biological control agent for *Acacia nilotica* (Mimosaceae). *Entomophaga,* 39(3–4): 335–340.

Martynov, A. (1924). Sur les organes facettiques des ailes des insectes. *C. Acad. Sc. Russie,* pp. 71–73.

Martynov, A. (1925). On the facettic organs in the wings of insects. *Trans. Soc. Nat. Leningrad,* 54(2): 3–24.

Maschwitz, U., Fiala, B., Lee, K.F., Chey, V. K., and Tan, F. L. (1989). New and little known myrmecophytic associations in Bornean rain forests. *Malayan Nature J.* 43: 106–115.

Maschwitz, U., Fiala, B., Moog, J., and Saw, L. G. (1991). Two new myrmecophytic associations from the Malay peninsula. *Insect Soc.* 38: 27–35.

Matile, L. (1993–1995). *Dipteres d' Europe Occidentals* (vols. I and II). Boubeé, Paris: 439 + 381 pp.

Matthews, E. G. and Kitching, R. L. (1984). *Insect Ecology.* University of Queensland Press, St. Lucia: 211 pp.

Mattiacci, L., Dicke, M., and Posthumus, M. A. (1995). B-glucosidase: an elicitor of herbivore-induced plant odor that attracts host-searching parasitic wasps. *Proc. Natl. Acad. Sci. USA,* 92: 2036–2040.

Mattson, W. J. and Addy, N. D. (1975). Phytophagous insects as regulators of forest primary production. *Science*, 190: 515–522.

Mattson, W. J., Levieux,J. L., and Bernard-Dagan, C., Eds. (1980). *Mechanisms of Woody Plant Defenses Against Insects*. Springer-Verlag, Berlin: 416 pp.

Maulik, S. (1947). Relationships between the assemblages of plants fed upon by different insects and between the assemblages of insects that feed upon different plants. *Nature*, 159(4034): 1491–1495.

Mbata, G. N. (1994). Sensory organs involved in egg distribution in *Callosobruchus subinnotatus* Pic. *J. Stored Prod. Res.* 30(4): 339–346.

McDade, L.A. (1992). Pollinator relationships, biogeography and phylogenetics. *BioScience*, 42(1): 21–26.

McDowell, P. G., Lwande, W., Deans, S. G., and Waterman, P. G. (1988). Volatile resin exudate from stem bark of *Commiphora rostrata*: potential role in plant defense. *Photochemistry*, 27(8): 2519–2521.

McKey, D. and Davidson, D. W. (1993). Ant-plant symbiosis in Africa and the Neotropics: history, biogeography, and diversity, in *Biological Relationships Between Africa and South America*, Goldblatt, P., Ed., Yale University Press, New Haven: pp. 568–605.

McMullen, C. K. (1993). Flower-visiting insects of the Galapagos Islands. *Pan-Pacific Entomol.* 69(1): 95–106.

Meeuse, B. J. D. (1961). *The Story of Pollination*. Ronald Press, New York: 243 pp.

Menken, S. B. J., Visser, J. H., and Harrewijn, P., Eds. (1992). *Proc. 8th Int. Symp. on Insect-Plant Relationships*. Kluwer Academic, Dordrecht: 424 pp.

Merrill, E. D. (1981). *Plant Life in the Pacific World*. Tuttle, Boston: 297 pp.

Mesquin, T. (1971). Competition for pollinators as a stimulus for the evolution of the flowering time. *Oikos*, 22: 398–402.

Metcalf, R. L. and Luckmann, R. H. (1975). *Introduction to Insect Pest Management*. John Wiley & Sons, New York: 587 pp.

Metcalf, R. L. and Metcalf, E. R. (1990). *Plant Kairomones in Insect Ecology and Control*. Chapman and Hall, New York: 168 pp.

Meyer, J. (1987). *Plant Galls and Gall Inducers*. Centre National de la Recherche Scientifique, Strasbourg: 291 pp.

Miller, J. R. and Miller, T. A., Eds. (1986). *Insect-Plant Interactions*. Springer-Verlag, Berlin, 342 pp.

Miller, J. S. (1987). Host-plant relationships in the Papilionidae (Lep.): parallel cladogenesis or colonization? *Cladistics*, 3: 105–120.

Miller, J. S. (1992). Host-plant associations among prominent moths. *BioScience*, 42: 50–57.

Mitchell, B. K. (1974). Behavioural and electrophysiological investigations in responses of larvae of the Colorado potato beetle (*L. decemlineata*) to amino acids. *Entomol. Exp. Appl.* 17: 255–264.

Mitchell, B. K. and Schoonhoven, L. M. (1974). Taste receptors in Colorado potato beetle larvae. *J. Insect. Physiol.* 20: 1787–1793.

Mitter, C. and Brooks, D. R. (1983). Phylogenetic aspects of convolution, in *Coevolution*, Futuyma, D. J. and Slatkin, M., Eds., Sinauer Associates, Sunderland: pp. 65–98.

Mitter, C. and Farrell, B. (1991). Macroevolutionary aspects of insect-plant relationships, in *Insect-Plant Relationships*, vol. 3, Bernays, E. A., Ed., CRC Press, Boca Raton, FL: pp. 35–78.

Mode, C. J. (1958). A mathematical model for the coevolution of obligate parasites and their hosts. *Evolution*, 12: 158–165.

Moeller, A. P. (1995). Bumblebee preference for symmetrical flowers. *Proc. Natl. Acad. Sci. USA*, 92(6): 2288–2292.

Moldenke, A. R. (1979). Pollination ecology within the Sierra Nevada. *Phytologia*, 42(3): 223–282.

Moldenke, A. R. (1979). Host plant coevolution and the diversity of bees in relation to the flora of North America. *Phytologia*, 43(4): 357–419.

Monod, T. and Schmitt, C. (1968). Contribution a l' étude des pseudogalles formicaires chez quelques acacias africains. *Bull. IFAN*, 30(A3): 953–1012.

Moody, S. (1978). Latitude, continental drift, and the percentage of alkaloid-bearing plants in floras. *Am. Nat.* 113: 965–968.

Moran, V. C., Hoffmann, J. H., Impson, F. A. C., and Jenkins, J. F. G. (1994). Herbivorous insects species in the tree canopy of a relict South African forest. *Ecol. Entomol.* 19: 147–154.

Moret, L. (1943). Manuel de Paléontologie Végitale. *Masson. Paris*, 216 pp.

Morris, M. G. (1991). *Weevils*. Richmond Publishers, Slough: 76 pp.

Mound, L. A. (1993). The first *Thrips* species inhabiting leaf domatia: *Domatiathrips cunninghamii* gen sp. nov. (Thys. Palaeothripidae). *J. N.Y. Entomol. Soc.* 101(3): 424–430.

Mound, L. A. and Waloff, N. (1978). Diversity of insect faunas. *Symp. Roy. Entomol. Soc.* No. 9: 204 pp.

Müller, H. J. (1958). The behaviour of *Aphis fabae* in selecting its host plants, especially different varieties of *Vicia faba*. *P. Entomol. Exp. Appl.* 1: 66–72.

Muller, V. G. and Wolf-Muller, B. (1991). Epiphyll deterrence to the leafcutter ant *Atta cephalotes*. *Oecologa*, 86: 36–39.

Naeem, S. (1968). Resource heterogeneity fosters coexistence of a mite and a midge in pitcher plants. *Ecol. Monogr.* 58(3): 215–227.

Nahrstedt, A. (1989). The significance of secondary metabolites for interactions between plants and insects. *Planta Medica*, 55(4): 333–338.

NAS (1969). *Insect-Plant Interactions*. National Academy of Sciences, Washington, D.C.: 98 pp.

Neal, L., Ed., *Pollination Biology*. Academic Press, New York: 338 pp.

New, T. R. (1990). *Associations Between Insects and Plants*. New South Wales University Press, Kensington: 113 pp.

Nichols-Orians, C. (1992). The acceptability of young and mature leaves to leaf-cutter ants varies with light environment. *Biotropica*, 24(2a): 211–214.

Nielsen, J. K. (1978). Host-plant discrimination within Cruciferae: feeding responses of four leaf beetles (Col. Chrys.) to glucosinolates, cucurbitacins and cardenolides. *Entomol. Exp. Appl.* 24(1): 41–54.

Nielsen, J. K., Kirkeby-Thomsen, A. M., and Petersen, M. K. (1989). Host-plant recognition in monophagous weevils: specificity in feeding responses of *Ceutorhynchus constrictus* and the variable effect of sinigrin. *Entomol. Exp. Appl.* 53: 157–166

Nikritin, L. M. and Ponomarenko, A. G. (1991). Fossil Coleoptera of the USSR; their evolution and distribution, in *Advances in Coleopterology*, Zunino, M., Belles, X., and Blas, M., Eds., University of Barcelona: pp. 29–34.

Nilsson, L. A. (1988). The evolution of flowers with deep corolla tubes. *Nature* 334(6177): 147–149.

Nilsson, L. A. (1992). Orchid pollination biology. *Trends Ecol. Evol. (TREE)*, 7(8): 255–259.

Norstog, K. J. and Fawcett, P. K. (1989). Insect-cycad symbiosis and its relation to the pollination of *Zamia furfuracea* (Zamiaceae) by *Rhopalotria mollis* (Curculionidae). *Am. J. Botany*, 78(9): 1380–1394.

Novak, V., Hrozinka, F., and Stary, B. (1976). *Atlas of Insects Harmful to Forest Trees.* Elsevier Science, Amsterdam: 126 pp.

Ochoa, M. G., Lavin, M. C., Ayala, F. C., and Perez, A. J. (1993). Arthropods associated with *Bromelia hemisphaerica* (Brom.) in Morelos, Mexico. *Florida Entomol.* 76(4): 616–621.

O'Dowd, D. J., Crew, C. R., Christophel, D. C., and Norton, R. A. (1991). Mite-plant associations from the Eocene of southern Australia. *Science*, 252: 99–101.

Olesen, J. M. and Balsley, H. (1990). Flower biology and pollinators of the Amazonian monoecious palm *Geonoma macrostachys*: a case of Bakerian mimicry. *Principes* 34(43): 181–190.

Olesen, J. M. and Knudsen, J. T. (1994). Scent profiles of flower colour morphs of *Corydalis cava* (Fumariaceae) in relation to foraging behaviour of bumble bee queens (*Bombus terrestris*). *Biochem. Syst. Ecol.* 22(3): 231–237.

Opler, P. A. (1974). Oaks as evolutionary islands for leaf-mining insects. *Am. Sci.* 62: 67–73.

Ostrofsky, M. L. and Zettler, E. R. (1986). Chemical defenses in aquatic plants. *J. Ecology*, 74: 279–287.

Painter, R. H. (1936). The food of insects and its relation to resistance of plants to insect attack. *Am. Nat.* 70: 547–566.

Painter, R. H. (1951). *Insect Resistance in Crop Plants.* Macmillan, New York: 520 pp.

Painter, R. H. (1958). Resistance of plants to insects. *Ann. Rev. Entomol.* 3: 267–290.

Palaniswamy, P. and Bodnaryk, R. P. (1994). A wild *Brassica* from Sicily provides trichome-based resistance against flea beetles, *Phyllotreta cruciferae* (Goeze). *Can. Entomol.* 126(5): 1119–1130.

Panda, N. and Khush, G. S. (1995). Host plants resistance to insects. *CAB Intern.* 444 pp.

Papaj, D. R. (1986). Conditioning of leaf-shape discrimination by chemical clues in the butterfly, *Battus philenor*. *Animal Behav.* 34(5): 1281–1288.

Papaj, D. R. and Rausher, M. D. (1983). Individual variation in host location by phytophagous insects, in *Herbivorous Insects: Host-Seeking Behaviour and Mechanisms*, Ahmad, S., Ed., Academic Press, New York: pp. 77–124.

Papaj,D. R. and Prokopy, R. J. (1989). Ecological and evolutionary aspects of learning in phytophagous insects. *Ann. Rev. Entomol.* 34: 315–350.

Papaj, D. R. and Lewis, A. C., Eds. (1992). *Insect Learning: Ecological and Evolutionary Perspectives.* Chapman and. Hall, New York: 412 pp.

Pasteels, J. M. and Daloze, D. (1977). Cardiac glycosides in the defensive secretion of Chrysomelid beetles: evidence for their production by the insects. *Science*, 197: 70–72.

Paulian, R. (1988). *Biologie des Coléoptères.* Lechevallier, Paris: 719 pp.

Peakall, R. (1984). Observations on the pollination of *Leporella fimbriata* (Lidl.). *Orchadian*, 8(2): 44–45.

Peakall, R. (1989a). Pollination ecology of *Leporella fimbriata*, the only pseudo-copulatory orchid known to mimic female ants, in *Symposium on Ant-Plants*, Huxley, C. R. and Cutler, D. F., Eds., Oxford University Press, London: 320 pp

Peakall, R. (1989b). The unique pollination of *Leporella fimbriata*: pollination by pseudocopulating male ants (*Myrmecia urens*). *Plant Syst. Evol.* 167: 137–148.

Peakall, R. (1994). Interactions between orchids and ants, in *Orchid Biology: Reviews and Perspectives*, V, Arditti, J., Ed., John Wiley & Sons, New York: pp. 103–134.

Peakall, R. and Beattie, A. J. (1989). Pollination of the orchid *Microtis parviflora* R. Br. by flightless worker ants. *Funct. Ecol.* 3(5): 515–522.

Peakall, R. and James, S. H. (1989). Outcrossing in an ant-pollinated clonal orchid. *Heredity*, 62: 161–167.

Peakall, R. and Beattie, A. J. (1991). The genetic consequences of worker ant pollination in a self-compatible, clonal orchid. *Evolution*, 45(8): 1837–1848.

Peakall, R. and Handel, S. N. (1994). Pollinators discriminate among floral heights of a sexually deceptive orchid: implications for selection. *Evolution*, 47(6): 1681–1687.

Peakall, R., Beattie, A. J., and James, S. H. (1987). Pseudocopulation of an orchid by male ants: a test of two hypotheses accounting for the rarity of ant pollination. *Oecologia* 73(4): 522–524.

Peakall, R., Angus, C. J., and Beattie, A. J. (1990). The significance of ant and plant traits for ant pollination in *Leporella fimbriata*. *Oecologia*, 84(4): 457–460.

Peakall, R., Handel, S. N., and Beattie, A. J. (1991). The evidence for and importance of ant pollination, in *Ant-Plant Interactions*, Huxley, C. R. and Cutler, D. F., Eds., Oxford University Press, London: pp. 421–429.

Peakall, R., Oliver, I., Turnbull, C. L., and Beattie, A. J. (1993). Genetic diversity in an ant-dispersed chenopod *Sclerolasna diacantha*. *Australian J. Ecology*, 18: 71–179.

Pellmyr, O. (1992). Evolution of insect pollination and angiosperm diversification. *Trends Ecol. Evol. (TREE)*, 7(2): 46–49.

Pemberton, R. H. and Vandenberg, N. J. (1993). Extrafloral nectar feeding by ladybird beetles. *Proc. Entomol. Soc. Wash.* 95(2): 139–151.

Perry, F. (1977). *Flowers of the World*. Hamlyn Publishers, London: 320 pp.

Pesson, P. and Louveaux, J. (1984). *Pollinisation et Production Vegetales*. INRA, Paris: 663 pp.

Phillips, W. M. (1977). Modification of feeding preference in the flea beetle, *Haltica lythri*. *Entomol. Exp. Appl.*: 71–80.

Pillemer, E. A. and Tingey, W. M. (1976). Hooked trichomes: a physical barrier to a major agricultural pest. *Science*, 193: 482–484.

Poinar, G. O. (1992a). Fossil evidence of resin utilization by insects. *Biotropica*, 24(3): 466–468.

Poinar, G. O. (1992b). *Life in Amber*. Stanford University Press, CA: 350 pp.

Poinar, G. O. and Poinar, R. (1994) *The Quest for Life in Amber*. Addison-Wesley, Reading, PA: 219 pp.

Ponomarenko, A. G. (1976). A new insect from the Cretaceous period of Tranabaikalia, a possible parasite of pterosaurians. *Paleont. Zhur.* 3: 102–106.

Potter, D. A. and Kimmerer, T. W. (1988). Do holly leaf spines really deter herbivory? *Oecologia*, 75: 216–221.

Potter, D. A. and Kimmerer, T. W. (1989). Inhibition of herbivory on young holly leaves: evidence for the defensive role of saponins. *Oecologia* 78: 322–329.

Powell, J. A. (1980). Evolution of larval food preferences in Microlepidoptera. *Ann. Rev. Entomol.* 25: 133–159.

Prance, G. T. and Lovejoy, T. E., Eds. (1985). *Amazonia*. Pergamon Press, Oxford: 442 pp.

Price, P. W. (1985). *Insect Ecology*, 2nd ed. John Wiley & Sons, New York: 607 pp.

Price, P. W., Lewinsohn, T. M., Fernandes, G. W., and Benson, W. W. (1991). *Plant-Animal Interactions: Evolutionary Ecology in Tropical and Temperate Regions*. John Wiley & Sons, New York: 693 pp.

Proctor, M. and Yeo, P. (1979). *The Pollination of Flowers*. Collins, London: 418 pp.

Proença, C. E. (1992). Buzz pollination — older and more widespread than we think? *J. Trop. Ecol.* 8: 115–120.

Prokopy, R. J., Averill, A. L., Cooley, S. S., and Rotberg, C. R. (1982). Associative learning in egg-laying site selection by apple maggot flies. *Science*, 218: 76–77.

Raman, A. (1994). Adaptational integration between gall-inducing insects and their host plants, in *Functional Dynamics of Phytophagous Insects*, Ananthakrishnan, T. N., Ed., Science Publishers, Lebanon, NM: pp. 249–275.

Raupp, M. J. (1985). Effects of leaf toughness on mandibular wear of the leaf beetle, *Plagiodera versicolora*. *Ecol. Entomol.* 10(1): 73–80.

Raupp, M. J. and Sadof, C. S. (1991). Responses of leaf beetles to injury related changes in their salicaceous hosts, in *Phytochemical Induction by Herbivores*, Tallamy, D. W. and Raupp, M. J., Eds., John Wiley & Sons, New York: 183–204.

Rausher, M. D. (1983a). Conditioning and genetic variation as causes of individual variation in the oviposition behaviour of the tortoise beetle *Deloyala guttata*. *Animal Behav.* 31: 743–747.

Rausher, M. D. (1983b). Ecology of host-selection behavior in phytophagous insects, in *Variable Plants and Herbivores in Natural and Managed Systems*, Denno, R. F. and McClure, M. S., Eds., Academic Press, New York: pp. 223–257.

Rausher, M. D. (1988). Is coevolution dead? *Ecology*, 69(4): 898–901.

Real, L., Ed. (1983). *Pollination Biology*. Academic Press, Orlando, FL: 338 pp.

Redfern, M. and Askew, R. R. (1992). *Plant Galls*. Richmond Publishers, Slough: 99 pp.

Rees, C. J. (1969). Chemoreceptor specificity associated with choice of feeding site by the beetle *Chrysolina brunsvicensis* on its food plant, *Hypericum hirsutum*. *Entomol. Exp. Appl.* 12: 565–583.

Rhoades, D. F. (1983). Responses of alder and willow to attack by tent caterpillars and webworms: evidence for pheromonal sensitivity of willows, in *Plant Resistance to Insects*, Hedin, P. A., Ed., American Chemical Society, Washington, D.C.: pp. 55–68.

Richard, A. J. (1978). *The Pollination of Flowers by Insects*. Academic Press, London: 213 pp.

Rickson, F. R., Cresti, M., and Beach, J. H. (1990). Plant cells which aid in pollen digestion within a beetle's gut. *Oecologia*, 82: 424–426.

Ridley, H.N. (1930). *The Dispersal of Plants Throughout the World*. Reeve Publishers, London: 744 pp.

Riek, E. (1970). Lower Cretaceous fleas. *Nature*, 227: 746–747.

Risley, L. S. and Crossley, D. A. (1988). Herbivore-caused greenfall in the southern Appalachians. *Ecology*, 69(4): 1118–1127.

Rodriguez, J. G. (1972). *Insect and Mite Nutrition*. North Holland, Amsterdam: 702 pp.

Rogers, C. E. (1985). Extrafloral nectar: entomological implications. *Bull. Entomol. Soc. Am.* 31(3): 15–20.

Rohdendorf, B. B. and Paznitsin, A. P. (1980). The historical development of the class Insecta. *Trudy Paleontol. Inst., Moscow*, 175: 1–268.

Rohfritsch, O. (1992). Patterns in gall development, in *Biology of Insect-Induced Galls*, Shorthouse, J. and Rohfritsch, O., Eds., Oxford University Press, London: pp. 80–86.

Roitberg, B. D. and Isman, M., Eds. (1992). *Insect Chemical Ecology: An Evolutionary Approach*. Chapman and Hall, New York: 320 pp.

Roitberg, B. D. and Mangel, M. (1993). Parent-offspring conflict and life history consequences in herbivorous insects. *Am. Nat.* 142(3): 443–456.

Rolfe, W. D. T. (1950). Early invertebrate terrestrial faunas, in *The Terrestrial Environment and the Origin of Land Vertebrates*, Panichen, A. L., Ed., Academic Press, London: pp. 117–157.

Rolfe, W. D. T. and Ingham, J. K. (1967). Limb structure, affinity, and diet of the Carboniferous "centipede". *Arthropleura. Scot. J. Geol.* 3: 118–124.

Roques, A. (1987). Interaction between visual and olfactory signals in cone recognition by insect pests, in *Insects-Plants*, Labeyrie, V., Fabres, C., and Lachaise, D., Eds., Junk, Dordrecht: pp. 153–160.

Rosenthal, G. A. and Janzen, D. H., Eds. (1979). *Herbivores: Their Interaction with Secondary Plant Metabolites*. Academic Press, New York: 718 pp.

Rosenthal, G. A. and Berenbaum, M. R (1991–1992). *Herbivores: Their Interactions with Secondary Plant Metabolites*, 2nd ed., Academic Press, New York: pp. 468–493.

Roskam, J. C. (1992). Evolution of gall-inducing guild, in *Biology of Insect-Induced Galls*, Shorthouse, J. and Rohfritsch, O., Eds., Oxford University Press, London: 34–49.

Rotheray, G. (1994). *Insect Life on Plants*. Chapman and Hall, New York: 330 pp.

Rothschild, M. (1984). Aide-mémoire mimicry. *Ecol. Entomol.* 9: 311–319.

Rothschild, M. and Schoonhoven, L. M. (1977). Assessment of egg load by *Pieris brassicae* (Lept. Pieridae). *Nature*, 266(5600): 352–355.

Rothschild, M., Nash, R. J., and Bell, E. A. (1986). Cycasin, in the endangered butterfly *Eumaeus atala floridana. Phytochemistry*, 25: 1853–1854.

Roubik, D.W. (1989). *Ecology and Natural History of Tropical Bees*. Cambridge University Press, London: 514 pp.

Rozario, S. A. (1995). Association between mites and leaf domatias. Evidence from Bangladesh, South Asia. *J. Trop. Ecol.* 11(1): 99–108.

Rumphius, G. E. (1750). *Herbarium Amboinense*. Changnion and Hermann Utywerf, Amsterdam.

Ryan, C. A. (1983). Insect-induced chemical signals regulating natural plant protection responses, in *Variable Plants and Herbivores in Natural and Managed Systems*, Denno, R. F. and McClure, M. S., Eds., Academic Press, New York: pp. 43–60.

Sacchi, C. F. (1988). Ricerche sulla strttura degli ecosistemi: invito al "Cenone". *Thalassia Salentina*, 18: 187–276.

Sagers, C. L. (1992). Manipulation of host plant quality: herbivores keep leaves in the dark. *Funct. Ecol.* 6: 741–743.

Saito, N. and Harborne, J. B. (1992). Correlations between anthocyanin type, pollinator and flower colour in Labiatae. *Phytochemistry*, 31(9): 3001–3015.

Samuelson, G. A. (1966). Epizoic symbiosis: a new Papuan colydiid beetle with epicuticular growth of cryptogamic plants. *Pacific Insects*, 8(1): 250–293.

Samuelson, G. A. (1994). Pollen consumption and digestion by leaf-beetles, in *Novel Aspects of the Biology of Chrysomelidae*, Jolivet, P., Cox, M. L., and Petitpierre, E., Eds., Kluwer Academic, Dordrecht: pp. 179–183.

Saxena, K. N. and Goyal, S. (1978). Host-plant relations of the citrus butterfly, *Papilio demoleus* L.: orientation and ovipositional responses. *Entomol. Exp. Appl.* 24(1): 1–10.

Saxena, K. N. and Schoonhoven, L. M. (1978). Induction of orientational and feeding preferences in *Manduca sexta* larvae for an artificial diet containing citral. *Entomol. Exp. Appl.* 23: 72–78.

Saxena, K. N. and Schoonhoven, L. M. (1982). Induction of orientation and feeding preferences in *Manduca sexta* larvae for different food sources. *Entomol. Exp. Appl.* 32: 173–180.

Saxena, O. P., Koul, O., Tikru, K., and Atal, C. K. (1977). A new insect chemosterilant isolated from *Acorus calamus* L. *Nature*, 270: 512–513.

Schillinger, J. A. and Gallon, R. L. (1968). Leaf pubescence of wheat as a deterrent to the cereal leaf beetle, *Oulema melanopus*. *Ann. Entomol. Soc. Am.* 61: 900–903.

Schimper, A. F. (1888). Die Wechselbeziehungen Zwischen Pflanzen und Ameisen im Tropischen Amerika. *Bot. Mitt. Tropen, Iena*.1: 1–95.

Schlegtendal, A. (1934). Beitrag zum Farbensinn der Arthropoden. *Z. Vergl. Physiol.* 20: 545–583.

Schnell, D. E. (1936). *Carnivorous Plants of the United States and Canada*. Blair, Winston-Salem, NC: 125 pp.

Schnell, P. (1970). *Introduction a la Phytogéographie des Pays Tropicaux* (2 vols.). Gauthier-Villars, Paris: 950 pp.

Schnell, P. and de Beaufort, G. (1966). Contribution a l'étude des plantes a myrmécodomaties de l'Afrique intertropicale. *Mem. IFAN*, 75: 1–66.

Schnell, R. (1963). Le probleme des Acarodomaties. *Marcellia*, 31(2): 95–107.

Schnell, R. (1966). Remarques morphologiques sur les myrmécophytes. *Bull. Soc. Bot. Fr. Mem.*: pp. 121–132.

Schoener, T. W. (1987). Leaf pubescence in buttonwood: community variation in a putative defense against defoliation. *Proc. Natl. Acad. Sci.* 84: 7992–7995.

Schoener, T. W. (1988). Leaf damage in island buttonwood, *Conocarpus erectus*: correlations with pubescece, island area, isolation and the distribution of major carnivores. *Oikos*, 53: 253–266.

Scholler, M. (1996). Oekologie mitteleuropaischer Blattkafer, Semenkafer, and Breitrussler. Chrysomelidae, Bruchinae, and Anthribidae. Brandstetter, Bürs: 65 pp.

Schoonhoven, L. M. (1967a). Les aspects chimiosensoriels de la reconnaissance des plantes-hotes par les insectes. *Mededel. Rijksfac. Landbouw Wetensch. Gent*, 32(3–4): 286–290.

Schoonhoven, L. M. (1967b). Chemoreception of mustard oil glucosides in larvae of *Pieris brassicae*. *Proc. Kon. Neder. Akad. Wetensch. Amsterdam C.* 70: 556–568.

Schoonhoven, L. M. (1967c). Loss of host plant specificity by *Manduca sexta* after rearing on an artificial diet. *Entomol. Exp. Appl.* 10: 270–272

Schoonhoven, L. M. (1968). Chemosensory bases of host plant selection. *Ann. Rev. Entomol.* 13: 115–136.

Schoonhoven, L. M. (1969a). Amino-acid receptor in larvae of *Pieris brassicae*. *Nature*, 221(5187): 1268–1269.

Schoonhoven, L. M. (1969b). Sensitivity changes in some insect chemoreceptors and their effect on food selection behaviour. *Proc. Kon. Neder. Akad. Wetensch. Amsterdam C.* 72(4): 491–498.

Schoonhoven, L. M. (1969c). Gustation and food plant selection in some lepidopterous larvae. *Entomol. Exp. Appl.* 12: 55–56.

Schoonhoven, L. M. (1970). Hoe herkennen Insekten hun voedselplant? *Overdruk uit Vakblad v. Biol.* 129–135.

Schoonhoven, L. M. (1972a). Some aspects of host selection and feeding in phytophagous insects, in *Insect and Mite Nutrition*, Rodriguez, J. G., Ed., North-Holland, Amsterdam: pp. 557–566.

Schoonhoven, L. M. (1972b). Plant recognition by lepidopterous larvae, in *Symp. Roy. Entomol. Soc. London 6*, Van Emden, H. F., Ed., Blackwell Scientific, Oxford: pp. 87–99.

Schoonhoven, L.M. (1974). Studies on the shootborer *Hypsila grandella* (Zeller). 23. Electroantennograms (EAG) as a tool in the analysis of insect attractants. *Turrialba*, 24(1): 24–28.

Schoonhoven, L. M. (1976a). On the variability of chemosensory information. *Symp. Biol. Hung.* 16: 261–266

Schoonhoven, L. M. (1976b). Feeding behaviour in phytophagous insects: on the complexity of the stimulus situation. *Coll. Int. CNRS: Comportement des Insectes en Milieu Trophique*, 265: 391–398.

Schoonhoven, L. M. (1977). On the individuality of insect feeding behaviour. *Proc. Kon. Neder. Akad. Wetensch. Amsterdam C.* 80(4): 341–350.

Schoonhoven, L. M. (1980a). Perception of azadirachtin by some lepidopterous larvae, in *Proc. 1st Int. Neem Conf. Rottach-Egern*: pp. 105–108.

Schoonhoven, L. M. (1980b). First International Neem Conference: Afterword, in *Proc. 1st Int. Neem Conf. Rottach-Egern*: pp. 291–298.

Schoonhoven, L. M. (1982). Biological aspects of antifeedants. *Entomol. Exp. Appl.* 31: 57–69.

Schoonhoven, L. M. (1983). Second International Neem conference: Afterword, in *Proc. 2nd Int. Neem Conf. Raulscholzhausen*: pp. 581–588.

Schoonhoven, L. M. (1987). What makes a caterpillar eat? The sensory code underlying feeding behavior, in *Perspectives in Chemoreception and Behavior*, Chapman, R. F., Bernays, E. R., and Stoffolano, J. G., Eds. Springer-Verlag, New York: pp. 69–97.

Schoonhoven, L. M. (1988). Stereoselective perception of antifeedants in insects, in *Stereoselectivity of Pesticides: Biological and Chemical Problems*, Ariens, E. J. et al., Eds., Elsevier Science, Amsterdam: pp. 289–302.

Schoonhoven, L. M. (1990a). Insects and host-plants: 100 years of botanical instinct. *Symp. Biol. Hung.* 39: 3–14.

Schoonhoven, L. M. (1990b). Insects in a chemical world, in *Handbook of Natural Pesticides*, Morgan, H. F. D. and Mandava, K. B., Eds., CRC Press, Boca Raton, FL: pp. 1–21.

Schoonhoven, L. M. (1991). The sense of distate in plant-feeding insects: a reflection on its evolution. *Phytoparasitica*, 19(1): 3–7.

Schoonhoven, L. M. and Henstra, S. (1967). Morphology of some rostrum receptors in *Dysdercus* spp. *Netherlands J. Zool.* 22(3): 343–346.

Schoonhoven, L. M. and Derksen-Koppers, I. (1973). Effects of secondary plant substances on drinking behaviour in some Heteroptera. *Entomol. Exp. Appl.* 16: 141–145.

Schoonhoven, L. M. and Derksen-Koppers, I. (1976). Effects of some allelochemics on food uptake and survival of a polyphagous aphid, *Myzus persicae*. *Entomol. Exp. Appl.* 19: 52–56.

Schoonhoven, L. M. and Jermy, T. (1977). A behavioural and electrophysiological analysis of insect feeding deterrents, in *Crop Protection Agents — Their Biological Evaluation*, McFarlane, N. R., Ed., Academic Press, New York: pp. 133–146.

Schoonhoven, L. M. and Meepman, J. (1978). Metabolic cost of changes in diet and neutralization of allelochemics. *Entomol. Exp. Appl.* 24: 689–693.

Schoonhoven, L.M. and Blum, F. (1988). Chemoreception and feeding behaviour in a caterpillar: towards a model of brain functioning in insects. *Entomol. Exp. Appl.* 49(5): 123–129.

Schoonhoven, L. M. and Liner, L. (1994). Multiple mode of action of the feeding deterrent, toosendanin, on the sense of taste of *Pieris brassicae* larvae. *J. Comp. Physiol.* A 175: 519–524.

Schoonhoven, L. M., Tramper, N. M., and van Drongelen, W. (1977). Functional diversity in gustatory receptors in some closely related *Yponomeuta* species. *Netherlands J. Zool.* 27(3): 287–291.

Schoonhoven, L. M., Sparnary, T., van Wissen, W., and Meerman, J. (1981). Seven-week persistence of an oviposition deterrent pheromone. *J. Chem. Ecology*, 7(3): 583–588.

Schoonhoven, L. M., Blaney, W. M., and Simmonds, M. S. J. (1987). Inconsistencies of chemoreceptor sensitivities, in *Insects-Plants*, Labeyrie, V., Fabres, C., and Lachaise, D., Eds., Junk, Dordrecht: pp. 141–145.

Schoonhoven, L.M., Blaney, W. M., and Simmonds, M. S. J. (1992). Sensory coding of feeding deterrents in phytophagous insects, in *Insect-Plant Relationships*, vol. 4, Bernays, E. A., Ed., CRC Press, Boca Raton, FL: pp. 59–79.

Schoonhoven, L. M., Jermy, T., and van Loon, J. J. A. (1998). *Insect-Plant Biology*. Chapman and Hall, London: 409 pp.

Schremmer, F. (1976). Other interactions of animals and plants, in *Encyclopedia of Ecology*, Grzimek, B., Ed., Von Nostrand-Reinhold, New York: pp. 125–133.

Schultz, J. C. (1988). Many factors influence the evolution of herbivore diets, but plant chemistry is central. *Ecology*, 69(4): 896–897.

Schupp, E. W. and Feener, D. H. (1991). Phylogeny, life form and habitat dependence of ant-defended plants in a Panamanian forest, in *Ant-Plant Interactions*, Huxley, C. R. and Cutler, D. F., Eds., Oxford University Press, London: pp. 175–197.

Schwartz, R. (1974). *Carnivorous Plants*. Praeger Publishers, New York: 128 pp.

Schwert, D. P. and Ashworth, A. C. (1990). Ice age beetles. *Nat. Hist.* 1: 10–14.

Scoble, M. J. (1992). *The Lepidoptera: Form, Function, and Diversity*. Oxford University Press, London: 404 pp.

Scott, A. C. and Taylor, T. N. (1983). Plant-animal interactions during the Upper Carboniferous. *Bot. Rev.* 49(3): 259–307.

Scott, A. C., Chaloner, W. G., and Paterson, S. (1985). Evidence of pteridophyte-arthropod interactions in the fossil record. *Proc. Roy. Soc. Edinburg*, 86: 133–140.

Scott, G. D. (1969). *Plant Symbiosis.* Edward Arnold, London: 58 pp.

Scriber, J. M. (1984). Larval foodplant utilization by the world Papilionidae (Lep.): latitudinal gradients reappraised. *Tokurana*, 6–7: 1–49.

Scriber, J. M. (1988). Tale of the tiger: Beringial biogeography, binomial classification, and breakfast choices in the *Papilio glaucus* complex of butterflies, in *Chemical Mediation of Convolution*, Spencer, K. A., Ed., Academic Press, San Diego: pp. 241–301.

Scriber, J. M. and Ayres, M. P. (1988). Leaf chemistry as a defense against insects. *ISI Atlas Sci. Plants Animals*, 1: 117–123.

Sedlag, O. (1978). *Salvia glutinosa*: an effective insect trap. *Entomol. Nach. Dresden*, 22(1): 1–6.

Seguy, E. (1950). *La Biologie des Diptères.* Lechevalier, Paris: 609 pp.

Seidel, J. L., Epstein, W. W., and Davidson, D. W. (1990). Neotropical ant gardens. I. Chemical constituents. *J. Chem. Ecol.* 16(6): 1791–1816.

Shapiro, A. M. (1992). Why are there so few butterflies in the high Andes? *J. Res. Lepid.* 31(1–2): 35–56.

Shear, W. A. and Kukalova-Peck, J. (1990). The ecology of Paleozoic terrestrial arthropods: the fossil evidence. *Can. J. Zool.* 68: 1807–1834.

Shear, W. A., Bonamo, P. M., Rolfe, W. D. I., Smith, E. L., and Norton, R. A. (1984). Early land animals in North America: evidence from Devonian age arthropods from Gikboa, New York. *Science*, 224: 492–494.

Shear, W. A., Palmer, J. M., Coddington, J. A., and Bonamo, P. M. (1989). A Devonian spinneret: early evidence of spiders and silk use. *Science*, 246: 479–481.

Shorthouse, J. and Rohfritsch, O., Eds. *Biology of Insect-Induced Galls.* Oxford University Press, London: 296 pp.

Shykoff, J. A. and Buchell, E. (1995). Pollinator visitation patterns, floral rewards, and the probability of transmission of *Microbotryum violaceum*, a venereal disease of plants. *J. Ecol.* 83(2): 189–198.

Sigal, L. L. (1984). Of lichens and lepidopterons. *Bryologist*, 87(1): 66–68.

Singer, M. C. (1984). Butterfly-host plant relationships: host quality, adult choice and larval success, in *The Biology of Butterflies*, Vane-Wright, R. I. and Ackery, P. R., Eds., Academic Press, London: pp. 81–88.

Singer, M. C. and Parmesan, C. (1993). Sources of variations in patterns of plant-insect association. *Nature*, 361: 251–253.

Sivaramakrishnan, K. G. (1990). Systematics and speciation in relation to insect-plant interactions. *Proc. Indian Acad. Sci. (Animal Sci.)* 99(3): 267–276.

Slack, A. (1979). *Carnivorous Plants.* Ebury Press, London: 240 pp.

Slansky, F. (1972). Latitudinal gradients in species diversity of the New World swallowtail butterflies. *J. Res. Lepid.* 2(4): 201–217.

Slansky, F. (1978). Phagism relationships among butterflies. *J. N.Y. Entomol. Soc.* 86: 15 pp.

Slansky, F. and Rodriguez, J. G., Eds. (1987). *Nutritional Ecology of Insects, Mites, Spiders and Related Invertebrates.* John Wiley & Sons, New York: 1016 pp.

Slocum, R. D. and Lawrey, J. D. (1976). Viability of the epizoic lichen flora carried and dispersed by green lacewing (*Nodita pavida*). *Can. J. Bot.* 54(15): 1827–1831.

Smiley, J. (1978). Plant chemistry and evolution of host specificity: new evidence from *Heliconius* and *Passiflora*. *Science*, 201: 745–747.

Smith, L. B. and Downs, R. J. (1974). Bromeliaceae, in *Flora Neotropica*. 14. Hafner Press, New York: 660 pp.

Smith, M. A. and Cornell, H. V. (1979). Hopkins host selection in *Nasonia vitripennis* and its implications for sympatric speciation. *Animal Behav.* 27: 365–370.

Smith, P. M. (1976). *The Chemotaxonomy of Plants*. Edward Arnold, London: 313 pp.

Sondheimer, E. and Simeone, J. B. (1970). *Chemical Ecology*. Academic Press, New York: 336 pp.

Souchon, C. (1965). *Les Insectes at les Plantes*. Presses Univ. France, Paris: 123 pp.

Southwood, T. R. E. (1961). The number of species of insect associated with various trees. *J. Animal Ecol.* 30: 1–8.

Southwood, T. R. E. (1985). Interactions of plants and animals: patterns and processes. *Oikos*, 44(1): 5–11.

Southwood, T. R. E., Ed. (1968). Insect abundance, *Symp. Roy. Entomol. Soc. Lond.* 4: 160 pp.

Southwood, T. R. E., Ed. (1973). The insect/plant relationship — an evolutionary prospective. *Symp. Roy. Entomol. Soc. Lond.* 6: 3–30.

Spencer, K. A. (1973). *Agromyzidae (Diptera) of Economic Importance*. Junk, The Hague: 418 pp.

Spencer, K. A. (1990). *Host Specialization in the World of Agromyzidae (Diptera)*. Kluwer Academic, Dordrecht: 444 pp.

Spencer, K. C., Ed. (1988). *Chemical Mediation of Coevolution*. Academic Press, New York: 609 pp.

Spruce, R. (1908). *Notes of a Botanist on the Amazon and Andes*. Macmillan, London: 524 pp.

Stadler, E. (1986). Oviposition and feeding stimuli in leaf surface waxes, in *Insects and Plant Surface*, Juniper, B. E. and Southwood, R., Eds., Edward Arnold, London: pp. 105–121.

Stamp, N. E. and Casey, T. M., Eds. (1993). *Caterpillars: Ecological and Evolutionary Constraints on Foraging*. Chapman and Hall, New York: 587 pp.

Stebbins, G. L. (1981). Why are there so many species of flowering plants? *BioScience*, 31(8): 573–577.

Stehr, F. W. (1987). Order Lepidoptera, in *Immature Insects*, Stehr, F. W., Ed., Kendall/Hunt, Dubuque, IA: pp. 288–305.

Stein, B. A. (1992). Sicklebill hummingbirds, ants, and flowers. *BioScience*, 42(1): 27–33.

Sterck, F., van der Meer, P. and Bongers, F. (1992). Herbivory in two rain forest canopies in French Guyana. *Biotropica*, 24(1): 97–99.

Stewart, W. N. and Rothwell, G. W. (1993). *Paleobotany and the Evolution of Plants*, 2nd ed. Cambridge University Press, New York: 521 pp.

Steyermark, J. A. (1986). *Speciation and Endemism in the Flora of the Venezuelan Tepuis: High Altitude Subtropics*. Oxford University Press, London: pp. 317–373.

Stidd, B. M. and Phillips, T. L. (1982). *Johnhallia lacunosa* gen. et sp. n.: a new pteridosperm from the Middle Pennsylvanian of Indiana. *J. Paleontol.* 56: 1093–1102.

Stiling, P. and Simberloff, D. (1989). Leaf abscission: induced defense against pests or response to damage? *Oikos*, 55: 43–49.

Stork, N. E. (1987a). Guild structure of arthropods from Bornean rain forest trees. *Ecol. Entomol.* 12(1): 69–80.

Stork, N. E. (1987b). Arthropod faunal similarity of Bornean forest trees. *Ecol. Entomol.* 12(3): 219–225.

Stork, N. E. (1988). Insect diversity: facts, fiction and speculation. *Biol. J. Linn. Soc.* 35: 321–337.

Stork, N. E. (1991). The composition of the arthropod fauna of Bornean lowland rain forest trees. *J. Trop. Ecol.* 7: 161–180.

Straatman, R. (1962). Notes on certain Lepidoptera ovipositing on plants which are toxic to their larvae. *J. Lepid. Soc.* 18: 99–103.

Strong, D. R. (1974). The insects of British trees: community, equilibration in ecological time. *Ann. Miss. Bot. Garden*, 61: 692–701.

Strong, D. R. (1979). Biogeographic dynamics of insect-host plant communities. *Ann. Rev. Entomol.* 24: 89–119.

Strong, D. R. (1988). Special feature: insect host range. *Ecology*, 69(4): 885–907.

Strong, D. R., Lawton, J. H., and Southwood, T. R. E. (1984). *Insects on Plants: Community Patterns and Mechanisms*. Blackwell Scientific, Oxford: 313 pp.

Sudd, J. H. and Franks, N. R. (1987). *The Behavioural Ecology of Ants*. Blackie and Sons, Glascow: 206 pp.

Swain, T. (1963). *Chemical Plant Taxonomy*. Academic Press, London: 543 pp.

Swanton, E. W. (1912). *British Galls*. Methuen Co., London: 287 pp.

Tahvanainen, J. O. and Root, R. B. (1972). The influence of vegetational diversity on the population ecology of a specialized herbivore, *Phyllotreta cruciferae*. *Oecologia*, 10: 321–346.

Takhtajan, A. (1969). *Flowering Plants: Origin and Dispersal*. Oliver Boyd, Edinburgh: 310 pp.

Takhtajan, A. (1991). *Evolutionary Trends in Flowering Plants*. Columbia University Press, New York: 241 pp.

Tallamy, D. W. (1985). Squash beetle feeding behavior: an adaptation against induced cucurbit defenses. *Ecology*, 66(5): 1574–1579.

Tang, W. (1987). Insect pollination in the cycad *Zamia pumila* (Zamiaceae). *Am. J. Botany*, 74(1): 90–99.

Tavares, I. I. (1985). Laboulbeniales (Fungi: Ascomycetes). *Mycologia Mem.* 9: 1–627.

Taylor, E. L. (1987). Tree ring structure in woody axes from the central Transantarctic Mountains, Antarctica, in *Proc. Int. Symp. Anarctic Res.*, China Ocean Press, Tianjin: pp. 109–113.

Taylor, P. (1964). The genus *Utricularia* (Utriculariaceae) in Africa (south of Sahara) and Madagascar. *Kew Bull.* 18(1): 1–245.

Taylor, T. N. (1978). The ultrastructure and reproductive significance of *Monoletes* (Pteridospermales) pollen. *Can. J. Bot.* 56: 3105–3118.

Taylor, T. N. (1981). *Paleobotany: An Introduction to Fossil Plant Biology*. McGraw-Hill, New York: 589 pp.

Taylor, T. N. and Scott, A. C. (1983). Interactions of plants and animals during the Carboniferous. *BioScience*, 33(8): 488–493.

Taylor, T. N. and Taylor, E. L. (1993). *The Biology and Evolution of Fossil Plants*. Prentice-Hall, Englewood Cliffs, NJ: 982 pp.

Teuscher, H. (1956). *Myrmecodia* and *Hydnophytum*. *Nat. Hort. Mag.* 35: 49–51.

Thaxter, R. (1926). Contribution towards a monograph of the Laboulbeniaceae, Part IV. *Mem. Am. Acad. Arts Sci.* 15: 427–580.

Theobald, N. (1937). *Les Insectes Fossiles des Terrains Oligocenes de France*. Georges Thomas, Nancy: 473 pp.

Thiery, D. and Visser, J. H. (1986). Masking of host plant odour in the olfactory orientation of the Colorado potato beetle. *Entomol. Exp. Appl.* 41: 165–172.

Thiery, D. and Visser, J. H. (1995a). Les messages du vent. *Pour La Sci.* 211: 27.

Thiery, D. and Visser, J. H. (1995b). Satiation effects on olfactory orientation patterns of Colorado potato beetle females. *C.R. Acad. Sci. Paris*, 318: 105–111.

Thomas, B. (1981). *The Evolution of Plants and Flowers.* Eurobook: 116 pp.

Thomas, W. W. (1984). The systematics of *Rhynchospora* section *Dichromena*. *Mem. N.Y. Bot. Garden*, 37: 1–116.

Thomasson, J. R. (1982). Fossil grass antoecia and other plant fossils from arthropod burrows in the Miocene of western Nebraska. *J. Paleontol.* 56: 1011–1017.

Thompson, J. N. (1981). Reversed plant-animal interactions. The evolution of insectivorous and ant-fed plants. *Biol. J. Linn. Soc.* 16: 147–155.

Thompson, J. N. (1982). *Interaction and Coevolution.* John Wiley & Sons, New York: 179 pp.

Thompson, J. N. (1988a). Variation in preference and specificity in monophagous and oligophagous swallowtail butterflies. *Evolution*, 42(1): 118–128.

Thompson, J. N. (1988b). Coevolution and alternative hypotheses on insect-plant interactions. *Ecology*, 69(4): 893–895.

Thompson, J. N. (1988c). Evolutionary ecology of the relationship between oviposition preference and performance of offspring in phytophagous insects. *Entomol. Exp. Appl.* 47: 3–14.

Thompson, J. N. (1989). Concepts of coevolution. *Trends Ecol. Evol. (TREE)*, 6: 179–183.

Thorsteinson, A. J. (1953a). The role of host selection in the ecology of phytophagous insects. *Can. Entomol.* 85: 276–282.

Thorsteinson, A. J. (1953b). The chemical sense in phytophagous insects. *Redia*, 38: 369–374.

Thorsteinson, A. J. (1960). Host plant selection by phytophagous insects, *Ann. Rev. Entomol.* 5: 193–218.

Torma, M. et al. (1958). First symposium on insect and food plants. *Entomol. Exp. Appl.* 1(1–2): 1–152.

Torre-Bueno, J. R. (1989). *The Torre-Bueno Glossary of Entomology* (revised by Stephen W. Nichols). American Museum of Natural History, New York: 840 pp.

Tower, W. L. (1906). *An Investigation of Evolution in Chrysomelid Beetles of the Genus* Leptinotarsa. Carnegie Institute, Washington, D.C.: 321 pp.

Tower, W. L. (1918). *The Mechanism of Evolution in* Leptinotarsa. Carnegie Institute, Washingon, D.C.: 384 pp.

Traynier, R. M. M. (1986). Visual learning in assays of sinigrin solution as an ovipositor releaser for the cabbage butterfly, *Pieris rapae. Entomol. Exp. Appl.* 40: 25–33.

Tremblay, R. L. (1992). Trends in the pollination ecology of the Orchidaceae: evolution and systematics. *Can. J. Bot.* 70: 642–650.

Treseder, K. K., Davidson, D. W., and Ehleringer, J. R., Jr. (1995). Absorption of ant-provided carbon dioxide and nitrogen by a tropical epiphyte. *Nature*, 375(6527): 137–139.

Tschudy, R. H., Pillmore, C. L., Orth, C. L., Gilmore, J. S., and Knight, J. D. (1984). Disruption of the terrestrial plant ecosystem in the Cretaceous period Tertiary boundary, Western Interior. *Science*, 225(4666): 1030–1032.

Turner, J. R. G. (1986). Drinking crocodile tears: the only use for a butterfly? *Antenna*, 10(3): 119–120.

Ule, E. (1902). Ameisengarten im Amazonasgebietes. *Bot. Jahrb. Syst.* 30(68): 46–52.

van de Water, T. P. M. (1982). Allopatric speciation in the small ermine moth *Eponomeuta padellus*, in *Proc. 5th Int. Symp. Insect-Plant Relationships*, Visser, J. H. and Minks, H. K., Eds., Pudoc, Wageningen: pp. 405–406.

van der Ent, L. J. and Visser, J. H. (1991). The visual world of the Colorado potato beetle. *Proc. Exp. Appl. Entomol. N.E.V. Amsterdam*, 2: 80–85.

van der Pijl, L. (1982). *Principles of Dispersal of Higher Plants*. Springer-Verlag, New York: 215 pp.

van der Pijl, L. and Dobson, C. M. (1966). *Orchid Flowers: Their Pollination and Evolution*. University of Miami Press, Coral Gables, FL: 214 pp.

van Emden, H. F. (1960). Plant-insect relationships and pest control. *World Rev. Pest Controls*: 115–123.

van Emden, H. F., Ed. (1972a). Insect/plant relationships. *Symp. Roy. Entomol. Soc. Lond.* 6: 1–215.

van Emden, H. F., Ed. (1972b). *Aphid Technology*. Academic Press, London: 344 pp.

van Emden, H. F. (1972c). Aphids as phytochemists, in *Phytochemical Ecology*, Harborne, J. B., Ed., Academic Press, London: pp. 25–43.

van Emden, H. F., Ed. (1973). Insect-plant relationships. *Symp. Roy. Entomol. Soc. Lond.* 6: 215 pp.

Vane-Wright, R. I. and Ackery, P. R., Eds. (1984). *The Biology of Butterflies*. Academic Press, London: 429 pp.

Vasconcellos-Neto, J. and Jolivet, P. (1994). Cycloalexy among chrysomelid larvae, in *Novel Aspects of the Biology of Chrysomelidae*, Jolivet, P., Cox, M. L., and Petitpierre, E., Eds., Kluwer Academic, Dordrecht: pp. 303–309.

Verschaffelt, E. (1910). The cause determining the selection of food in some herbivorous insects. *Proc. Kon. Neder. Akad. Wetensch. Amsterdam C.* 13: 536–542.

Visser, J. H. (1983). Differential sensory perceptions of plant compounds by insects, in *Plant Resistance to Insects*, Hedin, P. A., Ed., American Chemical Society, Washington, D.C.: pp. 215–230.

Visser, J. H. (1986). Host odor perception. in phytophagous insects. *Ann. Rev. Entomol.* 31: 121–144.

Visser, J. H. and Minks, A. K., Eds. (1982). *Insect-Plant Relationships*. PUDOC, Wageningen: 464 pp.

Waldbauer, G. P. and Friedman, S. (1988). Dietary self-selection by insects, in *Endocrinological Frontiers in Physiological Insect Ecology*, Sehnal, F., Zabza, A., and Denlinger, D. L., Eds., Wroclaw Technical University Press, Poland: pp. 403–422.

Wallace, J. M. and Mansell, R. L. (1976). *Biochemical Interaction Between Plants and Insects*. Plenum Press, New York: 425 pp.

Ward, D. B. and Fish, D. (1982). Threatened powdery *Catopsis*, in *Rare and Endangered Biota of Florida*, Pritchard, Ed., University of Florida Press, Gainesville: pp. 74–75.

Ward, L. K. (1988). The validity and interpretation of insect food-plant records. *Br. J. Entomol. Nat. Hist.* 1: 153–162.

Ware, A. B., Kaye, P. T., Compton, S. G., and Noort, S. (1993). Fig volatiles, their role in attracting pollinators and maintaining pollinator specificity. *Plant Syst. Evol.* 186(3–4): 147–156.

Waterlot, G. (1949). Ordre des Arthropleurides, in *Traite de Zoologie*, VI, Grassé, P. P., Ed., Masson et Cie, Paris: pp. 211–216.

Weber, W. A. (1974). Two lichen-arthropod associations in Australia and New Guinea. *Lichenologist*, 6: 168.

Wei-Chun, M. (1969). Some properties of gustation in the larva of *Pieris brassicae*. *Entomol. Exp. Appl.* 12: 584–590.

Weinstein, P. (1990). Leaf petiole chewing and the sabotage of induced defences. *Oikos*, 67: 58.

Weir, A. and Beakes, G. W. (1996). Biology and identification of species of *Laboulbenia* Mont. and C. P. Robin (Fungi Ascom.) parasitic on Alticine Chrysomelidae, in *Chrysomelidae Biology*, Jolivet, P. and Cox, M. L., Eds., SPB Academic Publishers, Amsterdam: pp. 117–139.

Weiss, M. R. (1991). Floral colour change as cues for pollinators. *Nature*, 354: 227–229.

Weiss, M. R. (1995). Associative colour learning in a nymphalid butterfly. *Ecol. Entomol.* 20: 298–301.

Welch, R. C. and Greatorex-Davies, J. N. (1993). Colonization of two *Nothofaqus* species by Lepidoptera in Southern Britain. *Forestry (Oxford)*, 66(2): 181–203.

Whalley, P. (1986). A review of the current fossil evidence of Lepidoptera in the Mesozoic. *Biol. J. Linn. Soc.* 28: 253–271.

Whalley, P. (1987). Insects and Cretaceous mass extinction. *Nature*, 327: 562.

Whalley, P. (1988). Insect evolution during the extinction of the Dinosauria. *Entomol. Gen.* 13(1–2): 119–124.

Wheeler, W. M. (1913). Observations on Central American *Acacia* ants. *Trans. Congr. Entomol. Oxford*, 2: 109–139.

Wheeler, W. M. (1925). *Ants: Their Structure, Development and Behavior*. Columbia University Press, New York: 633 pp.

Wheelwright, N. T. (1991). Frugivory and seed disposal: la coevolucion ha muerto — viva la coevolucion! *Trends Ecol. Evol. (TREE)*, 6(10): 312–313.

White, R. E. (1996). Leaf beetles as biological control agents against injurious plants in North America, in *Chrysomelidae Biology*, Jolivet, P. and Cox, M. L., Eds., SPB Academic Publishers, Amsterdam: pp. 373–432.

White, R. E. and Day, W. H. (1979). Taxonomy and biology of *Lema trivittata* Say, a valid species with notes on *L. trilineata* (Oliv.). *Entomol. News*, 90(5): 209–217.

Whitman, D. H. (1994). Plant body guards: mutualistic interactions between plants and the third trophic level, in *Functional Dynamics of Phytophagous Insects*, Ananthakrishnan, T. N., Ed., Science Publishers, Lebanon, NM: pp. 207–248.

Whitman, D. H., Blum, M. S., and Slansky, F. (1994). Carnivory in phytophagous insects, in *Functional Dynamics of Phytophagous Insects*, Ananthakrishnan, T. N., Ed., Science Publishers, Lebanon, NM: pp. 161–205.

Whittaker, R. H. and Feeny, P. P. (1971). Allelochemics chemical interactions between species. *Science*, 171: 757–770.

Wiebes, J. T. (1984). Fig wasp-fig coevolution. *Antenna*, 8(3): 122–126.

Wieczorek, H. (1976). The glycoside receptor of the larva of *Mamestra brassicae* L. (Lept. Noct.). *J. Comp. Physiol.* 106: 153–176.

Wiens, D. and Prourke, J. (1978). Rodent pollination in southern African *Protea* spp. *Nature*, 276: 71–73.

Wilding, N., Collins, N. M., Hammond, P. M., and Webber, J. E., Eds. (1989). *Insect-Fungus Interactions*. Academic Press, London: 344 pp.

Willemstein, S. C. (1987). An evolutionary basis for pollination ecology. *Bull. Leiden Bot. Ser.* 10: 425 pp.

Williams, C. E. (1991). Host plant latex and the feeding behavior of *Chrysochus auratus* (Col. Chrys.). *Col. Bull.* 45(2): 195–196.

Williams, G. and Adam, P. (1994). A review of rainforest pollination and plant-pollinator interactions with particular reference to Australian subtropical rainforests. *Australian Zool.* 29: 177–212.

Williams, K. S. and Gilbert, L. E. (1981). Insects as selective agents on plant vegetative morphology: egg mimicry reduces egg laying by butterflies. *Science*, 212: 467–469.

Williams, S. E. (1976). Comparative sensory physiology of the Droseraceae. The evolution of a plant sensory system. *Proc. Am. Philos. Soc.* 120(3): 187–204.

Willis, J. C. and Airy-Shaw, H. K. (1973). *A Dictionary of the Flowering Plants and Ferns.* Cambridge University Press, London: 1245 pp.

Willson, M. F. (1979). Sexual selection in plants. *Am. Nat.* 113(6): 777–790.

Wilson, C. L. and Graham, C. L. (1983). *Exotic Plant Pests and North American Agriculture*, Academic Press, New York: 528 pp.

Windsor, D. M. and Jolivet, P. (1996). Aspects of the morphology and ecology of two Panamanian ant-plants, *Hoffmannia vesciculifera* (Rubiaceae) and *Besleria formicaria* (Gesneriaceae). *J. Trop. Ecol. Cambridge*, 12: 835–842.

Woodhead, S. and Bernays, E. (1977). Changes in release rates of cyanide in relation to palatability of sorghum to insects. *Nature*, 270: 235–236.

Wootton, P. J. (1981). Palaeozoic insects. *Ann. Rev. Entomol.* 26: 319–344.

Wratten, S. D. (1974). Aggregation in the birch aphid, *Euceraphis punctipennis* (Zett.) in relation to food quality. *J. Animal Ecol.* 43: 191–198.

Wratten, S. D. and Edwards, P. J. (1984). Wound-induced defences against insect grazing. *Antenna*, 8(1): 26–29.

Wratten, S. D., Edwards, P. J., and Winder, L. (1988). Insect herbivory in relation to dynamic changes in host plant quality. *Biol. J. Linn. Soc.* 35: 339–350.

Yamaoka, R. (1990). Chemical approach to understanding interactions among organisms. *Physiol. Ecol. Jpn.* 27: 31–52.

Zamora, R. (1996). The trapping success of a carnivorous plant, *Pinguicula vallisneriifolia*: the cumulative effects of availability, attraction, retention, and robbery of prey. *Oikos* 73(3): 309–322.

Zamora, R. and Gomez, J. M. (1996). Carnivorous plant-slug interaction: a trip from herbivory to kleptoparasitism. *J. Animal Ecol.* 65(2): 154–160.

Zhang,Z. Q. and McEvoy, P. B. (1996). Factors affecting the response of *Longitarsus jacobaea* (Col. Chrys.) to upwind plant odours. *Bull. Entomol. Res.* 86: 307–313.

Zinov'yev, A. G. (1994). Geographic variation of *Pontania acutifoliae* (Hym. Tenthr.) and possibility of parallel evolution of the gall maker and its host plant. *Entomol. Rev.* 73(1): 142–155.

Zwolfer, H. (1970). Regional change of foodplant of phytophagous insects as an evolutionary problem. *Z. Angew. Entomol.* 55: 233–239.

Zwolfer, H., Paulus, H., Regenfuss, H., and von Wahlert, G., Eds. (1978) *Coevolution: Phylogenetics Symposium.* Paul Parey Publishers, Berlin: 125 pp.

Plant index

A

Acacia, 3, 65, 125, 127, 128, 129, 131, 132, 133, 134, 135, 136, 158, 189, 210
  A. sphaerocephala, 124
Acanthaceae, 159
Acarus calamus, 96
Acer, 158
  A. campestris, 152
Aconitum, 189
Actinidia, 178
Adansonia, 174
Adenium obesum, 42
Adenostemma, 36, 217
Aegle marmelos, 208
Agaricus, 68
Agavaceae, 182
Agave, 171
Aglaophyton, 18
Aldrovanda, 110, 119
  A. vesicularia, 111
Aldrovandra, 105, 106
Alnus parvifolia, 12
Alnus, 44, 170
Aloeaceae, 190
Alternanthera
  A. philoxeroides, 170, 172
Amaranthaceae, 39
Ambrosia artemissifolia, 48
Amentaceae, 38
Amentiferae, 44, 45
Amoebidium parasiticum, 61
Amorphophallus, 183, 185
  A. johnsonii, 208
  A. titanus, 176
Anemone, 35

Angiospermae, 104, 173
Angraecum sequipedale, 188
Anguria, 34, 82, 197
Anthorrhiza, 130
Anthyllis, 39
Anurosperma, 107
Apiaceae, 15, 30, 40, 41, 44, 72, 85, 86, 186
Apocynaceae, 42, 76, 94, 95, 211
Apocynum, 51
Aquifoliaceae, 73
Aquilegia, 205
  A. formosa, 205
  A. pubescens, 205
Araceae, 85, 96, 116, 139, 176, 208
Araliaceae, 175
Araucaria, 4, 19, 20, 33
Aristolochia, 44, 82, 83, 98
  A. elegans, 46, 84
  A. praevenesa, 84
  A. reticulata, 82
  A. serpentaria, 82
Arthopyrenia halodytes, 144
Arum, 116, 183, 192
  A. conophalloides, 186
  A. maculatum, 185
  A. orientale, 187
Asclepiadaceae, 37, 42, 51, 52, 76, 90, 94, 95, 96, 116, 130, 138, 185, 189, 211
Asclepias, 37, 51, 94
  A. curassavica, 196
  A. syriaca, 51, 94
Ascomycetes, 56, 60, 62, 66, 67, 69, 146
Aspergillus, 57
  A. flavus, 57
Aspidistra, 174

Insect index

General index

A

Acari, 18, 37
acarocecidia, 137, 151; *See also* galls
acarodomatia, 23, 154, 155
acarophytes, 24, 123, 124, 136–137
acids, 90
Agromyzid mining species, 15
Albany pitcher plant, 109
alfalfa, 179
algae, 11, 17, 19, 43, 60, 61, 66, 68, 77, 147, 148
alkaloid receptors, 87, 91
alkaloids, 30, 42, 71, 77, 85, 86, 90, 91, 92, 95, 98, 188, 196, 205
  of aquatic plants, 84
alkyl isothio-cyanates, 48
allelochemicals, 6, 93, 205
allelopathy, 93, 99
allomones, 93, 94
allotrophy, 31, 43–45
alticines, 78
amber, 20–21
  Baltic fauna, 20
  DNA sequencing, 20
amino acid receptors, 87
amino acids, 30, 90, 157, 184
  in nectar, 188
ammonites, extinction of, 21
anabasine, 83
anemogamy, 174
anemophily, 23, 199
anemotropism, 83
angiosperms, 4, 14, 34, 58, 104, 120, 128, 137, 158, 173, 180, 199, 202, 203, 205, 207, 210, 212; *See also* flowering plants

and beetle pollination, 186
appearance during Triassic, 15
as sister group of Benettitales, 13
as sister group of Gnetales, 13
dinosaurs feeding upon, 22
evolution of, 197
pollination in, 174
reestablishment of dominance, 15
Animalia kingdom, 55
ant-plants, 123–142, 160, 163
antherozoa, 196
anthocyanin receptors, 87
anthocyanins, 23
antifeedants, 50, 72, 83, 206
antigonadal agent, 96
antioxidizers, 188
ants, 28, 52, 92, 166, 178, 189, 194; *See also* leafcutter ants
  Acromyrmex, 66
  and Cephalotaceae, 109
  and fungus, 63–66
  as protection against herbivores, 127
  as wood-borers, 6
  Atta, 2, 64, 127, 139
  Attini, 62–68
    and fungus gardens, 63–64
  Azteca, 6, 124, 128, 188
  "biting", 137
  Camponotus, 129
    C. femoratus, 139
  Cataulacus mackei, 132
  Crematogaster, 130, 132
    C. africana, 132
    C. parabiotica, 139
  Dolichoderus, 130
  Engramma kohli, 133

297